AF681346

ROCKPECKER

by

T.D. LaRUNE

Copyright (c) 1993

A MINERAL PROSPECTOR'S PRIMER

SECRETS, CLUES, TIPS, HINTS, DATA

&

TRICKS OF THE TRADE

Skill//Quest Co.
675 Fairview Drive, # 246
Carson City, Nevada 89701 USA

DEDICATION

To the legendary one blanket, one jackass prospector,
and the political foresight necessary
to enable his successors to continue the search.

COPYRIGHT 1993 BY T.D. LaRUNE.
All rights reserved

COPYRIGHT: BERNE COPYRIGHT CONVENTION, UNIVERSAL COPYRIGHT CONVENTION, AND PANAMERICAN COPYRIGHT CONVENTION.

No part of this publication may be copied or distributed, transmitted, transcribed, stored in a retrieval system, or translated into any human or computer language, in any form or by any means, electronic, mechanical, magnetic, manual, or otherwise, or disclosed to third parties without express written permission of:

T. D. LaRune, P.O. Box 620995, San Diego, California 92162 U.S.A.

This publication is designed to provide accurate and authoritative information in regard to the subject matter covered. It is sold with the understanding that neither the author nor contributing authors nor the publisher is engaged in rendering legal, accounting, or other professional service. If legal advice or other expert assistance is required, the services of a competent professional should be sought.

From a Declaration of Principles jointly adopted by a Committee of the American Bar Association and a Committee of Publishers.

PUBLISHED BY: Skill//Ware Books, AN IMPRINT OF: Skill//Quest Co., NEVADA, U.S.A.

Sponsoring editor: Premium General Agency

Project editor: MS Worth Evaluations

Jacket designer: T. D. LaRune

Cover illustrations: Fanci Boyd

LIBRARY OF CONGRESS CATALOG CARD NUMBER: 94-67809

ISBN 1-886499-00-4 - Hardcover

ISBN 1-886499-01-2 - Paperbound

FIRST EDITION

CONTENTS

ROCKPECKER — PART I

GOLDPECKER — PART II

PART III
277 MINERAL RECOGNITION NOTES

CHEMICAL TESTING - PART IV

PARTS 3 & 4 AVAILABLE IN PC DOS COMPUTERIZED FORMAT

See page 504

ABOUT THE AUTHOR:

T.D. LaRune, B.A., G.se., J.D., seeks his recreation with a goldpan. In 1982, while prospecting near an old, long-abandoned gold mine, he glimpsed something odd in a small water source flowing from a shale overlying slate formation. In an attempt to improve the water volume, he used his hunting knife to pry some of the shale loose. The flow did not speed-up, but the knife blade dislodged the oddity.

It measured two inches long by three-eights inches wide. It had the thickness of a twist-off strip seal, such as you find on a tin of coffee or corned beef. This shiny, silvery, yellow-toned strip of can sealer was out of place. Curious, he thought, how did it get there? Why wasn't it corroded? Since it was not rusty, he put it in safekeeping, then promptly forgot it.

Four days later he was visiting the local mining museum. Lo and behold, there in the display case, was the answer to his ignorance. He now knew he had some silver-gold electrum.

The shale-slate-gold association [*] demanded research for the *"WHY"*. Over the years he had filled many note books with clues and answers to similar, perplexing, prospecting questions. This book is the distillation of that research. [* See Rockpecker Tip # 53, page 130]

"ANYTHING WORTH KNOWING, CANNOT BE TAUGHT IN THE CLASSROOM."

Law & Order 11/91

PREFACE TO FIRST EDITION

A mineral specimen is the same composition of chemicals wherever in the world it is found. Proper identification of your mineral find requires some knowledge of the conditions under which it was formed. That is, the environment in which it is usually found.

Visual appearance is the first identification method. To that end, rock and mineral familiarity can be gained by a trip to the local museum of natural history, many visits to rock shops, and a diligent review of the many color picture books published for the rockhounding community. Avail yourself of this visual recognition handshake at each opportunity. In time, you will reach a comfort level as your confidence matures.

Simple chemical tests and blowpipe tests can direct you on a path that is often neglected and overlooked. This book uses the streak test followed by a specific gravity measurement to eliminate possibilities. Follow the mineral I.D. Chart clues as closely as you can to narrow your probables. When you are dealing with thousands of possible minerals, eliminating all but a few probables is a giant step toward recognition.

Since this is intended as a practical book, to be of value to the amateur and seasoned old-timer alike, reference citations have been omitted. The author will be pleased to hear from the reader who can input observations and experiences that will enhance later *ROCKPECKER* editions. It is the author's fondest hope that some of the joy and excitement he found in this field of activity will be contagious.

The author has deliberately omitted references to specific ore locations. This is to foster a spirit of *open mindedness.*

The writer wishes to thank the many who have helped in the preparation of this book.

INTRODUCTION

"We must grope our way around the truth to remove the veil that hides nature. It is not enough to be satisfied with a superficial view." *-SENECA-*

The ore deposit is the goal that urges the prospector towards its discovery. The geologist directs attention to understanding the deposit so as to aid in the discovery of additional sources. The Rockpecker who understands and applies the principles of ore deposit formation is using sets of clues found over time in general earth history.

Fundamentals of Geology, Mineralogy and Geochemistry encountered in this work are explained in simplified terms. I trust that I have reached both the pilgrim and pro Rockpecker equally.

Let us define the word *"deposit"* as a concentration of useful substances of economic importance that can be gathered and processed at a profit.

Ore deposits, besides being mineral associations of useful compositions, are geological genesis snapshots. They are inseparable from their earthly birthplace. It is this kinship or *PARENTAL RELATIONSHIP* that give us exploration clues.

Ore is a mixture of mineral grains that permits extraction of one or more useful metals or non-metals. For example: *PYRITE,* for its sulfur, selenium and thallium; *SPHALERITE,* for its zinc, cadmium, gallium and indium; *LEPIDOLITE,* for its lithium, rubidium and cesium. An ore's non-economic accessory minerals, such as quartz in gold ore, is called *GANGUE.*

ROCKPECKER TIP # 1. Be aware that certain gangue minerals can be indicators of specific ore. *BE ALERT FOR THE INDICATORS!*

ROCKPECKER TIP # 2. SMOKY QUARTZ AS AN INDICATOR OF URANIUM. Radioactivity is a property of uranium's daughters. Crystal quartz, when bombarded, turns smoky. Therefore, the quartz smoky color effect can reasonably indicate the presence of uranium.

ROCKPECKER TIP # 3. Calcite that FLUORESCES deep pink or red under ultraviolet light has been linked to manganese and lead ores. Only the calcite IN or NEAR the mineralized area fluoresces.

ROCKPECKER TIP # 4. White quartz, sometimes called BULL-QUARTZ, is especially associated with gold. All bull-quartz should be tested for gold values. Remember, one troy ounce (31.1 grams) per ton is SUPER high grade ore.

ROCKPECKER TIP # 5. Oftentimes, BARITE and FLUORITE are gangue materials. Where they are CONCENTRATED, they may well have ore value. In those cases the associated calcite or limestone may well be the gangue.

ROCKPECKER TIP # 6. Does your quartz carry trace metals such as copper? You may have a marketable copper smelting flux product. Copper smelters use silicon to produce a slag melt to remove unwanted iron impurities. Here, the quartz is the ore.

ROCKPECKER TIP # 7. Concentrations of uranium minerals are strongly associated with DEEP PURPLE FLUORITE. Again, radiation caused the color change.

ROCKPECKER TIP # 8. WHEN YOU KNOW THE COMMON GENESIS OR ORIGIN OF AN ORE AND ITS GANGUE, PARTICULARLY IF BOTH FORMED THE SAME WAY, YOU HAVE AN IMPORTANT GUIDE TO OTHER ORE-BEARING STRUCTURES AND FAVORABLE SITES.

ROCKPECKER TIP # 9. TRACE METALS, detected by assay as impurities in gangue minerals, may reach ore grades. An example would be gold-bearing pyrites.

ROCKPECKER TIP # 10. Other common ore-gangue associations have been:

1. Magnetite with apatite.
2. Fluorite with lead and zinc.
3. Barite with silver, copper and lead.

4. Tourmaline and topaz with cassiterite.
5. Arsenopyrites with gold and tungsten.

ROCKPECKER TIP # 11. MICA and CLAYS with PASTEL COLORS are a good indicator or guide to ore.

The percent of economic mineral concentration in a deposit is called *TENOR.* Thus, the sulfur concentration in pyrites is theoretically 53%. If your deposit carries 45% sulfur, that is the tenor of your deposit.

You can have economic ore concentrations and still have nothing profitable. For example, the ore is inaccessible because of an uncontrollably high water table; the size of the deposit does not justify the expense to extract it; the expense of dressing or processing the ore exceeds its marketable value.

As a rule, the larger the reserves of a deposit, the more can be invested in technical extraction installations.

The genesis of any mineral deposit is not only of theoretical interest, but can become a legal issue to be resolved by the courts. The California borate deposits illustrate this point.

The Kramer Borate District is located in Kern County, California. This case wound up in litigation on the issue of whether the deposit fell under the FEDERAL SODIUM LEASING ACT OF 1920. If so, the sodium on PUBLIC LAND could only be mined by leasing and the payment of royalties. The stakes were high. The borate deposits were estimated to contain a minimum of 100 million tons with a thickness of 250 feet.

The commercial beds of BORAX and KERNITE aggregates lie in a matrix of BLUE CLAY. COLEMANITE and ULEXITE is scattered intermittently throughout the BLUE SHALE. The kernite was derived from borax by water loss during recrystalization. There is evidence that the process is reversible, in that borax has been formed from kernite.

The 1920 Act was limited to sodium minerals *"Dissolved in*

and soluble in water, AND ACCUMULATED BY CONCENTRATION." (Ancient lake deposits)

If it could be shown that the deposit originated in some way other than *accumulated concentration*, the deposit could be claimed under the GENERAL MINING LAWS and be ROYALTY-FREE for the claimholder. In this event, there were *big bucks* to be saved.

This case was heard in 1929. After several eminent geologists testified that the deposits were formed through *REPLACEMENT* of shales by hot borax solutions *RISING FROM BELOW*, the court ruled that the deposits had not been *accumulated by concentration"*. It also ruled that claims could be filed, patented and mined under the General Mining Laws.

Thereafter, the 1920 Act was revised. The clause relating to *"accumulation"* was deleted.

The expert witnesses in the 1929 case were subsequently discredited in favor of a *LAKE BED or PLAYA LAKE* concentration concept.

How this mineral deposit was originally discovered has an interesting background. In 1913 a water well was being drilled on a homestead claim. At 370 feet the mineral COLEMANITE was discovered. Further exploratory drilling encountered the thick borax bed, together with an unknown mineral later named KERNITE. Now, over 200 BORON derivatives are produced from this deposit.

ROCKPECKER TIP # 12. The secret of successful prospecting is based on two GUIDING PRINCIPLES. Both are so simple that spelling them out appears to belabor the obvious. To overlook their direct application at every step of your search is to court failure.

The principle of *ELIMINATION* is the FIRST, which is directly based on the SECOND, the principle of *ANALOGY*. Simply stated: To dig random holes in the ground, in the hope of finding a relatively rare ore body, is a hopeless and futile endeavor. You must eliminate futility by applying the principle of analogy.

That is, to know what rock associations in a particular district have yielded ore deposits in the past. Restated: Look for elephants where elephants have been reported in the past.

The search clues, hints, tips and ore theories presented in this book are your association guide lines. They help you develop the principle of analogy for search applications.

You will find that the explanatory format in this book covers the headings of the MINERAL CHART. The intent is to explain and provide background information to clarify and show the purpose and utility of the various headings. In this way you can use the mineral data more fruitfully. For example, the chart heading *"MOST PROBABLE OCCURRENCE ENVIRONMENT"* appears as:

	MINERAL	*MOST PROBABLE OCCURRENCE ENVIRONMENT*
5.	ALBITE	PLUTONIC IGNEOUS ROCKS; MICA SCHISTS & GNEISS
124.	GOLD	HITEMP HYDROTHERMAL QUARTZ VEINS. PLACERS.
230.	SILVER	SULFIDE REDUCTION IN LOWER LEAD, ZINC ORES.

For clarity, we will explore and explain ore genesis theory involving *igneous rocks, hydrothermal veins and sulfide reduction.* Refer to other pertinent chapters and the index when more than one occurrence environment is shown. Remember too, that *most probable* is simply the *most obvious.* Gold has been found with at least 33 different gangue materials. Quartz is simply the most obvious association. For other associations you would check the chart column *"ASSOCIATED WITH"* for more detail. However, in the case of gold, there is a special chapter dealing with gangue associations.

USE THIS BOOK AS A GUIDE. STUDY THE MINERAL CHARTS. DO NOT CLOSE YOUR MIND TO OTHER POSSIBILITIES. USE YOUR INITIATIVE!

HAPPY PROSPECTING!

T.D. LaRUNE - 1991

CHAPTER 1

GANGUE

An ore's non-economic accessory minerals, such as quartz in gold ore, is called gangue. Gangue minerals can be indicators of specific ore, particularly if both were formed the same way. Some gangue materials, if of sufficient concentration, may well be the ore. What you consider to be gangue, could be of ore value, simply because you lack some bit of expertise. What is truly gangue today, may be an economic mineral tomorrow. A case in point involves the element uranium.

In 1789, Martin Klaproth, a pioneer in true chemistry, was in great demand for his expertise in separating a single element from natural ores. That year he received an ore sample from Cornwall, England. He recognized it as similar to a Bohemian ore; black nodules in a matrix of green-yellow and ochre. Some men called it "pitchblende" or "iron pitch ore". It was thought to be a natural compound of zinc, tungsten and iron.

He had great difficulty testing the ore. His final results were a yellow powder which yielded a gray metallic button. It looked like a metal, but to a degree it was different. He called it a "half metal" or "metalloid", then pronounced it a new element which he named "uranium'. In 1840 uranium was proven to be a true metal.

In 1895, Roentgen called the emanations from pitchblende "X rays"; Bacquerel, the next year, identified their source as from uranium. In 1898, a Polish girl working in Paris, probed the emissions from pitchblende for her doctoral thesis. She named the emissive phenomenon "radioactivity". We know her as Marie Curie, the discoverer of radium and its sister element polonium. It was later chemically proven that the new elements were derived from uranium.

In 1915 a prospector for a large Belgium mining company was searching for copper in Katanga Province, Africa. In this mostly flat terrain he approached a small hill, more like a big mound, looking for outcrops. At the top of the mound he found old scars from earlier digging activities. It did not look like much. As he waited for his native boys to catch up, he literally kicked rocks, more in boredom, than in prospecting. A flash of yellow caught his eye. He picked up a heavy, hand-sized rock laced with green streaks and ochre in a matrix of bright yellow. The heft of the rock was his first indication of value; he had seen the same ore in a sample collection.

It was that useless uranium ore, good only because it contained a trace of radium. His ore might be worthless for it contained much more yellow than the sample he remembered. Yet, that yellow rock had plenty of brothers nearby.

Playing it safe, he staked the mound, fixed the position, and wrote his notes. Was there really some worthwhile radium in his discovery ore?

On a piece of zinc he scratched "radium", mounted it on a pole, and claimed a 1,000 meter radius circle from the pole as his company's claim. Instructions to the company to file the paperwork followed. Then he went off to World War 1.

Medical demands for radium as a magic bullet to fight cancer caused great uranium prospecting activity. There was only one supply source, and that held less than a 1% uranium tenor.

In 1919, with the end of the war, the prospector returned to Africa shocked to find that the claim had never been filed. His company had no claim! He immediately visited the ore site, saw his pole intact, then pushed for the paper filings.

Geologists called the deposit a "spur, wedged in a fold fault". It had squeezed to the surface in reefs, much like gold. The ore averaged 68% uranium. For every 1000 tons mined, 680 tons of uranium ore were to be had. The uranium was the gangue.

For every 3 tons of hand-cobbed uranium ore refined, one single gram of radium was recovered. In 1920 the going price of radium was 3 million dollars the gram.

At the refinery in Belgium, tons and tons of yellow, uranium cakes were being dumped in the slagheap. From time to time a ton or so was sold as a coloring agent for glass, the only other known use for uranium. By 1936 more than 10,000 tons had been processed. Radium was radioactive gold dust; uranium was the refinery dross.

The Katanga mine yielded over 40,000 tons of the richest uranium ore. Uranium, as an element, has an atomic number of 92, atomic weight of 238.03. For that reason it is known as U 238.

With the separation of U 235 as an isotope of U 238, fission, fusion, and the atomic age was upon the world. The story of how the Belgian uranium gangue dump was obtained for the U.S. atomic bomb project is a tale of intrigue in itself.

To end this chapter of the most worthless gangue ever to become valuable, suffice it to say that the Katanga uranium ore deposit showed cobalt, nickel, native silver, uranite, phosphates, carbonates, molybdates, until the list totaled 67 different minerals. The Katanga deposit was a geological freak because of its remarkable richness.

Legions of prospectors with their Geiger counters proved that uranium was as common as copper. In the U.S. Midwest, for example, every square mile of soil to 12 inches deep, contains 3 tons of uranium. Thus, there is no shortage of uranium as a raw material. Most discoveries are simply not concentrated enough to make the deposit economic. To that extent, uranium is gangue.

CHAPTER 2

GENESIS – MINERAL BIRTH THEORIES

"THE DISCOVERY DIFFICULTY FACING THE MINERALS INDUSTRY TODAY IS NOT SO MUCH LACK OF MINERAL DEPOSITS TO BE FOUND AS IT IS THE LACK OF KNOWLEDGE HOW TO FIND THEM."
H. M. BANNERMAN, 1957 MIN. ENG., V.9, P. 1108

Most valuable mineral discoveries have been made by prospectors rather than geologists. Up to 1960, 85% of Canadian mineral prospects were located by Rockpeckers. They were observant outdoorsmen, with a fair level of common sense, initiative and curiosity. They had very little money, and nearly as much formal education. They did have hope, expectations and a sense of adventuring. More than anything, they stood as a breed apart. In the romance stories of early mining days they were described as *one blanket, one jackass* prospectors.

Did he know that the tree root, flaming an eerie green in his campfire, was a test clue for copper? Probably not!

This work, diligently studied, should help you become a practical exploration person. You should see geological structures in a new and intriguing way. If it gives you a *leg up,* I will have accomplished my task.

The *ECONOMIC MINERAL PROSPECTOR (who I call a Rockpecker)* must successfully develop his exploration thought patterns to include, not only the optimism needed to forge ahead, but also imaginative use of fundamental structural geology. The successful search will rest on all the input and clue analysis available to him right up to the moment of discovery. In this sense, the Rockpecker is a field scientist *applying geological theories and testing them in the field.*

As an aid in bringing you up to speed for your good-results search, we will deal with fundamental general principles.

Historically, we pick up three basic ideas from AGRICOLA (1556) and STENSON (1669). They are:

1. ORE CHANNELS ARE YOUNGER THAN THE ROCKS THEY TRESPASS UPON.

2. CIRCULATING MINERAL SOLUTIONS DEPOSIT THEIR ORES IN THESE CHANNELS.

3. MINERAL-PREGNANT VAPORS RISING THROUGH FISSURES CONDENSE ALONG THE FISSURE WALLS TO FORM ORE DEPOSITS.

At the time forked twigs, divining rods and hazel sticks were generally popular prospecting tools, these principles were truly revolutionary.

Agricola wrote his essays on prospecting in Latin. His work was largely unknown, until in 1912, it was translated into English by Hoover and Hoover. (Lou H. Hoover was Stanford's first female geology graduate in 1898. Her husband, Herbert, was educated as a mining engineer. He became the 31st U.S. president serving from 1929-1933.)

In the EARLY 1700's the importance of *HYDROTHERMAL SOLUTIONS* bearing dissolved rock substances and ore deposition by *REPLACEMENT* was recognized.

In the MID 1700's distinctions were made between *VEINS* and *BEDDED* deposits. Veins being filled open fissures; bedded deposits were sediment strata.

In the LATE 1700's veins were considered open fissures filled with ore deposits leached from wall rocks by water. One hundred years later, the *VEIN LEACH THEORY* was expanded to in clude *RISING* liquids transported from deep magma injecting ore materials *UP* and *INTO* the fissure.

When it was shown that minerals, including the metallics, were soluble in water combined with dissolving agents, *mineral stream solution transport and vein deposit* was an easy next step.

Where *ore materials are leached* from wall rocks, then carried by solution into open fissures and filling them, the theory of *LATERAL SECRETION* was developed.

Later observations and field data showed a relationship between ore deposits and *INTRUSIVE ROCKS,* which has become the basis for the modern theory of *HYDROTHERMAL* ores.

THE FINDINGS WERE:

(a) Large ore deposits were *LOCALIZED* near and around *IGNEOUS* zones.

(b) Ore-bearing fluids spread from *DEEP MAGMA* chambers.

(c) *RISING MINERAL STREAMS* spread through the rocks and deposit in favorable environments.

Still later observations recognized *MAGMATIC SEGREGATION and REPLACEMENT* ores formed at zones in *CONTACT and ADJACENT to the INTRUSIVE IGNEOUS ROCKS.*

The next theory stage led to the view that *ROCK STRUCTURE* was vital in ore deposition and localization. *GUIDE CHANNELS* and favorable settling sites for the hydrothermal solutions were essential. For example, disseminated copper ores commonly found in *STOCKS* of monzonite and quartz monzonite.

These associations prompted the assumption that a *PARENTAL RELATIONSHIP* existed between *IGNEOUS ROCKS* and many ore deposits. Distinctions were accepted between *MAGMATIC and SEDIMENTARY ore deposit ORIGINS.* The *BIRTH* of minerals in a particular association, at around the *SAME TIME, CONDITION and PLACE,* is called

GENESIS. Genesis can be strictly divided into deposits of *MAGMATIC (igneous), SEDIMENTARY and METAMORPHIC origin.*

IT MUST BE CLEARLY UNDERSTOOD THAT NO SINGLE THEORY EXPLAINS THE BIRTH OR GENESIS OF ALL ORE DEPOSITS.

Geology grew around the study of ore deposits. Problems remain to be unraveled about ore material transportation and deposition. The Rockpecker must apply all known theories to maximize successful exploration. These theories will become clearer as we encounter them again.

NOW, ABOUT THEORIES AND THE SCIENTIFIC METHOD.

The truth or falsity of an idea does *NOT* depend on a majority consensus. *Theories*, like all scientific knowledge gained by science, evolve through a process called *THE SCIENTIFIC METHOD.*

If you start with the premise that the universe is orderly; that facts about it can be objectively analyzed; then its workings can be known. These facts are called *DATA.*

A probable explanation or solution of a known fact results in a *HYPOTHESES.* Predictions, based on the hypotheses, are then tested for correctness over the widest possible range of investigation. A hypotheses that passes multiple tests for correctness can then become a *THEORY.*

A theory is merely a hypotheses now regarded as having an excellent chance of being true. It is still open to question. It is not an article of faith. Since nothing in science is considered as absolutely proven, any theory is always open to question. Consider too, is the data you are applying really a *fact*, or is it only wishful thinking? Remember, a theory only gives you a mental track to follow. Think of it as a guiding pathway.

In the next chapters we will examine the genesis of igneous (magmatic), metamorphic and sedimentary rocks. They are the *GENESIS-CLUES* to ore deposits! The ore or mineral we seek is hidden in a cloaking device called *rock.* But where and in which rock? Shall we dig an endless series of holes as they did in the Middle Ages? What about using a forked twig? How about using a blind, random search method? Wouldn't it be more efficient if we knew more about nature's *rock vault?* Can we improve the odds through the use of some theory as to how the mineral or ore was originally deposited? How it was born?

****

NEXT LET'S LOOK AT A FEW ANCIENT ORE-BIRTH THEORIES.

CHAPTER 3

DISCREDITED ORE THEORIES

Ancient learning, advanced by philosophers, assigned the origin of ore bodies within the earth's crust, to the influence exerted by the sun and stars. This theory has been totally exploded.

One early theory, generally accepted at the time, involved the growth of metals from metallic seeds. It was based on the discovery of copper and silver growths on the timbers of exhausted mines. Subsequent research showed that the deposits were caused by waters containing silver and copper in solution.

Another early theory held that ores and metals came into existence through transformation of water into metal. The true explanation involved metal-bearing solutions traveling a long distance to redeposit elsewhere under favorable conditions.

The *mother gives birth theory* was based on ground potters clay, mixed with water, then left to stand. Tiny lumps of silicate formed concretions. Cassiterite, in exhausted tin mine tailings apparently regenerated the tin, so that it could be reworked again. The explanation that finely divided tin concentrations, over time, recombined to be of sufficient size for later recovery, seemed logical.

Heap leaching is based on the premise that earlier extraction methods were inefficient. That ore recovery with new methods could be both efficient and profitable is now a fact.

A theological theory states that all ore was deposited in the earth's crust at the "beginning". It has been shown that in many parts of the world, many ore types are presently being deposited. Fumaroles, depositing metals in their exhaust stream are a specific contradiction.

For generations alchemists tried to change or transmute base metals into gold or silver. This was based on the theory that gradual and progressive change turned lower order metals into gold in a *ripening* process. The maturity into gold has long been discarded. However, in a sense the theory has a reverse truth. We now know that uranium and thorium undergo change, over time, to produce lowly lead.

Had the ancients known of an igneous intrusion such as a stock forming a gossan or iron hat, the transmutation of iron sulfide into the higher metals of zinc, lead and silver would have been cited as proof that natures heap leaching had transformed iron into silver in a zonal arrangement downward. It would have been called a metal ripening or maturing.

These older theories had no basis in observed fact to support them and now receive a different explanation.

The ancient theories of lateral secretion and magmatic segregation, in the broadest sense, have a large measure of confirmation in present time. Lateral secretion describes the formation of ore deposits by concentration of solutions percolating through country rock. The metal-bearing solutions disseminate the ore particles from deep sources. Magmatic segregation holds that the origin of certain ore bodies was caused by segregation of metallic elements from a fused magma forcing the separated metallic constituents upward into the deposit. This theory is still being debated.

Modern advances as to genesis of alluvial deposits from surface alteration of an ore body by weathering and transportation with concentration by specific gravity were unknown by the ancients. They believed that gold-bearing sands of alluvial deposits came from the interior of the mountain, carried by water from caves.

Most of these discredited armchair theories were refuted by observable facts discovered in the field.

CHAPTER 4

GEOLOGIC TIME CLOCK

THE RECORD OF THE ROCKS

When we talk about *earth time* as eons, eras, and epochs, confusion can easily creep into these ideas. We can divide earth time into two main points of view. We can look at it as the evolution of LIFE on the earth; or as NATURAL EVENTS that have shaped the crust of the earth. The record of both views is read in the rocks.

Our focus will be on the events that have produced changes in the crust. This is the area that most concerns the Rockpecker.

Geologists divide the span of earth history into two periods called EONS. The *hidden life* eon includes the earliest known rocks up to 600 million years ago. The *visible life* eon covers from that time to the present.

The hidden life eon is a rare find; the core of ancient mountains in eroded exposure; deep gorges cut in high plateaus, or huge *shields* exposed on the earth's surface, such as the Canadian Shield.

This 4.5 billion year span is measured by the organic carbon content of the rocks. This really measures former living things or *fossil life.* Very little is known of this eon, that's why it's called the hidden life.

The visible life eon of the past 600 million years has much more to tell us about earth conditions.

We can divide the visible life eon into *ancient life* or PALEOZOIC; *in-between-life* or MESOZOIC; *recent life* or CENOZOIC.

These are the great subdivisions we call *ERAS*.

TIME SCALES OF NORTH AMERICA

The three eras are further subdivided into PERIODS defining fossil life finds.

PALEOZOIC ERA - ANCIENT LIFE

The ancient life era is the PRECAMBRIAN PERIOD with little evidence of fossil life. The seas and landmasses appear with the formation of the earth's crust.

CAMBRIAN PERIOD - Shallow seas and low-lying land masses form sediments which originate the basis for the APPALACHIANS, ROCKIES, and other mountain ranges.

ORDOVICIAN PERIOD - There is much geologic disturbance. 70% of America is covered by shallow seas. North America was a group of islands. Volcanic activity forms rising eastern land masses running from Newfoundland to Alabama as a huge mountain chain.

SILURIAN PERIOD - The eastern United States is covered by landlocked inland seas caused by coastal uplift. Volcanoes are active in Maine.

DEVONIAN PERIOD - Volcanoes are very active with the beginnings of many mountain ranges. Vast forest swamps form the basis for later coal deposits. Coastal landmass from Canada to North Carolina rises from the sea.

CARBONIFEROUS PERIOD - Earlier plant life forms coal. Crustal activity raises many mountain ranges. The seas withdraw to expose landmasses.

PERMIAN PERIOD - Violent mountain building of the Appalachians. Climate changes cause severe ice ages with glaciers. Large deserts form to cover much of the continent.

MESOZOIC ERA - IN-BETWEEN LIFE

TRIASSIC PERIOD - The Appalachians, once as tall as the ALPS begin to erode. This was the time of the dinosaur's first appearance.

JURASSIC PERIOD - Dinosaurs are in dominance. Disturbances thrust up the Sierra Nevada's from Alaska to Southern California.

CRETACEOUS PERIOD - The sea inundates much of North America for the last time. The Rocky Mountains rise out of sedimentary troughs from Alaska to Central America.

CENOZOIC ERA - RECENT LIFE

TERTIARY PERIOD - Extensive volcanic activity in the Northwestern United States. The Cascade Range and Columbia Plateau rises. The Rocky Mountains reach their present height.

PLEISTOCENE PERIOD - Four Ice Ages send glaciers across the North American continent. Life forms retreat, adapt or die due to resultant habitat erasure.

Less than one million years ago, sheets of mile-thick ice ground across North America. Ice-dug hollows formed the basins for the Great Lakes with the retreat and melting of the ice pack.

ROCKPECKER TIP # 13 - QUIZ: WHAT DO THE GEOLOGICAL CLOCK PERIODS HAVE TO DO WITH PROSPECTING?

LET'S TAKE A HYPOTHETICAL FACT SITUATION.

You have discovered a mineral. Assume it is gold. As you look at the rock strata above your find, you notice a layer of limestone overlain by a pumice-like rock expanse. The next upward layer reveals a cement-like rock. It looks like slate, but you know it is till.

The strata you are examining lies at a 45 degree angle pointing upward toward the sky. Off, in the distance about two miles away, you make out the shape of layers pointing upward and towards you.

ANALYZE THE FACTS AS TO THE PROBABILITIES. THERE IS NO RIGHT OR WRONG ANSWER, ONLY A VARIATION AS TO FACT INTERPRETATION.

QUESTION: Determine each geologic period working from the topmost layer down toward your gold location?

The limestone is overlain by pumice rock which is overlain with a fine-grained slate-like rock, probably till.

Top #1 stratum: The fine-grained rock may be rock flour ground by passing glaciers. (Scratches on nearby rock faces could verify this conclusion.) You are now in the Ice Age or Pleistocene.

Top #2 stratum: The next clue is the pumice rock. Volcanic activity of the Tertiary blew extrusive volcanic ash over the site of your find. Tertiary Period.

Top #3 stratum: The limestone was precipitated from a non-moving water source. You are in the Cretaceous Period.

Since your find lies UNDER the limestone stratum, the prior geologic period seems the earliest logical time. This would make you gold stratum the Jurassic Period.

The strata lying at 45 degrees to the sky, indicates a severe tectonic disturbance. Did the two mile distance to the mating ridges result from a landmass drop? Did an upthrust separate them? All we really know on these facts is that a severe disturbance caused a crustal change.

We can now date our gold find to early Jurassic Period, with a later Jurassic upthrust or center drop. WHAT GOOD IS ALL THIS ANALYSIS?

First, you want to date the ridge two miles distant for the same strata layering. Will most of your gold-find lie beneath the limestone in the two mile corridor? If the corridor was NOT disturbance effected, then your gold claim is best made at the discovery location. Can you be sure? Not without core drilling and more strata investigation! **NOW WHAT?**

YOU NEED TO DO MUCH MORE SEARCH-WORK, BUT ALL WE ARE LOOKING AT HERE IS HOW GEOLOGIC PERIODS CAN BE KNOWLEDGE-FURTHERING CLUES. BUT ONLY IF YOU USE THEM !

CHAPTER 5

ENDANGERED SPECIES

FOSSIL MYTH OR FOSSIL REALITY?

THE BURGESS SHALE STORY

We appeal to the human compassion present in all of us to protect certain pristine areas from human interference. We even set up National Parks to control population misuse; to save nature for our children's, children future enjoyment. There is little general disagreement with that idea. National Parks are a national commitment, a national treasure. That same compassionate appeal commits us to "ecology". That is, in its political meaning, to "leave nature alone".... at whatever price.

The spotted owl, snail darter, gnat catcher are thought to have their extinction avoided by "protecting their habitat". That is, by eliminating human interference in a particular area occupied by the endangered species.

Marine habitats are politically endorsed and set up to vacate the offending public from beachlands, where the hidden agenda is really to privatize public lands for the benefit of beach frontage owners. Even boating and surfing in a marine habitat area is forbidden. This illustrates the power of emotional appeal an ecological argument swings in the voting public's mind. It also illustrates the political connections and willingness to spend freely to gain privacy. Needless to say, boating and surfing were never mentioned in pre-vote publicity which led to the political approval.

Spotted owl compassion has closed off vast woodland tracts to the detriment of logging jobs. Gnat Catcher compassion has stopped or altered residential development plans. The Desert Tortoise appeal has closed off potentially valuable mining resource areas. Wilderness set-aside areas have closed off much worthless desert land to recreational wheels and mineral resource exploitation.

Only a footfall is allowed, otherwise that environment will be harmed.

The Snail Darter appeal delayed beneficial improvements for years until it was shown that there existed other habitat areas. Too much was taken on faith, almost as a matter of religion. People want to believe that they are compassionate; to feel good about endorsing a "save the planet" argument.

Compared to the western U.S., where millions of acres are owned by the Federal Government, and are thereby "PUBLIC LANDS", consider the "Endangered Species Act" as it applies in the East to PRIVATE PROPERTY owners of small valuable forest tracts.

Timber supply in the East is effected by clean water and wetlands issues PLUS habitat protection pressures. Today, when private land is included in 'critical habitat", or when "land management guidelines" are imposed on PRIVATE LAND, the private property owner becomes responsible for species recovery.... with NO HEARING and NO COMPENSATION.

"Critical habitat designation" at once renders many lands close to valueless for commercial purposes.

The small landowner-forester has two options:

1. He can spend money for habitat conservation by reducing or ceasing the tree harvest on some or all of his land.

2. He can cut the trees before anyone discovers that the endangered species exists on his land.

THE ALABAMA RED HILLS SALAMANDER

This salamander is a few inches long, lives underground, and sneaks out only at night from holes the diameter of a 20 gauge shotshell.

One large forest products company solved the Red Hills Salamander problem by agreeing to a habitat conservation plan that avoided costly litigation. It privately agreed, (in the nature of a plea bargain) to protect the salamander by withdrawing from forestry operations, 4500 acres out of 500,000 privately owned Alabama acres.

Of course, Fish and Wildlife, environmental organizations, public review, the Interior Department, and the Corporation had to meet minds in the solution process. You can bet that this solution was costly to that forestry corporation. It had to lay out substantial sums over two years to pay for three sets of outside experts that were needed to combat and overcome objections.

This causes one to wonder about the value of the Red Hills Salamander in the scheme of nature. What next? The endangered insect species whose entire and sole habitat is confined to a single tree in your front yard?

Our culture subtly teaches compassion in schools, then deeply ingrains and reinforces the idea of delicate plant and animal communities in nature studies and public television save-the-planet documentaries. The usual theme promotes the idea that any interference, impact or imbalance upsets "nature's way"; that through millions of years of adaptation each species has its place and role in "nature's community". Even the idea of "community" itself, is taught to be an evolving organism, as if it had a life of its own. Any negative impact is destructive to the "community", and therefore harmful to plant and animal species within it. THIS IS TOTAL EXAGGERATION !

Yes, there are many interdependencies, such as a bee's need for the blossom, but this network of nature does not collapse, like a house of cards, if one or some impacts are added or removed.

IT MUST BE CARRIED IN MIND THAT EXTINCTION OF A SPECIES IS ULTIMATELY A BIOLOGICAL EVENT. IF A SPECIES IS FRAGILE AND ALWAYS VULNERABLE TO EXTINCTION, THEN THE MILDEST IMPACT CAN ERASE ITS EXISTENCE.

Fragility of species is always a problem of MINIMUM animal population, wherein a very small impact can cause erasure. The specific impact I have in mind is a low birthrate caused by a geographic area so large that the potential mates have trouble finding each other.

The conservation biologist grounds his claim for employment on the size of an area and the number of species it can sustain. This is called the SPECIES AREA EFFECT. This theory is actively promoted in spite of studies that conclude:

1. Extinction is impartial in choosing victims; RARITY proves to be the best index of vulnerability.

2. To double a geographic area, the number of species increases by 20%; cutting the geographic area by half, reduces the number of species by the same 20%.

BUT REGARDLESS OF AREA SIZE, AN ENDANGERED SPECIES (RARITY) MIGHT NOT SURVIVE IN EITHER A LARGER OR SMALLER GEOGRAPHIC AREA.

Still, the conservation biologist persists in the idea that species loss can be minimized by judicious control of the amount and distribution of habitat placed in reserves. By that they mean: Create artificial islands of haven to slow down rarity in species. YET THE SPECIES TO BE CONFINED IS ALREADY ENDANGERED, AND THUS ALREADY A RARITY. We already know this is an exercise in futility... BUT A JOB IS A JOB! ANY END JUSTIFIES THE MEANS TO PROTECT IT.

To a paleontologist, 1000 years is trivial; to a conservation biologist, survival for 100 years is immortality.

FORTUNATELY, MOST SPECIES ARE RESILIENT. THEY MUST BE IMPACTED BY SEVERE STRESSES AND CONDITIONS TO CAUSE EXTINCTION.

DOESN'T IT SEEM STRANGE THAT ON BALANCE, HABITAT PROTECTION CAN BE ABUSED TO THE POINT OF INTERFERING WITH HUMAN OBJECTIVES. ISN'T MAN A PART OF NATURE TOO?

IMPACTS

As a matter of compassion, no one wants to wipe out or erase a life species. But what if "IMPACTS", other than habitat, will erase a species anyway? Should we extend medical benefits to animals and plants, such as the condor or astrogalus, to thwart extinction? Even if we do, "NATURAL IMPACTS" will overcome these pseudo-extinction-delaying tactics. This is truly man's interference with nature and the natural order.

IS MAN AN ENDANGERED SPECIES?

If man is an endangered species, is one species more valuable than another? If so, which species? Little by little, over time, "impacts" will erase human life. Consider these "impacts" that threaten extinction of the human species:

INTERNAL IMPACTS

1. ABORTION - Early termination of a new life.

2. BIRTH CONTROL - Prevention of new life.

3. HOMOSEXUALITY - No conception possible.

4. AIDS - Lifestyle termination of young life.

5. CANCER - Early termination of mid-life.

6. DRUG ABUSE - Termination of life at all ages.

7. MURDERS - Selective termination of life.

8. GENETIC DEFECTS - Inborn defects that favor certain victims such as Tay-Sachs in eastern European Jewish victims; sickle cell anemia in Negroid victims.

9. POLLUTION OF THE ENVIRONMENT - To shorten all lifespans.

THESE ARE ONLY A FEW INTERNAL "IMPACTS" THAT ENDANGER HUMAN EXISTENCE.

THEN THERE ARE "EXTERNAL IMPACTS":

1. WARS

2. PESTILENCE

3. FAMINE

4. NATURAL DISASTERS

THAT ERASE HUMAN LIFE EN MASSE.

Does anyone favor "ghetto clearance" as "habitat protection"? It is, you know! Pollution control too!

FINAL IMPACT:

What if our oxygen atmosphere is altered to methane? Which human or other oxygen-breathing species will survive? You are right. None!

THE POINT IS: OVER TIME, "MULTIPLE IMPACTS" OPERATE TO TERMINATE A SPECIES OF LIFE TO THE POINT OF EXTINCTION.

Survival of the fittest is a circular argument. Fittest is defined in terms of survival; survival is defined in terms of fittest. The real-world idea is "adaptation to environment". For example: If man learned to breath methane instead of oxygen, the human race will survive in a methane environment.

PLANTS AND BUGS

The medusaweed, a spiky plant with barbs that lacerate the mouths of cattle, sheep and other grazers, is a hardy plant. It is slow to rot and so piles up debris to fuel wild fires; its seed is heat resistant. After a wild fire, the medusaweed expands its growing range at the expense of other plants. To date it has covered eleven western states.

Herbicides are in restricted use on public lands so medusaweed freely grows to fuel rangeland fires. But there is hope! Soil fungi in the medusaweed habitat keeps the weeds in check. Fungal enzymes destroy the plant's passageways to stop water from the roots and thereby food storage within the plant.

The California Poppy is considered a plant indicator for copper deposits in that it favors a copper-rich soil. Yet, it thrives in non-copper soils. Either it has adapted from copper to normal soils, or from normal to copper soils. It will survive.

The Astrogalus is an endangered plant that favors selenium-rich soil. This member of the loco-weed or rattleweed family is poisonous to man and grazing animals. Astrogalus and serpentine rocks are promoted as associates. Serpentine soils are a mineral associate of chromite, the main ore for chromium. If you can stop serpentine disturbances, you can stop chromium exploration and mining. But what if Astrogalus is susceptible to fungal enzyme destruction and has adapted to serpentine soils? What if selenium does not make a happy home for fungi? Has anyone tried to raise Astrogalus from seed in a hothouse environment, or would they rather keep that information in the dark? Is this plant worth saving? Is extinction the destiny of Astrogalus?

WOULD YOU BELIEVE THAT AN ENDANGERED INSECT IS NOW BEING PROMOTED FOR THE ENDANGERED LIST? IF THE MEDITERRANEAN FRUIT FLY OR MEDFLY BECOMES ENDANGERED, SHOULD THIS DESTRUCTIVE INSECT BE HABITAT PROTECTED?

ANIMALS

The cat family is particularly susceptible to F.I.T. virus. In domestic felines it is common and fatal in the very young under age one and in cats over eleven years old. F.I.T. virus is known to have wiped out entire colonies of cheetahs.

The Tasmanian devil, small with a badger-like fierceness, is endangered in part, because it attacks the tires of crushing 18 wheelers rolling along Tasmanian highways. Now confined to isolated Tasmanian areas, U.S. conservation biologists would seriously propose:

1. Make the entire island of 26000 square miles a protected habitat.
2. Ban all 18 wheelers from the roads.
3. Eliminate all roads in Tasmania.
4. All of the above.

This solution is deliberately exaggerated to illustrate a point: To some human victims of "habitat protection" the effect is the same, i.e. road closures in millions of Wilderness Area acres. "Leave nature alone... at whatever price".

THE BURGESS SHALE STORY

The theory of evolution, and thus endangerment of species, is depicted as a pine tree pictograph. Each change of branch extension is an adaptation to change. Each limb end is a species extinction with an overall 90% survival rate.

The Burgess Shale discovery upset that theory, and reversed the pictograph so that the pine tree is now roots in the air. It tells us that extinction of species is the rule, not the exception. Only ten percent of species survive, ninety percent will not.

Bear in mind that the thrust of this story involves EXTINCTION OVER A TIME FRAME OF MILLIONS OF YEARS. As a matter of proven life history 90% of ALL species will disappear, not merely the 10% that are thought endangered. The 10% that do continue existence are the adaptable species, 90% are not FULLY adaptable. In other words, through the operation of INTERNAL and EXTERNAL IMPACTS, a species will erase, as a matter of historical fact, whether currently considered as endangered or not.

Geologists divide the span of earth history into two periods called eons. The hidden life eon includes the earliest rocks up to 600 million years ago. The visible life eon picks up from that point to the present. The hidden life eon is also known as the Precambrian, when some single cell life first appeared.

600 million years ago an ancient sea formed a small limestone quarry in the Canadian Rockies. This rock quarry, found at the 8000 foot level, measured 10 feet high and covered a city block in area. It was called the Burgess Shale.

Discovered in 1909 by Charles Walcott, America's greatest paleontologist and head of the Smithsonian Institute, these Canadian fossils located in Yoho National Park, are precious because they preserve, in detail down to the gill of a Trilobite, or the components of the last meal in a worm's gut, the SOFT anatomy of organisms. Prior to the Burgess discovery, the fossil record was exclusively the story of fossil HARD parts.

THE BURGESS SHALE IS OUR ONLY WINDOW INTO THE HISTORY OF ANIMAL LIFE KNOWN AS THE CAMBRIAN EXPLOSION OF MULTI-CELLULAR LIFE. THE BURGESS TEACHES US THAT WITH FAR FEWER SPECIES, A DISPARITY IN DESIGN ANATOMY, EXCEEDS THE WORLD'S MODERN RANGE. Before I lose you with that last line ..... let me explain....

In the early nineteenth century geologists discovered that fossil species of SHORT DURATION provided the best means of arranging geologic events in a time sequence. By identifying fossils, a rock can be located with great precision in a time sequence. ORGANIC EVOLUTION IS CENTRAL TO ALMOST ALL ASPECTS OF PALEONTOLOGY, BIOLOGY AND ECOLOGY.

Since virtually ALL species found in rocks are extinct, few geologists and paleontologists have taken an active part in today's concern about endangered species and PREDICTION OF FUTURE EXTINCTIONS.

Biologists classify living organisms much as paleontologists classify fossils: That is, on the external appearance. Those benchmarks include anatomy, color, biochemistry, breeding systems, and sometimes behavior. A classifier of the organic world is known as a TAXONOMIST. His job is to recognize and distinguish natural species. A SPECIES IS A SPECIES IF A COMPETENT TAXONOMIST SAYS IT IS. Because the biological world is divided into natural units, it works. Geographical variants of a species are called SUBSPECIES, VARIETIES or RACES, indicating they could INTERBREED if they lived together. By separation over long, long time, a SUBSPECIES CAN BECOME A SPECIES. Therefore, the SPOTTED OWL, IF PHYSICALLY MOVED TO A NEW GEOGRAPHIC LOCATION, would interbreed and thus enhance survival and not suffer immediate endangerment.

BUT THE BURGESS SHALE STORY REVEALS A SERIOUS TAXONOMIC CLASSIFICATION ERROR. Dr. Walcott accommodated unclassifiable forms of fossil life within the standard system. The fact is, Walcott misinterpreted many of the peculiar fossils and misclassified them so as not to throw traditional evolution into confusion. WHERE IN THE STANDARD SYSTEM, CAN YOU CLASSIFY A SOFT-BODIED ANIMAL WITH 5 SEPARATE AND DISTINCT EYES ?

Forty years later H.B. Whittington reinterpreted fossils that had lain in Smithsonian drawers, protected from further inspection by Walcott and later, his surviving wife. The conclusion seems to be that Walcott may have been protecting his professional standing from possible embarrassment.

If you step on a cockroach or other soft-bodied organism like a caterpillar, it gets squashed flat. In a fossil rock it looks like a film showing only one side. Whittington realized that if you peel off the top layer of the fossil you could see the lower fossil section, such as hidden legs, fins, backbone and even gut contents. A fossil rock is known as a PART and COUNTER-PART when split to expose the fossil. Traditionally, the fossil in the rock face is the PART, the impression of the top view is the COUNTERPART.

The Burgess investigators called the top view the PART, and the impression under the fossil the COUNTERPART. This departure from normal investigative practice revealed features of external carapace, eyes, antennae of soft bodies on the counterpart, with legs and internal anatomy on the part.

With this fossil investigative method-change, it was proven that serious fossil classification errors were earlier made by other investigators; that a new interpretation of life's history, as to extinction of species, was mandated. The misconception that soft-bodied fossils are usually preserved as flat films of carbon on the surface of rocks was discarded; THAT UNKNOWN LIFE FORMS EXISTED IN THE CAMBRIAN THAT HAVE NO CURRENT LIVING RELATIVES.

In 1986 Simon Conway (Whittington's graduate student) published an ecological analysis of the entire Burgess fossil findings. With about 74,000 specimens on 34,000 slabs, (90% reside in Walcott's Washington Smithsonian collection) 88% are animals, of which 86% are soft-bodied.... the others were algae.

Conway's conclusion as to the Burgess Cambrian animals: A relatively few species survived at the expense of the extinction of many other groups. Over 600 million years, IMPACTS DECIMATED AND ERASED 20 BURGESS SPECIES OR GROUPS, WITH ONLY 4 SPECIES SURVIVING TO THE PRESENT.

It was always known that the phenomenon of decimation and extinction applied to hard-bodied groups of animals, such as dinosaurs in later geological times. The Burgess story proves that extinction existed during the Cambrian, the earliest known multi-cellular life period.

THE REALITY OF LIFE HISTORY SHOWS US THAT MANY SPECIES PROLIFERATE, THEN OVER TIME, DECIMATE TO EXTINCTION. THAT'S NATURE'S WAY.

It can be reasonably argued that habitat protection ALONE will NOT save and preserve, or alone delay, the extinction of an endangered species that is destined for decimation as a rule of nature.

RESTATED: An endangered species is destined for decimation and extinction....AND NOTHING CAN STOP THAT EVENT. IT IS A MYTH TO BELIEVE OTHERWISE. It is sad, but true. It is as certain as death at the end of an individual's life span.

FOR ADDITIONAL READINGS SEE:

1. WONDERFUL LIFE, S. J. GOULD, NORTON, 1989

2. EXTINCTION, BAD GENES OR BAD LUCK?, D. M. RAUP, NORTON, 1991

CHAPTER 6

TECTONICS — BELIEVE IT OR NOT

Generally, *tectonics* refers to the forces or conditions within the earth that cause deformed crustal surface structures such as continents, mountain building, folds and faults. Another name for the same concept is *diastrophism.*

Can you believe? The summit of Mt. Everest is marine limestone? Limestone usually forms in an ocean or sea environment. Logic tells us that the rocks on the summit of Everest once rested on the bottom of a sea. 200 million years ago, the landmass that we know as the Rocky Mountains was part of a continental shelf. The continental divide was once the Western shoreline. The Appalachian Mountains were formed and folded by multiple plate collisions with the Eastern continental shelf. The Sierra Nevada and Appalachian Mountains were once 10,000 feet higher in elevation than they stand today.

Can you visualize? The earth with a single overall crust that broke into five major pieces called *plates.* Through seafloor spreading, these plates move away from and into each other in push and shove collisions. When a plate collides with a continental mass, the continent is imperceptibly moved and raised, as the plate digs into the continental trench. This subduction raises the continental shelf to form even higher land masses that eventually will become a coastal mountain range. This movement is possible because the earth's crust floats on a heated taffy-like rock mantle that is kept soft by the molten core of the earth.

On the East side of North America the plate crashed into the continental plate to raise the Appalachian Mountains. Continued plate pressure, over time, bent the strata layers into roller-coaster undulations called *folds.* The law of superposition was upset. The oldest rock strata that should rest on the bottom were reversed. They now rested atop the later deposited, thereby younger strata, to create an *overthrust belt* over 1000 miles long. That belt runs from the Carolinas to Newfoundland.

On the West side of North America, an island arc array collided with the continent. Some blocks slid onto the continent, others slipped North along the San Andreas strike-slip fault. In time, the North-bound blocks welded themselves onto the continent to form the Eastern Oregon to Alaska landmass.

Rocks of the old continental shelf were compressed by multiple collisions into an accordion fold. When the folds became overstressed, they broke into fault blocks and piled over each other. Where plates pinch, crash, grind, separate and rotate into one another, grand scale geologic activity such as volcanism, earthquakes and mountain building operate to deform the crustal surface. For example, a plate, subducting under a continental shelf, could crack the crust to produce a faulted structure. The cracked crust then provided avenues, passages and channelways for molten rock to rise and exit the surface, or to form underground structures such as sills and dikes filled with metal and mineral-rich igneous rock mixtures.

This brings us to the thrust of this chapter.

The penetration of the earth's crust by hot molten magma rising through fault-formed passageways forms the basis of igneous rocks that rise *extrusively* and eruptively to the surface, and of those *intrusive* igneous rocks that stay deep within the earths crust. These magma melts carry within their structures the minerals that will form specific ore deposits that will be associated with a given rock type. The hot rising vapors and thermal waters carry dissolved metals, and other mineral streams, along with igneous rocks, for deposit in veins, sills and dikes. Thermal waters can effectively deposit their dissolved ores in country rock that lays miles from the originating magmatic source.

The deep seated molten intrusive rocks can take a thousand years to cool in their insulating blanket of country rocks.

Where the cooling magma touches the country rock, mineral transfers, and alterations occur to form new distinct mineral forms in rocks that once were igneous, sedimentary or even other metamorphic rocks. For miles, over the region of heat transfer, country rocks undergo internal change. This roasting, baking and pressure cooking of rocks forms the basis of *metamorphic* rock divisions. Metamorphic rocks can become plastically deformed just short of melting; can soak up mineral and vapor streams to recombine their internals into economic ore deposits and unusual mineral compositions.

ROCKPECKER TIP # 14 – From specific mineral distributions in a metamorphosed region, it is possible to determine the general direction and location of the intrusive that caused those changes and alterations.

When mountains erode to become mere stumps of their former grandeur, the general process is called weathering. Along with stream movements they transport and separate rock products into differing densities. Where these sediments deposit, the more resistant metals fall to rest in favorable sites to form economic sedimentary ores. Over time, compaction and cementation will form the stratified layers into sedimentary rocks that act as vaults for their valuable, disseminated, mineral inclusions.

ROCKPECKER TIP # 15 – It is said that sedimentary structures hide more economic minerals in disseminated form than have been found since the beginning of mining.

****

In CHAPTER 24 we will examine sedimentary processes and mineral depositions that begin with mechanical weathering, which simply makes smaller pieces out of boulders, without changing the chemical content of the individual pieces.

CHAPTER 7

STREAMS AND LANDSCAPES

THIRTYFIVE QUESTIONS

The two most important agents of erosion are running water aided by mass wasting. Nearly every landscape shows the results of stream erosion and sediment deposition. Landforms such as valleys, terraces, deltas, alluvial fans are all children of stream action.

If bread is the staff of life, water is its lifeblood. Thanks to the *water cycle*, water evaporating from land and sea, condenses to form clouds, only to fall again as rain or snow. Water falling on land runs off over the surface as a stream, or percolates into the ground to become ground water. Evaporation and transpiration return it to the atmosphere to renew the cycle.

Mountains are changed by water action. Streams and rivers wear down rocks to transport the particles. Stream erosion carves its valley-signature into the landmass.

ROCKPECKER TIP # 16 - For the Rockpecker this water action opens the crust to visual inspection; for the goldpanner, it deposits its metal load for easy gleaning.

1. WHAT IS THE DEFINITION OF A STREAM ?

A stream is a body of channelized water that moves downstream under the influence of gravity.

2. WHAT IS SHEETWASH?

Sheetwash is a thin layer of UNCHANNELED water flowing downhill.

3. WHAT IS SHEET EROSION?

It is the intermediate process between mass wasting and stream erosion.

5. WHAT WORK IS DONE BY A STREAM ?

1. Streams erode the landscape.
2. Transport the sediments downstream.
3. Deposit the sediments in favorable deposition locations.

6. WHAT IS A DRAINAGE BASIN?

The total area drained by a steam and its tributaries. (Like water flowing down the inside of a cup.)

7. WHAT IS A DRAINAGE DIVIDE?

It is a high ground ridge or strip dividing a drainage basin. (Like the lip of the cup forcing the water to flow either down the inside or outside of the cup.)

9. WHAT FACTORS AFFECT STREAM EROSION AND DEPOSITION?

1. Velocity of water in streambed - the faster the flow, the greater the erosion.
2. Stream gradient - the steeper the streambed slope, the greater the erosion.
3. Channel shape and roughness - the more youthful a stream, the greater the erosion. - The rougher the streambed, the slower the erosion...because it impedes velocity.
4. Volume of water discharge over a given point - the greater the volume, the greater the erosion.

10. WHAT ARE THE METHODS OF STREAM EROSION?

1. Hydraulic action - the force of water digging at the stream course.
2. Solution - chemical weathering dissolves minerals to weaken the materials.
3. Abrasion - grinding and rubbing of streamcourse fragments. (THIS IS THE MOST EFFECTIVE METHOD)
4. Grinding depressions and potholes in bedrock.

11. WHAT ARE THE METHODS OF STREAM TRANSPORTATION?

1. SOLUTION- dissolved minerals as an invisible water load.
2. SUSPENSION - light sediment is actually visible and remains above bed indefinitely.
3. SALTATION - lifting, leaping, hopping and bouncing along the streambed.
4. TRACTION - a rolling and sliding along in contact with the streambed.

12. WHAT IS STREAM CAPACITY?

It is the total load of sediment the stream can carry. The more water in the stream - the more sediment it can carry.

13. WHAT IS STREAM COMPETENCE ?

It is the measurement of the largest particle the stream can carry. The faster the velocity, the greater the size of particle it can carry.

14. WHAT ARE THE FORMS OF STREAM DEPOSITION?

1. SORTING OF PARTICLES - Heavier particles drop earlier than the lighter ones.
2. BAR FORMATION - a temporary ridge of sediment along the middle or at the banks of the stream, or as flood plain deposits.
3. DELTA FAN FORMATION - A permanent deposit at the mouth of a river, lake or sea.
4. ALLUVIAL FAN FORMATION - sediment deposited in a fan shape at the end of the stream usually within the continent.

15. WHAT IS A BRAIDED STREAM ?

It is a series of mid-channel bars in a wide shallow channel that widen the stream through water diversion along the banks.

16. WHAT IS A MEANDER ?

It is a formation of a streambed, usually deep and narrow, carrying sand and silt in suspension that eats away the outside curve of a streambed to form a snake-like steam appearance. (The eating action is called *lateral corrasion.*)

17. WHAT IS A POINT BAR?

It is a sand bar deposited on the inside of a curve or bend due to the decreased velocity of the stream that aids deposition.

18. WHAT IS A MEANDER CUT-OFF?

Where the stream takes a shortcut across the neck of a meander in the form of a shorter channel.

19. WHAT IS AN OXBOW LAKE ?

It is the old position of a river meander that has been cut off to form a lake or pond.

20. WHAT IS A DELTA?

It is a stream flowing into standing water such as a lake or sea.

21. WHAT IS A DISTRIBUTARY?

They are small, shifting channels that carry water away from the main river channel to distribute it over the surface of the delta. The surface of most deltas is marked by distributaries.

22. WHAT TYPE OF DEPOSIT COMPRISE THE INTERNAL CONSTRUCTION OF A SMALL DELTA?

Small deltas, particularly freshwater lake deltas are built up of 3 types of deposits called foreset beds, topset beds, & bottomset beds. Foreset beds are the main body of the delta usually deposited at an angle. Topset beds are laid horizontal (as strata) on top of the foreset beds. They deposit as varying grain sizes. Bottomset beds are in front of and as a leading edge of the foreset beds. They comprise the finest silt & clay.

23. WHAT IS AN ALLUVIAL FAN?

It is a large fan or cone shaped pile of sediment where a stream emerges from a narrow to wider condition. The sudden decrease in velocity causes load to drop. Oftentimes, found in desert areas and within continents.

24. WHAT IS THE MOST COMMON LANDFORM FOUND ON THE EARTH'S SURFACE?

Valleys are the most common.

25. HOW ARE VALLEYS FORMED?

They are usually cut by streams through a process of down-cutting to base-level. Downcutting is the process of deepening a valley through erosion of the streambed. The lowest level that a stream can cut is known as the base level, which is sea-level, or a lake or reservoir. Base level can also be the lowest point in a valley....if it is below sea level.

26. WHAT IS A GRADED STREAM.

In a side view, it has a smooth longitudinal profile. This smoothness is maintained by the cutting and filling in of streambed irregularities. It is said to have a delicate balance between transport capacity and sediment load.

27. WHAT IS THE ENERGY OF A STREAM USED FOR?

Energy is used for 2 things:
1. Transporting sediment. 2 Overcoming flow resistance.

28. WHAT IS LATERAL EROSION AND HEADWARD EROSION?

Lateral erosion undercuts the stream walls side to side as the stream swings its course. Headwater erosion is a slow UPHILL growing of a valley through gullying, mass wasting and sheet erosion.

29. IN REGIONAL EROSION, WHAT IS PARALLEL RETREAT ?

Parallel retreat is the theory that angle of slope is constant. Rather than becoming gentler, it laterally retreats along a region. This is opposed to erosion eventually producing a level plain. (The land becomes flat, but the slopes remain constant over time.)

30. WHAT FACTORS CONTROL SLOPE ANGLES?

1. Rock type - Solid rocks can hold a vertical cliff; talus holds a 45 degree angle; sand dunes pile up at 34 degree angles.
2. Climate - A wet humid climate produces rounded hills; an arid climate, due to low precipitation, produces angular topography.
3. Rock structure - folds offer varying resistance to erosion; faults or faulting can produce unequally sloped landscapes.

31. WHAT IS A DRAINAGE PATTERN?

In a map view, the arrangement of a river and its tributaries is a drainage pattern. It can reveal the underlying nature and structure of rocks.

32. DESCRIBE THESE DRAINAGE PATTERNS WHICH ARE CLUES TO THE TYPE OF ROCKS UNDERLYING THE DRAINAGE SYSTEM:

1- DENDRITIC. 2- RADIAL. 3- RECTANGULAR. 4- TRELLIS. 5- CENTRIPEDAL. 6- BRAIDED. 7- DERANGED. 8- INTERMITTENT. 9- YAZOO. 10- SUPERIMPOSED. 11- ANTECEDENT. 12- INCISED MEANDER.

1. Dendritic - resembles tree branches - develops on uniformly erodible rock such as layered sedimentary rock or unfractured igneous rock.

2. Radial - like outward wheel spokes - develop on high conical mountains, domes and composite volcanoes.

3. Rectangular - tributaries join main stream with frequent 90 degree angles - develops on uniformly fractured rock due to faulting cracks. - fracture network forms more easily erodible stream pathways.

4. Trellis - short tributaries join main stream at 90 degree angles - tilted layers due to folding as in Appalachian Mountains. Topography of parallel ridges and valleys.

5. Centripedal - streams converge at a central point in a depression - limestone sinkholes. (reverse of Radial)

6. Braided - numerous mid-channel bars split stream in multiple directions - usually as alluvial fans in a desert.

7. Deranged - Glacier-caused depressions formed lowland lakes to capture drainage.

8. Intermittent - No continuous flow upon the surface. - limestone regions, karst topography.

9. Yazoo - flows between bluffs and levee. Enters mainstream through break in levee.

10. Superimposed - Stream eats sediments overlying a buried range; cuts through the sediments into and through the buried range. - Flow slope increases by the cutting and so does the energy of the stream.

11. Antecedent - A pre-existing river slowly being up-folded cuts through the rising fold.

12. Incised meanders - shallow, old age, wide contoured stream forms a meandering valley as its original base level is regionally uplifted. (A case of land slowly rising and the stream slowly cutting.)

33. WHAT ARE STREAM TERRACES?

Step-like landforms found lying higher that the streams current flood plain.

34. WHAT ARE PAIRED TERRACES?

A thick section of flood plain deposits built up on each side of the river and lying at the same elevation.

35. WHAT ARE UNPAIRED TERRACES?

Terraces with unequal elevation caused by stream shifting back and forth across a valley with downcutting and lateral erosion operating at the same time.

ROCKPECKER TIP #17 - WHERE YOU CAN RECOGNIZE PRIOR CUTTING ACTION OF A STREAM, YOU CAN ALSO VISUALIZE EARLIER ALLUVIAL MINERAL CONCENTRATION POINTS.

CHAPTER 8

MASS WASTING or MASS MOVEMENT EIGHT QUESTIONS

To this point we have considered the slow, weathering, erosive breakdown of rocks, with transport of their soluble or insoluble minerals and subsequent deposition. What about the bulk movement of rock materials? Soil and rock moving downslope in bulk, at the earth's surface, under the pull of gravity, is known as mass wasting. Landslides are the best known type of bulk movement.

ROCKPECKER TIP # 18 - Mass wasting can be helpful to mineral discovery. It can also present a safety risk. For example: In your travels you pick up an interesting, dark, heavy rock. When it is an isolated specimen, it is called *float.* Some questions to consider suggest a fall from up-slope, perhaps as the lowest piece of a mass movement. Did it arrive at this location because of earlier sheetwash or flooding? If a downslope bulk movement of rocks, did it bare or erase a valuable mineral deposit? If you start digging at the base of the slope, will factors involving rock instability, bury you? Let's examine some aspects of mass wasting.

1. *WHAT IS THE DEFINITION OF MASS WASTING?*

Mass wasting is a collective term to indicate various kinds of *down-slope movements* occurring under the *PULL OF GRAVITY.*

2. *WHAT FACTORS CAUSE MASS WASTING?*

GRAVITY - SLOPE - WATER AS WEIGHT AND LUBRICANT - AIR TO EASE FRICTION BETWEEN THE LOOSE PARTICLES.

3. *WHAT IS THE IMPORTANCE OF MASS WASTING?*

It is an erosional process that creates a new earth surface. It differs from erosion in that it is a *bulk* movement. Mass wasting causes more total damage than earthquakes and volcanic eruptions together.

4. *WHAT FACTORS CONTROL OR ENCOURAGE MASS WASTING?*

a. SLOPE as to whether steep or gentle.

b. RELIEF, that is the terrain whether high or low, but mainly the difference in elevation between the highest and lowest. High relief is more likely to encourage mass movement.

c. WATER & ICE - The flow of water-saturated material over an impermeable surface. This is known as *SOLIFLUCTION.* This is a freeze-thaw process where ice or water saturated soil particles glide over a surface that cannot pass excess water such as impenetrable bed rock or permafrost.

d. The AMOUNT of LOOSE rock and debris AVAILABLE for movement.

e. The LACK of VEGETATIVE COVER.

f. EARTHQUAKE PRONE AREAS to act a TRIGGERS.

g. ORIENTATION OF BEDDING PLANES - foliation of schist or gneiss strata. (How the rock strata lies)

h. CLIMATIC CONDITIONS: 1. Unusual rainfall amounts.
2. Nature of occurrence - Heavy rains after drought conditions.

i. RUMBLING MOVEMENTS OF TRUCKS, TRAINS OR EXPLOSION VIBRATIONS.

j. SONIC BOOM VIBRATIONS.

k. OPERATIONS INVOLVING EXCAVATIONS OR BLASTING VIBRATIONS.

l. ANIMAL OR PLANT ROOT UNDERMINING SURFACE STABILITY.

5. *WHAT ARE THE BASIC FACTORS USED IN THE CLASSIFICATION OF MASS WASTING?*

1. RATE OF MOVEMENT
2. TYPES OF MATERIALS (mud, shale, etc.)
3. NATURE OF MOVEMENT WHETHER FLOW, SLIP OR FALL.
   - A. FLOW is a slow movement
   - B. SLIP - a. SLIDE - *movement along the slope.*
     b. SLUMP - *vertical movement as a drop and move.*
   - C. FALL - *a fast movement.*

6A. *WHAT ARE COMMON TYPES OF SLOW MASS WASTING?*

1. SOIL CREEP - only observable in relation to other objects such as a fence which will appear wavy or leaning.

2. EARTHFLOW - AS IN SOLIFLUCTION AND PERMAFROST.

6B. *WHAT ARE COMMON TYPES OF FAST OR RAPID MASS WASTING?*

1. MUDFLOW

2. LANDSLIDE

3. ROCKFALL

4. ROCKSLIDE

5. AVALANCHE

7. *WHAT ARE THE WARNING SIGNS OR PREDICTIONS OF FORTHCOMING RAPID MASS MOVEMENTS?*

1. Behavior of animals or insects.
2. Rock fissure widening.
3. Coincident with particular patterns of rainfall.
4. When mules were used in the coal mines, they sensed and telegraphed an impending roof-fall with an obstinate refusal to walk

under that roof section.
5. In those same coal mines, when the rats scurried to leave, it was time for the miners to get out. A cave-in was coming.

8. WHAT PREVENTION STEPS CAN BE TAKEN TO MINIMIZE MASS MOVEMENT?

A. On slopes, install vegetative cover.
B. Install water drains.
C. Build retaining walls.
D. Stitch a rock slope along planes of weakness.

CHAPTER 9

EARTHQUAKES — 16 QUESTIONS

It is hard to realize that the crust of the earth is constantly changing. Gigantic forces reshape the rocks of the crust into enormous twisted folds and long cracks. While there can be other causes, those of most interest to prospectors are tectonic and volcanic. The tectonic earthquake prepares the ground, in the sense of creating fissures and other deformities in the crust, favorable for LATER ore-bearing depositions.

Volcanic earthquakes are due to explosions or to fractures occurring within the structure of the volcano. It is believed that the warning rumble, prior to eruption, is due to the magma moving up through the crust as molten rock. Of course, millions of years later, you and I are scratching the crust for clues that may lead us to the fruits of this activity. We call it ore!

1. WHAT CAUSES AN EARTHQUAKE?

1. ELASTIC REBOUND THEORY - Rocks stressed and strained beyond their elastic limits, snap-break like a rubber band to suddenly release stored rock energy as seismic waves, which causes movement along a fault and the trembling and shaking of the surface.

2. Plate interaction between plate boundaries either transform, divergent or convergent movements.

3. Explosive volcanic surface eruptions.

4. Intrusion of underground magma into overlying crustal areas.

2. WHAT ARE THE TWO WAYS TO MEASURE AN EARTHQUAKE?

1. Intensity - by using the Marcalli scale with a range of I to XII to measure the destruction caused by the quake. The theory of damage rests on the fact the damage is greatest at the epicenter of the quake and diminishes in intensity as the distance from the epicenter increases.

2. Magnitude - by using the Richter scale with a range of 1 to 9 to measure released faulting energy. Each number on the scale in 10 times stronger than the one below it. (a 4 is 100 times stronger than a 2)

3. WHAT ARE THE EFFECTS OF AN EARTHQUAKE?

1. Ground motion causes buildings to vibrate. Intensity greater on soft sediments than solid rock.

2. Fire - caused by broken, ignited gas lines in cities.

3. Landslides as mass wasting.

4. Permanent displacement of or changes in landforms.

5. Minor aftershocks.

6. Tsunami creation - seismic sea waves.

7. Flooding of coastal areas.

8. Liquification as foundation failures.

4. WHAT ARE THE DISTRIBUTION OF EARTHQUAKES?

1. Circum-Pacific belt = 80%.

2. Mediterranean Himalayan belt = 15%

3. Plate boundaries & isolated areas = 5%

5. WHAT ARE THE PRECURSORS, PREDICTABILITIES OR WARNINGS OF EARTHQUAKES?

1. Animal behavior - Yaks; snakes leave holes; bee activity.

2. Palmdale bulge - as bulging land.

3. Cluster theory - earthquakes occur in multiples....light leading to heavy then light aftershocks.

5. Seismic gaps - Sections along a known faultline that have been quiet for a long time, but both ends have recorded large recent earthquakes. Probability says such a gap is ripe for activity.

6. HOW IS A FAULT CAUSED?

When a rock breaks, one section slides past the other broken portion, usually in a horizontal or vertical movement. The crack between TWO rock masses is a fault.

7. WHY DO EARTHQUAKES TAKE PLACE?

The elastic rebound theory involves the sudden release of accumulated strain in rocks causing movement along a fault or rock break. The strain release energy is expended in 2 ways:

1. By rock mass movement.

2. By creating seismic waves.
The brittle behavior of rocks is a surface characteristic. At depth, temperature and pressure effects on rocks reduces brittleness and increases ductility. Earthquakes during explosive volcanic eruptions or during magma chamber filling seem to be exceptions to usual associated fault movement.

8. WHAT IS THE FOCUS OF AN E.Q. ?

It is the point of seismic wave origination.

9. WHAT IS THE EPICENTER OF AN E.Q. ?

That point on the earth's surface directly above the focus is the epicenter.

10. WHAT ARE THE TWO TYPES OF WAVES RADIATING OUTWARD FROM THE FOCUS?

1. Body waves - travel outward from the focus through the earth's interior.

2. Surface waves - travel the earth's surface outward from the epicenter.

11. WHAT ARE THE TWO MAIN TYPES OF BODY WAVES ?

1. 'P' waves = Primary, Particle motion, Push wave, Parallel - travels through solids and liquids.

2. 'S' waves = Secondary, Slow, Shake at right angles to the wave with Shearing motion. - travel through solids but not liquids.

SURFACE WAVES CAUSE MORE GROUND MOVEMENT AND LAST LONGER IN PASSING.

12. WHAT IS THE FUNCTION OF A SEISMOMETER?

It acts as a point of reference in determining the amount of ground motion. When a recording device called a seismograph is added, it produces a paper record of earth motion called a seismogram.

13. HOW DOES THE PAPER RECORD DISTINGUISH P, S, & SURFACE WAVES.?

The 3 types of waves travel at different speeds. First to arrive is the P wave, then the S wave and lastly the surface wave.

14. HOW IS THE FOCUS/EPICENTER LOCATION DETERMINED?

The time interval between the arrival of the first P wave and the first S wave is plotted on a travel time curve. This determines the distance, but NOT the direction of the focus. When 3 stations are employed to find the distance, they can draw 3 circles, one from each station as the center, and the distance as the radius. Where the circles intersect, that is the focus/epicenter of the quake.

15. WHAT IS THE BENIOFF ZONE?

Distinct earthquake zones that begin at the oceanic trench, slope landward and downward 30 to 60 degrees into the earth. Island arcs and andesitic volcanoes are associated with the Benioff Zone.

16. HOW ARE EARTHQUAKES AND PLATE TECTONIC CONCEPTS RELATED?

Earthquakes occur at the plate boundaries; plate edges are identified and defined by them. Plate tectonics suggests that interaction of plates causes earthquakes. HOW???

1. The plate moves from spreading center:

2. Toward the subduction zone at the ocean trench:

3. Plate plunges through the trench into the mantle:

a. Diverging boundaries = plates move away from each other

b. Transform boundaries = plates slide past each other.

c. Converging boundaries = plates move toward each other to collide or subduct when two continents override. A thick crust and a mountain is created at the suture zone.

Stresses and strains within the earth's crust are a long time building up. Any additional force can cause a rupture in the rock structure to cause movement along a fault. It is the up and down, back and forth movement along the fault, that fills the fault with breccia and finely divided rock powder we know as gouge. Years from now, as the magma from Mt. St. Helens cools, we may have a new storehouse of valuable mineral deposits.

CHAPTER 10

GEOSTRUCTURES — SEVEN QUESTIONS

GEOLOGIC STRUCTURES - THE CREATION OF FOLDS AND FAULTS.

This chapter deals with bed rock changes caused by tectonic forces deep within the earth, and how the rocks respond to these changes. Rocks in the earth's crust respond to stress and strain. Stress forces a change in size or shape; strain is the adjustment of the rock unit to the size and shape caused by stress.

DIRECTION OF FORCE IS SHOWN AS: >>>>>>>> <<<<<<<<<<<<<.

OR AS: >>>>>>>>>>>>>>
<<<<<<<<<<<<<<.

OR AS: <<<<<<<<< >>>>>>>>>.

OR AS: >>>>>>> <<<<<<<<<< TO SHOW INTENSITY PATH.

1. WHAT ARE THE 3 KINDS OF STRESS?

1. Compressive stress - it is a vice-like force that tends to squeeze and shorten the rock body. >>>>> <<<<<.

2. Tensional stress - the force that tends to stretch, pull apart or elongate the rock body. <<<<< >>>>>.

3. Shear stress - is a parallel force acting in opposite directions resulting in horizontal movement. >>>>>>>>>>
<<<<<<<<<<

SHEAR IS A TOUCHING---SLIDING---PASSING FORCE---MUCH LIKE SCISSOR ACTION.

2. WHAT ARE THE 3 TYPES OF STRAIN?

1. Plastic strain – the stress deforms the rock which thereafter retains the deformed shape. (ductility)

2. Elastic strain – the stress-deformed rock returns to the original shape after the stress is released. (a spring)

3. Fracture stress – rather than yield to the stress, the rock snaps and breaks after the elastic limit is reached. (brittle rocks)

3. HOW DO SEDIMENTARY ROCKS BECOME BENT?

Slow stress under confining pressure deform rock layers into bends. (folds)

When the stress exceeds the elastic limit, the rock unit breaks to create a fracture. (fault)

4. WHAT IS STRIKE AND ANGLE OF DIP?

Tilted sedimentary layers are measured by strike, as the line of layers compared to the horizontal. It is a compass direction. Angle of dip is the downward angle of the tilt measured from the horizontal plane downward to the bedding plane. It is measured in degrees from the horizontal. When measured in a compass direction, the angle of dip is the direction of dip. DIASTROPHISM IS A TERM FOR ALL TECTONIC ROCK MOVEMENTS THAT INVOLVE FOLDING, FAULTING OR WARPING.

5. WHAT IS ANTICLINE, SYNCLINE AND HINGE LINE?

Because a fold is an arc it has an upward and downward leg or limb. Anticline is the upward leg; syncline is the downward leg. The hingeline is the crest-point where the folds are bent. On an anticline the hinge is on top. On a syncline, the hinge is on the bottom. (Remember "sink" for the down leg.)

6. DESCRIBE FIVE KINDS OF FOLDS?

1. Symmetrical - equal compressive force forms limbs at the same angle. (like a "U") ››› ‹‹‹

2. Assymetrical - unequal pressure with unequal folds. (more like a "V") ››››› ‹‹‹

3. Isoclinal - intense compressive stress, legs are parallel to one another. (like a hairpin) ››››››› ‹‹‹‹‹‹‹

4. Overturned - Limbs dip in same direction. Compressive force overrides upper limbs in same direction. (like a leaning hairpin)

››››››››››

‹‹‹‹‹‹‹‹‹‹

5. Recumbent - same pressure points as overturned folds, but more intense. Limbs nearly horizontal. ("S" shape)

›››››››››››››››

‹‹‹‹‹‹‹‹‹‹‹‹‹‹‹

6. WHEN A FOLD ERODES, WHAT GEOLOGIC STRUCTURES ARE FORMED?

The anticline forms a valley; the syncline forms a low hill. This is a case of the top of the crest becoming the valley; the bottom of the crest turns into a low hill. (Due to erosion.)

7. WHEN A ROCK FRACTURES, IT PRODUCES A FAULT. DESCRIBE THE TYPES OF FAULTS.

TYPES OF FAULT:

1. Normal fault - tensional force pulls apart rock. ‹‹‹ v ›››. Hanging wall moves downward.

2. Reverse fault - Horizontal compressive stress ››››››› ^ ‹‹‹‹‹‹‹. Hanging wall squeezed up.

3. Strike-slip fault - Opposite wall movements as a horizontal sliding due to shear stress. (San Andreas) ›››››››

‹‹‹‹‹‹‹

4. Double fault line - A normal fault with two fractures with the

center section dropping. (Graben) <<<<<<< v >>>>>>>

– same as a graben, but both sides drop with center section rising. (Horst) <<<<<<< ^ >>>>>>>.

Structural changes imposed on rocks are the pathways of mineral streams. The most common are faults and folds, followed closely by breccia zones and pipes.

Many ore deposits are directly related to faults, which in their movement, produce breccia in shear zones, as the surfaces grind in passing. Clean, coarse breccia with a minimum of gouge, (a fine-grained rock material that can plug the fault) is most favorable for ore deposition.

When a fault occurs, the rising of one face, or the lowering of the opposite face can be visually identified by the strata layers. The line of strata at the bottom is a footwall. The same strata line on the opposing face is the hanging-wall.

PIPES AND CHIMNEYS – A pipe can descend like a chimney in a steep downward nearly vertical angle. You should be aware that a horizontal pipe is not unusual. A pipe filled with angular, broken rubble is known as a *breccia pipe.*

Folds are umbrella-like shaped landforms. They are roller coaster strata-lines that have bent either as an umbrella shedding rain, or one acting as a rain catcher.

The uplift stress on the bent, folded rocks cause cracks on the upward arc (anticline), and on the downward arc (syncline). This tectonic folding before a mineral stream enters, allows ore deposits to form in and around the stress cracks on BOTH anticline and syncline sides. The fractured rocks permit a permeability increase for easy fluid deposits to settle in the rocks themselves. In this case the resultant ore bodies are arched one above the other and are termed saddles. But why not? If you suppose ten separate strata layers, each bent layer received the same upward stress, the only variation would be in the flexibility of a particular strata. Theoretically, this fold should contain ore in ten strata layers.

ROCKPECKER TIP # 19 – Carbonate rocks underlying shale are favorable places to find ore where faulting has taken place near a shale contact. Folding and faulting of the carbonate rocks, with bedding slip in the more flexible shales, allows mineral streams to concentrate **BELOW** the shale.

CHAPTER 11

GROUND WATER — 21 QUESTIONS

HOW DOES WATER IN THE GROUND RELATE TO MINERAL PROSPECTING?

Ground water or *subsurface water* feeds springs, brooks, streams, wells, and rivers. It remains trapped in sedimentary rocks, and also originates deep in the earth as steam from molten rock materials. The water table plays a vital role in supergene enrichment precipitation. Streams transport alluvial gold. Every *valley* landform was made by a stream. Water acts as the transporting agent for dissolved minerals that later become the ore deposit you are now seeking.

1. WHAT IS GROUND WATER?

It is water that lies beneath the surface that collects in cracks, crevices, pore spaces between grains in clastic rocks.

2. WHAT IS THE SOURCE OF GROUND WATER?

Rain or snow that falls onto the land which soaks into the ground. How much soaks in depends on 1. climate. 2. land-slope. 3. soil. 4. rock type. 5 vegetation.

3. WHAT IS POROSITY AND PERMEABILITY?

Porosity is that percentage of a rock volume that consists of openings that can hold water.

High porosity rock is: sandstone, conglomerate, limestone

Low porosity rock is: granite, schist, and crystalline rocks.

Permeability is the relative ease of fluid flow through the rock openings.

4. WHAT IS THE WATER TABLE?

It is the upper surface of the saturated zone where all the rock openings are filled with water.

5. WHAT IS THE UNSATURATED ZONE?

It is the portion above the saturated zone where rock openings are partly filled with air & water.

6. WHAT IS THE CAPILLARY FRINGE?

It is the transitional zone above the saturated zone and below the unsaturated zone where capillary action causes water to rise.

7. WHAT IS A PERCHED WATER TABLE?

It is a body of ground water above and separated from the main water table, collected in a lens of shale within a permeable rock. The upper water basin is called a perched water table.

ROCKPECKER TIP # 20 - You should be aware of a top-lying water table as an isolated cup-like depression, that can hold lake-size water in the crust. Formed near the surface, this perched water table may be supergene enriched by precipitation. However, where the lower water table lies very deep, then instead of enrichment, disseminated deposits form above the primary water table. This is especially true in arid areas where the descending water is sponged-up before it can reach the true water table.

8. WHAT ARE THE CAUSES OF GROUND WATER MOVEMENT?

Velocity is influenced by both the slope of the water table and the permeability of the rock the water must trespass.

9. WHAT IS A SPRING?

It is a water discharge where the water table intersects the land surface, or from caves, fractures or faults.

10. HOW DOES CLIMATE INFLUENCE WATER MOVEMENT?

Gaining streams - in rainy climates, streams gain by addition of water.

Losing streams - in arid climates, streams lose water.

11. ESPECIALLY IN HUMID REGIONS, HOW DOES THE WATER TABLE LIE IN RELATION TO THE SURFACE?

Usually, the water table roughly parallels the land surface. Water will not move on flat land.

12. WHAT IS AN AQUIFER?

It is a body of rock that is BOTH porous and permeable. Permeable examples are: sandstone, conglomerate, well jointed limestone, sand and gravel mix, fractured volcanic rock.

Examples of relatively impermeable rocks are: shale, granite, gabbro, gneiss, schist and crystalline rocks in general.

13. WHAT IS A WELL?

It is a deep hole dug in the ground to penetrate the saturated zone within an aquifer. The water is then pumped to the surface.

14. WHAT IS THE ZONE OF DEPRESSION IN A WELL?

It is an inverted cone around the well pump inlet caused by draw-down during pumping. The water table flow direction is changed to recharge the cone.

15. WHAT IS AN ARTESIAN AQUIFER?

It is an aquifer lying between a permeable upper, and lower less permeable rock layer that has been bent (folded) by tectonic events, with a subsequent fault drop at the low end, (syncline) to seal the saturated zone from leakage, together with surface erosion at the high side to form a catchment basin. Water rises in wells drilled into the saturated zone without pumping, due to high end water pressure on the entire lower lying saturated zone.

16. WHAT IS THE EFFECT OF GROUND WATER ACTION?

Karst topography, underground streams, petrified wood, concretions and geodes.

17. WHAT IS KARST TOPOGRAPHY?

An area with many sinkholes and a cave system beneath the land surface, usually lacking a surface stream.

]18. DESCRIBE THE PROCESS IN FORMATION OF KARST TOPOGRAPHY?

Rain water plus atmospheric carbon dioxide (CO_2) forms a weak carbonic acid, which plus the calcite of limestone forms calcium bicarbonate which is easily soluble. Chemical weathering erodes the limestone rock layers at joint intersections. Erosional valleys lower the water table causing streams to run underground to form dissolved limestone caves. Solution collapse causes sinkholes to form. Stalactites, stalagmites, meet to form columns to support the surface; dripstone; and flowstone cover walls and floors of the cave; as the sinkholes enlarge to meet each other and merge, dolines are formed. Eventually the surface drainage reappears on insoluble bedrock as the underlying limestone is solution-removed leaving only shale and low hills of dolomitic limestone as remnants to form a new surface.

19. WHAT ARE HOT SPRINGS?

They are waters that are warmer than human body temperature. They are formed by water circulation near a magma chamber or a body of cooling igneous rock. They are also formed by deep circulation in the earth, probably along deep joints or faults. Hot water rises because it is lighter than cold water.

20. WHAT IS A GEYSER?

It is a type of hot spring which intermittently erupts both hot water and steam. The water is generally near boiling point.

21. DETAIL THE ADVANTAGES AND DISADVANTAGES OF GEOTHERMAL ENERGY?

ADVANTAGES:
1. No costly fuel needs to be burned to heat water.
2. Air pollution is minimal.
3. Eliminates radiation hazards of nuclear plants.

DISADVANTAGES:
1. Hazardous gases, salts and metals enter the air and streams.
2. Salts corrode generating equipment.
3. Surface can subside as waters are withdrawn.
4. Plant and wildlife can be harmed through salt pollution of streams.
5. Current technology shortfalls limit potential.

Water that soaks into the ground filling pore spaces between soil particles, rock fragments, and mineral grains becomes ground water. It may reenter the stream run-off days later, or it may remain underground for a thousand years.

It is groundwater that flows into mines, spurts out a geysers and provides the water vapor gases during volcanic activity. Groundwater produces dripstone in caves; in combination with weathering leaches and redeposits iron, copper and aluminum ores.

As groundwater seeps down-slope to settle at the water table, the aerated zone above the table contains air and water to form a highly active weathering area. Where the water table touches on the surface, springs may emerge. In its down-slope movement the water dissolves salt, gypsum and limestone to form a sinkhole type of karst topography. Ultimately, the saline water reaches the ocean to maintain the salt content.

Groundwater transports iron oxide in solution to cement sedimentary rocks such as sandstone. Dissolved calcium carbonate can form concretions in sandstone. Solution silica precipitates as opal, agate and chalcedony. Sometimes it produces petrified wood.

CHAPTER 12

ENVIRONMENTAL CONCERNS ACTION AND REACTION

ACTION: — *PAST SINS*

Who doesn't cringe when one thinks of:

1. Brown mine waters polluting rustic streams.
2. Trout floating belly-up in a mountain stream.
3. National forests denuded by clear-cutting loggers.
4. Landscapes scarred through strip mining.
5. Abandoned mines and pits turning into child traps.
6. Redneck-rape of land, leaving only spoilage behind.

REACTION:

RADICAL ENVIRONMENTALISTS

Who doesn't cringe when one thinks of:

1. 66% of Public Lands are now closed to mineral entry.

2. Designated Wilderness public-excluded areas arbitrarily set up as "barren of minerals". (Because prospecting is prohibited, the legal fiction becomes unrebuttable.)

3. The paper chokehold: Mineral extraction start-up permitting processes that mandate 327 different operating permits before extraction begins. Thereafter, filing obligatory, regular,

detailed, monitor reports with 25 various government agencies.

4. Closing of large Public Land tracts because of an endangered animal species. (Spotted Owl supposedly favors mature forests.)

5. Closing of large Public Land tracts because of endangered plant habitat. (Astragalus favors selenium in serpentine soils. We know this plant as one of the Rattleweeds in the Loco Weed family.)

6. Public Land Management Regulations that classify recreational gold panning and rockhounding as "MINING". (So as to exclude these recreational pursuits.)

7. Frivolous environmental law suits that purposely delay proposed extraction ventures for years.

8. The Comprehensive Environmental Response, Compensation and Liability Act and its amendments for cleanup of hazardous waste dump sites and dangerous toxic spills.

9. The Resources Conservation and Recovery Act - provides for cradle to grave documentation for handling, storage and disposal of hazardous waste.

10. The Toxic Substance Control Act - requires pre-market testing of potentially harmful chemicals.

11. The Clean Air Act - authorizes the Environmental Protection Agency to establish national ambient air quality standards for emissions that pose a risk to public health and safety.

12. The Clean Water Act - enables the Justice Department to bring enforcement actions against polluters. Requires States to establish water quality standards.

Each statute contains penalty provisions in the form of minimum or maximum fines and imprisonment terms. Historically, state and federal judges rarely imposed sentences sufficient to deter violations. Corporations knew this. They simply included the fines in the cost of doing business. In 1984 the Sentencing Reform Act tempered the judges discretion with objective and well-defined parameters. Now judges are compelled to impose prison sentences within those parameters. In 1990 more than 50% of those convicted served jail time.

The advantage of prosecuting corporate executives who violate environmental statutes is obvious. They can no longer hide behind the corporate veil. The "responsible corporate officer doctrine" will impose criminal liability. If that executive, by reason of his position, responsibility and authority in the corporation, could have prevented, or should have promptly corrected the violation, BUT FAILED TO DO SO, he WILL be guilty. Every responsible corporate executive should be aware of this trend and always remember: *POLLUTE THE ENVIRONMENT - GO TO JAIL.*

13. California voters approved a proposition ostensibly to stop gill net fishing. The fine print establishes a Marine Ecological Reserve that halts nude bathing, boating, swimming, fishing, scuba diving, and surfing in the set-aside areas. Was this preserve designed by oceanfront owners solely to exclude people from public beaches?

THE HIDDEN AGENDA OF THE RADICAL ENVIRONMENTAL MOVEMENT SEEMS TARGETED AS AN EMOTIONAL APPEAL TO RESTRICT MINERS, LOGGERS AND RANCHERS. IS IT THEIR COVERT OBJECTIVE TO RID THE PUBLIC LANDS OF ALL PEOPLE-USE? CLUE IN POINT: OVER 100 YEARS AGO, SIERRA CLUB FOUNDER, JOHN MUIR, DESCRIBED GRAZING SHEEP AS *"LOCUSTS WITH CLOVEN HOOFS".*

Nature is adaptable. Entrenched endangerment allegations, casually accepted as true, are mere rebuttable assumptions. How do we know that the Astralagus weed will not grow in other than serpentine soils; that the Spotted Owl and Gnat Catcher can not adapt to new habitat; that the Darter can live no other place on earth; that Wilderness areas are devoid of minerals? These are presumptions that can be overcome through a simple "reward for information" program as soon as specious allegations are born. If not, idea-entrenchment becomes public-mind certainty.

Vulnerable industries that make contributions, either voluntarily, so they can crow about environmental concern, or under a feeling of coercion as a ransom payment to be left alone, should examine the folly of their actions. Those funds are the fuel that will promote even more onerous and aggressive activity. The more outrageous the distortion, the greater the audience will be.

To curtail the radical environmentalist movement, the first order of business must be to starve them of funds. The second, and even more important thrust, must deprive them of credibility.

EXPOSE THE LIE. REDIRECT YOUR FUNDS TO RESEARCH THE TRUTH.

COUNTERACTION —

COUNTER—THRUST

SPECIMEN PRESERVATION

There are thousands and thousands of mineral collectors and Rockhounds who actively seek to build their collections. Most purchase desirable specimens within the constraints of their pocketbook. Others covet nature's glories without regard to cost, much like the compulsive gambler.

To service these hoards of avid mineral seekers, a host of rock-shop dealerships was born to cater to their wants and desires. Many are part-time in the sense that they display and sell only at gem and mineral shows sponsored by mineral society chapters. The last show that I attended drew over 4000 visitors in 3 days. Apart from entrance fees, I spent over ten dollars on a few specimens that I craved. Mostly, I enjoyed the feeling of camaraderie between the visitors and dealer network. It was a mutual sharing of profit and joy through a common bonding. There was something else I sensed: a desire to possess a mineral specimen that was priced out of the collectors financial reach because of relative scarcity. More than one dealer bemoaned that scarcity with words like "I could have sold many times more than I had available."

Unfortunately, the attitude of the general public toward mineral collecting is not shared by the mineral extractive industry. Official policy, based on the excessive liability risk, prohibits or discourages dump-ground scavenging. In our litigious society, that policy makes good management sense.

Consider, if you will, this alternative threat:

Those responsible for most mineral production must stop the wanton, deliberate smashing of magnificent crystal pockets encountered in day to day operations. The few cents worth of metal cannot be measured in the enmity that can result.

For example, the vandalizing of mine mineral specimens has reared a body of thought that characterizes that action as bordering on the criminal. Any sort of legal pressure to stop it justifies the end. This is not environmentalism, per se, but a genuine desire to preserve the unique wonders of nature.

The thinking of the activists is quite simple. A mining venture is not justified in destroying objects of beauty, merely by occupying the apex of a vein with the right to mine it. Take your economic minerals, but spare and preserve the specimens of artistic and scientific merit. Force mine operators to arrange for mineral specimen preservation one way or the other. Form pressure groups to apply all legal means possible. Unique mineral specimens belong to the public and not the mine owner.

If you think this is far-fetched, consider the tuna industry and the porpoise killing! Each tuna boat must carry and pay the expense of a government observer to monitor the porpoise kill.

How would you like to double the number of mine foremen without increasing your ore production one iota? How would you like to see production decrease as your expenses rise and profits dwindle? Is there a government mineral specimen protector in your future?

The important objective is to save specimens from the work

ing mine, particularly the shallower levels where secondary mineral pockets abound. Enlightened recognition of the mine owners public responsibility to preserve unique mineral specimens for future generations is the most cost-effective counter-thrust.

ROCKPECKER TIP # 21 – Consider the good will and potential profit in displaying and selling your mine finds. If the discovering miner and his foreman share in the sales revenue, you know wanton destruction will cease. By making working arrangements with the local mineral society chapter, they will clean-up, display and sell the specimens. Contacts through mineral dealer networks can establish a wholesale market.

ROCKPECKER TIP # 22 – It may sound overly simplistic, but all you need to do is provide a lidded plastic barrel, tag the specimen with the name of the mine, county and state, the working depth, and the name of the discovering miner. The specimen should remain in its natural matrix whenever possible, for this enhances its resale value. The contents in the barrel should be protected from travel shock through a sawdust-type of cushioning material. You then ship the barrel, truck freight FOB, to your distribution point, probably a trustworthy dealer, perhaps even on consignment. The sales price will be measured in pounds instead of tons, in good will in lieu of animosity.

Good will can be difficult to establish for those who are seen as mineral plunderers. Is it not better to formulate a plan where the wonders of the inanimate world are shared with the area communities? Can not smaller mining ventures combine to set up an area showcase display with a local merchant? Can not an association group form a specimen mineral arm to travel the mineral show circuit? Can not show sales defray those costs?

Without good public relations, one who works a wasting asset, such as a mine, can be at the mercy of anyone perceiving injury to the environment. In these changing times, doesn't the extractive industry need all the friends it can influence?

CHAPTER 13

U.S.A. IMPORTED MINERALS

Domestic self-sufficiency in mineral commodities makes for a strong national industrial foundation. Based on import figures, the U.S. is at least 60% deficient in the following minerals. Obviously, a ready-made domestic market is waiting for your ore. Target these minerals!

To give you a feel for birth origins of minerals, we begin to point at probable genetic sources. At this stage that information will sound unfamiliar. Be patient. Genesis is given here only to show its relevance to a mineral search.

CHROMIUM — The U.S. imports 100% of domestic use. Alloy steel uses 8 to 35% chromium in the manufacture of stainless and chrome steels. There has been no U.S. production since 1962.

CHROMIUM - FROM THE ELEMENT CHART:

Symbol: Cr.
Atomic #: 24
Melting Point: 1856.0 c.
Boiling Point: 2673.0 c.
Atomic crystal lattice: BCC - (Body centered cubic)
Atomic Weight: 52.00
Hazards: Toxic, Irritant.
Natural Occurrence: Chromite.

CHROMITE - FROM MINERAL IDENTIFICATION CHARTS: Mineral # 71 --

Chemical Name: Iron chromium oxide.
Crystal System: Isometric.
Most Probable Occurrence Environment: Magmatic crystallization in mafic rocks.

Streak: Brown.
Mineral color: Black.
Luster: Opaque, submetallic.
Hardness: 5.5.
Specific Gravity: 4.5-4.8.
Ore of Cr.
Looks Like: Magnetite. (But weakly magnetic)
Alters to: Goethite (Limonite).
Associated With: Serpentine, Uvarovite, Magnetite, Pyrrhotite (magnetic pyrites), Corundum.

The ore mineral chromite carries a theoretical 68% chromium oxide. Your ore should contain at least 45% chromium oxide with a 3% or greater iron content for metallurgical ore grades. In January, 1992 chrome ore was quoted at $ 3.70 the pound.

Chromite is a rock-forming mineral closely associated with nickel and magnesium in ultramafic igneous rocks such as gabbro, peridotite and dunite. Rock-forming means the chromite is deposited at the same time as the associated igneous rocks. The Chromium content for an economic deposit would run 2000 to 4000 parts per million. (Abbreviated as ppm from now on.)

Chromite is heavy and resists chemical weathering. Therefore, placer deposits can contain secondary deposits when the host rock physically disintegrates. Chromite is one of the weakly magnetic black sands you can find in your gold pan.

In wet, hot, tropical conditions leaching of magnesium from the host chromite-bearing rock can leave a 2 to 4% Cr residue in lateritic soils. Laterites can be identified by the deep red color in the upper soil strata, called the zone of alluviation. Measured from the surface downward, this zone of accumulation starts at about 9 inches down to the weathered bedrock. The red color is caused by leached iron, usually hematite (135), which sometimes may be mined as iron ore. More than likely, the hematite has been hydrated to limonite (158), which will not be rich enough to mine economically. Aluminum-rich igneous rocks that have weathered to aluminum oxides may be found in laterites as nearly pure layers of bauxite (037).

ROCKPECKER TIP # 23 - Simply because the Rockpecking area is now arid, does not mean laterites are not present. This area may have been tropical 50 million years ago. Witness, as an example, the Arkansas bauxite deposits. Of course, any laterites may now rest under hundreds of feet of overburden, or have been folded into hills and valleys by later tectonic events.

Most chromite deposits are found as layered deposits. A few are in pod-like sills. The ultramafic host rock is multi-jointed at very close intervals like simple fractures. These fractures are often filled with gabbro. Fault-filled gabbro dikes, show a coarser grain when chromite is present.

GENESIS - There is much unexplained in present theories of origin. The layering is said to be caused by magmatic separation and differentiation that flowed off to form the rock layers of chromite, dunite and peridotite. The pods may have been formed by magmatic liquid injection as an intrusion between layers.

GENESIS - A magmatic segregation deposit is a direct crystallization product of the magma, formed in the magma chamber as a crystal mush, then pressure driven into the magma chamber walls and roof to form ore bodies such as intrusive sills and dikes. The ore and host rock are deposited at about the same time.

ROCKPECKER TIP # 24 - Peridotite and dunite host rocks or as their serpentine alteration forms are strong chromite associations. Foliated peridotite is an exceptionally strong indicator.

ROCKPECKER TIP # 25 - Peridotite and dunite rock associations are also a guide to nickel, diamonds, ilmenite and platinum group elements.

ROCKPECKER CLUES - SEE MINERAL #(071) - IDENTIFICATION CHART "D".

ROCKPECKER TIP # 26 - TEST FOR CHROMIUM - First powder your unknown mineral. A simple borax bead test in a reducing flame will show a green flame color when hot; a green residue on the bead when cold. In an oxidizing flame, the flame will color yellow; but a leave a green residue on the bead upon cooling.

IN THE U.S. CHROMITE HAS BEEN FOUND IN MARYLAND, MONTANA, CALIFORNIA, OREGON, NORTH CAROLINA AND ALASKA.

Chromite was identification-detailed to show the information value of the element and mineral identification charts. USE THEM!

MANGANESE — The U.S. imports 100% of domestic use. Its chief use is in steel making. About 20 pounds is used for each ton of carbon steel. It acts as a deoxidizer to remove oxygen and sulfur from the steel. There is no substitute for it.

Symbol: Mn.
Atomic # 25
Melting point: 1244.0 c.
Boiling point: 1962.0 c.
Atomic crystal lattice: Complex cubic.
Atomic weight: 54.94.
Hazards: Chronic poison.
Natural occurrences: Pyrolusite, rhodochrosite.

Some manganese ore minerals:

Pyrolusite (208) carries a theoretical 63% manganese oxide. Found in sedimentary deposits, it also forms as surface alteration of manganese deposits.

Manganite (163) - 62% hydrated manganese oxide.
Found in hydrothermal veins and sedimentary deposits.

Psilomelane (204) - 60% hydrous manganese oxide.
Found: The dendritic type forms in limestone fractures.

Hausmannite (132) - 72% manganese oxide.
Found in metamorphic contact zones in limestone and skarn.

Rhodochrosite (215) - 47% manganese carbonate.

Found as a sedimentary mineral in the oxidation zone of sulfide ores, and in caves as pink stalactites.

Rhodonite (216) - 41% manganese silicate.
Found in metamorphosed impure limestone; may be confused with rhodochrosite, but is softer. Black manganese oxide inclusions are visible to the eye. Makes an attractive pink gem material which is much sought after by collectors.

Your ore for carbon steel should contain a minimum 46% manganese. If your ore-find grades 80% iron and 20% manganese, you then have a natural pig iron ore. Pig iron is a crystalline form of cast iron often called spiegeleisen.

Sedimentary manganese deposits are the result of chemical weathering and the affinity of iron for manganese. Most soluble iron compounds found in nature are less soluble than manganese. Therefore, when chemical weathering operates, iron being less soluble, goes into solution last, but precipitates first, naturally separating the manganese. The deposits appear as thin banded layers of manganese distributed throughout iron oxide layers. This layering has puzzled geologists. Since all streams eventually empty into a pond, bog, lake or ocean, it is thought that seasonal water runoffs cause the layering, with the dissolved manganese being solution-retained longer in acidic waters, then precipitating as the acidity drops, to form its own sedimentary layer. But this is just a theory.

Layered sedimentary deposits are sometimes referred to as ribbon-shaped deposits or as banded deposits. As a rule, manganese and iron accumulate under similar conditions and environments.

GENESIS - There are three types of manganese deposits.

Sedimentary- which we have just reviewed.

Volcanogenic - where manganese-bearing hot gases or thermal springs percolated through volcanic tuffs, breccias or other extrusives to act as a manganese rock cement, much as iron or silica acts as a rock-forming binder.

Hydrothermal - about the same as volcanogenic, except the

percolation of manganese-bearing fluids is at greater depth, usually through sedimentary material.

ROCKPECKER TIP # 27 - Fine, brown, ribbon-shaped layers, like laminations on green or red andesite tuff or clay localized along faults, are worth a closer examination for manganese.

IN THE U.S., MANGANESE HAS BEEN FOUND IN MONTANA, MINNESOTA, NEVADA, ARKANSAS, GEORGIA, VIRGINIA, SOUTH DAKOTA, NEW MEXICO, ARIZONA, CALIFORNIA, WISCONSIN AND MICHIGAN.

ROCKPECKER TIP # 28 - Chemical leaching has produced known sedimentary deposits of manganese. It would be a real ground-breaker to find the leach-source of those deposits.

ROCKPECKER TIP # 29 - TEST FOR MANGANESE - First powder your unknown mineral. A simple borax bead test in a reducing flame will be colorless in both the flame and cool bead. In an oxidizing flame, the flame will color violet, leaving a reddish-violet residue on the cool bead.

ROCKPECKER CLUES - SEE MINERAL # 208, 163, 204, 132, 215, 216 IN MINERAL IDENTIFICATION CHART "D" .

COBALT — The U.S. imports 100% of domestic use. In specialty steels and temperature-resisting alloys, cobalt has the ability to maintain strength at high temperatures. It is found as the mineral *cobaltite* #078, cobalt arsenic sulfide, in copper, nickel and iron sulfide deposits associated with mafic rocks, and as disseminated grains in metamorphic rocks associated with cobalt-nickel arsenides.

ROCKPECKER TIP # 30 - Erythrite #108, hydrated cobalt arsenate, is a common alteration product of cobalt arsenides as they weather. With its pink color and dusty appearance, it is a distinct guidepost to primary cobalt deposits. Think of it as *cobalt blush*.

ROCKPECKER TIP # 31 - TEST FOR COBALT - First powder your unknown mineral. A simple borax bead test will show a blue flame color and a blue bead in both the reducing and oxidizing flame test.

IN THE U.S., COBALT HAS BEEN FOUND IN PENNSYLVANIA, IDAHO, NEW MEXICO, CALIFORNIA, ARIZONA.

TANTALUM — NIOBIUM — The U.S. imports 100% tantalum for domestic use. Both are used in metallurgy for heat resisting alloys and rust-proofing stainless steel. Important for electromagnetic superconductors.

Columbium is the old name for niobium. In rock types such as carbonatites and nepheline syenite, niobium is more abundant than tantalum. Many minerals contain both these elements.

For example:

Columbite-tantalite (080) (niobium tantalum oxide).
Fergusonite (111) (yttrium niobium oxide).
Microlite (169) (hydrous sodium calcium tantalum oxide).

Placer deposits have been important producers. On a lesser scale of production, pegmatites have small concentrations. In January, 1992 tantalum (spot ore) was quoted at $28.00 the pound.

ROCKPECKER CLUES - SEE MINERAL IDENTIFICATION CHART "D" FOR MINERAL # 80, 111, 169.

IN THE U.S., TANTALUM - NIOBIUM HAS BEEN FOUND IN IDAHO, COLORADO, CALIFORNIA AND ARKANSAS.

Large placer reserves have been found of manganocolumbite (niobium-rich) in recent times. Tantalum search efforts should be on-going for the tantalum-rich phase known as manganotantalite.

THORIUM — The U.S. imports 100% of domestic use. Small quantities are used in a nitrate conversion process to produce gas mantles such as the ones you use in your camping gas lamp. Uses now are for radioactive purposes.

Thorium is found as the mineral:

thorianite #250, thorium uranium oxide.

monazite #174, a rare earth metals phosphate with a 12% thorium oxide content. Monazite is the main ore for thorium and cerium.

GENESIS - Monazite is a common accessory mineral in S-type two mica granites; pegmatites - where hydrothermal fluids traveled along fractures to react with earlier emplaced pegmatite minerals to form a new mineral of replacement origin.

Monazite #174 is heavy and resists chemical weathering. Therefore, placer deposits can contain secondary deposits when the host rock physically disintegrates. Monazite is one of the non-magnetic black sands you can find in your gold pan. Beach placers in elevated sandbars have been worked but yield a very low mineral content.

Thorianite #250 has been found in metamorphosed limestones as a contact marble with serpentine in Pennsylvania .

IN THE U.S., MONAZITE HAS BEEN FOUND IN NORTH CAROLINA, FLORIDA, IDAHO, CALIFORNIA, MAINE, CONNECTICUT, VIRGINIA, NEW MEXICO, PENNSYLVANIA AND COLORADA.

ROCKPECKER TIP # 32 - This group of minerals in highly radioactive. A scintillation counter can be a helpful search-mate.

CHAPTER 14

CAUTION: THE MINING LAWS ARE UNDER SEVERE ATTACK. CHANGES ARE CERTAIN TO BE MADE. CHECK !

U.S. MINING LAWS

Staking of lode or placer claims, or acquiring a "land position" is the first step in the right to extract minerals. Dependent on where the ownership of minerals lie, this involves Federal, State or private lands.

One third of the United States is Federal land. Under the Mining Law of 1872, you can stake a claim on public land and mine it. That is, unless the mineral to be mined falls under the Leasing Act of 1920, or the Federal land has been withdrawn from mineral entry.

The Mining Law of 1872 was an American concept of mineral rights. It embodied District Miners Codes under which miners worked on Federal lands without permission.

The Leasing Act of 1920 brought coal, phosphate, sodium, oil, oil shale, gas and certain sulfur deposits under a leasing-royalty payment umbrella. Limited prospecting permits became available for areas with no known economic reserves of these leasable minerals.

Certain minerals fall under the "saleable minerals" limitation. These are sand, gravel, stone, pumice, clay, cinders and other materials without a distinct and special value. You cannot locate a claim or lease them. They are available by purchase only.

Trustee Indian Lands, designated scenic, recreational, military, and "Wilderness Areas" have also been **WITHDRAWN** from mineral entry.

There exists much controversy about Wilderness Areas because these areas are presupposed to contain "little mineral potential". What is particularly galling to mineral interests is the fact that saying there are no worthwhile minerals, by law, makes it so. A flat-out write-off.

State lands usually provide for state ownership of minerals. You need to check with the state you expect to prospect, for the mechanism they employ as to mineral search and extraction.

Private lands are called "Fee Lands". Fee Simple land titles include ownership of the subsurface minerals. You need to strike your arrangement with the owner of the land. In the case of "reserved mineral rights", someone other than the surface owner is involved as mineral owner. A land title search can reveal that interest.

The Mining Law of 1872 is under attack and critical review. The next few years could turn the old ways topsy-turvy. For that reason, this chapter must be read with that in mind.

CLAIM STAKING

Assuming your prospect is on lands open to mineral entry, and your claim target is not already filed upon by a previous locator, you can claim the mineral rights. REMEMBER, YOU ARE SUBJECT TO BOTH FEDERAL AND STATE MINING LAWS.

A valid mining location requires that a VALUABLE mineral be discovered on EACH of your claims. To establish legal ownership of the mineral rights, the mining laws require a *verifiable discovery of a valuable mineral.* Simple locating, staking and recording a claim is NOT enough. Sometimes, strong evidence that indicates the potential presence of a valuable deposit, can suffice. THE FEDERAL GOVERNMENT CAN CHALLENGE YOUR CLAIM ANYTIME!

Until you find your mineral in ECONOMIC amounts, the standards of "PRUDENT MAN" and MARKETABILITY are conditions that must be met. That is, would a prudent man expend money and effort to extract the mineral, and additionally, is there a reasonable market for the production of that mineral. Unless you meet those conditions, (in theory) your claim can be defeated by the Feds.

When your prospect includes a particular mineral deposit, you need to make an important claim-filing decision.

THE DISTINCTION BETWEEN A LODE CLAIM AND A PLACER CLAIM IS A MATTER OF LAW. SIMPLY STATED: LODE CLAIMS ARE CORRECT FOR VEINS OR LODES OF QUARTZ OR OTHER ROCK *IN PLACE,* BEARING GOLD, SILVER CINNABAR, LEAD, TIN, COPPER OR OTHER VALUABLE DEPOSITS. *PLACER CLAIMS ARE CORRECT FOR ALL FORMS OF DEPOSITS, EXCEPTING VEINS OF QUARTZ OR OTHER ROCK IN PLACE.*

As a matter of law, if it is not a lode, *in place,* it is a placer. The terms lode and vein are interchangeable. The geologic meaning is ignored. Nonmetallic minerals such as gypsum, limestone, feldspar can occur as legally defined lode deposits.

The U.S. Supreme Court has ruled that a placer discovery will not sustain a lode location. A lode discovery will not sustain a placer location. The general rule is: *A miner acts at his peril when he decides the type of claim to be located.*

If the mineral deposit is the result of weathering and erosional forces, you will file a *PLACER CLAIM.*

If, however, you have an ancient sedimentary fossil placer deposit in conglomerate, or a mineralized area in hardrock, (and in place) as in a vein or dike, you will file a *LODE CLAIM.*

These are important distinctions. A placer claim gives you NO lode mineral rights; a lode claim confers NO placer rights.

ROCKPECKER TIP # 33 – If there is any question about the type of claim to file. FILE BOTH A LODE AND OVERLAPPING PLACER CLAIM.

PLACER CLAIM LOCATING

Each person can locate one claim of 20 acres per claim. As long as there is an additional, valid discovery, that person can file as many additional claims of 20 acres each, as needed to cover the deposit area. EACH CLAIM WILL COST YOU A LOCATION FEE PLUS AN ANNUAL MAINTENANCE FEE IF YOU HOLD MORE THAN 10 CLAIMS.

LODE CLAIM LOCATING

The mining law allows a single lode claim to a maximum length of 1500 feet, with a maximum width of 600 feet. There is no limit to the number of claims an individual can file.

The easiest way to locate a lode claim is to post a location notice on the middle of the vein or dike outcrop. Then measure 750 feet along the strike or direction of the vein. At that point, measure 300 feet at a right angle on each side of the centerline to make the end CORNER lines. The opposing sidelines and endlines must be parallel.

Legal location and boundary marks vary by state law. Generally, mark your corners. Place the LOCATION NOTICE in the center of the claim. It is wise to mark the CENTER of your side and end lines. Each marker must contain a description of its position on the claim.

A rockpile three feet high, or a 4 x 4 post, marking and describing the corners is generally acceptable. But you must check state statutes to be sure. You can't use plastic pipe.

After you have POSTED your location notice, you may have as long as 20 days to return to erect your markers. (The time delay should be checked for each jurisdiction.)

Within 90 days from the date of your location notice, you must file and record a CERTIFICATE OF LOCATION with the state office of the Bureau of Land Management (BLM) and the County Recorders office. You will also need a CLAIM MAP defining the claim location. Again, check for state variations. If you fail to file these documents within the 90 days time frame, AND ALSO PAY THE CORRECT FEES, both lode and placer claims are VOID.

Placer claim descriptions should track on the public survey legal subdivision descriptions whenever possible.

After August 11, 1993 and before September 30, 1998 any mining claim located must pay a $25.00 location fee at the time of recording plus a $10.00 recording fee. There is an additional $100.00 maintenance fee due at recordation to pay for the current year ending August 31, and must be paid annually on or before August 31 under pain of claim forfeiture. If you hold 10 or fewer claims, you can escape the $100.00 per claim SECOND annual maintenance fee by filing an exemption waiver by AUGUST 31, PLUS the annual *PROOF OF LABOR* statement with the BLM BEFORE DECEMBER 30. THAT DATE IS CRITICAL! Failure to file proof of labor [DONE BY AUGUST 31] with the BLM and County Recorder is conclusive proof of claim abandonment. State requirements do vary, so check them.

ROCKPECKER TIP # 34 – If you miss the ANNUAL FILING deadline with the BLM, simply refile your claim anew. (And hope someone doesn't beat you to it.) YOU PAY $235.00 AS A NEW CLAIM FILING.

ROCKPECKER TIP # 35 – IF YOU HOLD MORE THAN 10 CLAIMS, YOU CAN'T DUCK THE MAINTENANCE FEE BY RECORDING IN YOUR WIFE'S OR DEPENDENT CHILDREN'S NAME TO SHOW 10 OR LESS, NOR WILL SETTING UP DUMMY CORPORATIONS OR PARTNERSHIPS ESCAPE THE FEE. IN ADDITION TO THE MAINTENANCE FEE, RELATED PARTIES WITH MORE THAN 10 CLAIMS MUST FILE A NOTICE OR PLAN OF OPERATIONS. Small miners with 10 or fewer claims must recertify their exemption waiver each year on or before August 31, and are exempt from operation plan rules.

The claim holding rules are simple: Pay the $100.00 maintenance fee or file the exemption waiver by August 31 each year or suffer a null and void claim by operation of law. NO APPEAL !

ROCKPECKER TIP # 36 – Claim filing, location claim maps, monument markers may get tricky. Secure professional help. Sleep better knowing you really own the mineral rights. At the very least, contact a licensed land surveyor to set your monuments exactly.

Thus far, we have been talking about UNPATENTED lode or placer claims. You may later decide to PATENT your claim.

PATENTING YOUR CLAIM

You have a valuable mineral discovery. Your unpatented mining claim vests the mineral rights in you as a possessory interest, but the Federal government still owns the paramount title to the land. WHY NOT BUY THE LAND AND CONVERT YOUR HOLDING TO A PATENTED CLAIM? You will then own the FEE land outright.

If you hired a registered, licensed land surveyor to mark your claim boundaries and draw your claim map, you are almost home free. You will need to post in a conspicuous place on your claim, a copy of the plat, together with a copy of application for patent. Next file the patent application, together with plat and field notes, which are made under the direction of the BLM. THIS MEANS MORE COSTS. NOW HOPE THERE IS NO PATENT MORATORIUM.

You will also need to file an affidavit from at least two persons who WITNESSED the claim posting. You can bet the BLM will look closely, at the economic mineral's marketability and the prudent man standard, in an attempt to defeat the patent application.

The land office registrar will publish a NEWSPAPER NOTICE of patent application for 60 days. During this advertising period, or after filing the application for patent, you must also register a certificate with the BLM, that $500. worth of labor or improvements have been made on each claim seeking the patent.

At the end of the 60 day advertising period, you file an affidavit that the claim was posted during that period. Next pay the five dollar per acre charge to be entitled to your patent. THIS SOUNDS LIKE A REAL ESTATE BARGAIN. Don't be surprised if your final costs total $500. to $1,000. the acre.

ROCKPECKER TIP # 37 - There is a strong push in Congress to change the mining laws, particularly as to the passing of public land into private ownership. When you patent your claim, you are no longer subject to the annual maintenance fee of $100. There seems to be a trend to make fees a revenue producer, and to stop the passing of Federal land into private ownership. Stay tuned !

CHAPTER 15

A GENERAL REVIEW

Petrographers have classed rocks into endless subdivisions. Never-the-less, they belong to three great rock groups: igneous, sedimentary, and metamorphic.

Igneous (magmatic) rocks were fire-formed from a once molten state. They are believed to come from deep in the earth. Various cooling rates form the rock varieties from basalt to granite.

Sedimentary rocks are composed of layers of sand and clay laid down into beds by the agency of water, ice, or wind. The weight of the upper tiers press and compress the lower layers into rocks we know as shale, sandstone, limestone that are then cemented with iron oxides, silica, even silt.

Metamorphic rocks are deep-lying igneous, sedimentary, or other metamorphic rocks, that undergo internal change without melting, through the action of pressure and heat that squeeze, bend, and deform their shapes.

It is important to remember that the changes in metamorphic rock occur deep in the crust. Just because you now see them on the surface, only indicates that weathering has removed the overburden to expose them. When you consider a heat treatment lasting a thousand years before the igneous heat source cools down, rocks termed metamorphic had a long time to bake. Then add the millions of years for the top cover to erode down to the exposed metamorphic rocks. This is how clay turned to shale and finally to slate; how sandstone became quartzite; how limestone became marble.

Generally, granite is considered an igneous rock. There are some geologists who believe it to be metamorphic in origin through a process called *granitization.*

When you think about it, no man has ever seen a rock form as a direct experience. There is but one exception, molten rock with a taffy-like consistency pouring from a volcano. This molten rock is the parent of all igneous rocks. When it blows or flows from a volcano, we call it an extrusive rock. We can tell by its small grain size that it received a sudden cooling shock. Slower cooling, over time, would cause the grains to grow larger.

Slow cooling rocks had to remain buried to account for their larger grain sizes. We call these rocks intrusive rocks. This is magma that never reached the surface

When magma cannot reach a surface outlet, it may thrust into cracks or earlier faulting breaks; even between sedimentary rock layers near the surface. We can also expect the molten magma to melt a path through other rocks.

Where a natural fissure or vertical crack in sedimentary rock is magma-filled and hardens, we call this rock wall a *dike.* The width of a dike can be a few inches to yards wide; it can run for miles in length. Where, like a dike the intrusion cross cuts strata, but looks like a whale's head breaking water, that structure is named a stock. Stocks can easily be a mile in diameter.

Where the magma penetrates between rock layers in a horizontal path, we call that structure a sill. If the sill has a mushroom shaped, domed head, it's called a laccolith.

Gigantic bodies of igneous rock, many miles in diameter, and extending downward to undetermined depths, are called *batholiths.* They are the main heat source for causing metamorphic rock alteration.

Typical batholiths, laccoliths, stocks, sills, and dikes are composed of a grainy, speckled rock called granite.

When you think about it, over the next 100 million years, the rocks we walk on today will be churned into the crust to be remelted, deformed, metamorphosed, uplifted to the surface to be eroded again. In that future time, another prospector will be looking over outcrops for whatever economic minerals will make him rich. His successful search will mean raw material availability. Civilization will be bettered through his efforts.

PROSPECTING IS AN INVISIBLE PROFESSION.

CHAPTER 16

THE COMPASS

It would be a shame, if your bones were found, ten years after you dropped from exhaustion. Your pack, still intact, holds clues to a remarkable gold mine. Ore specimen assay values run six to ten ounces per ton. This is the grist for a "lost mine" story. The shame would be equal, should you survive, but not be able to relocate the "lost mine" site. For these reasons, it is wise to carry and use your compass on any exploration trek.

COMPASS THEORY

The pivoted, magnetized, steel needle points to the MAGNETIC NORTH. The MAGNETIC North Pole is located about 1000 miles from the TRUE GEOGRAPHIC North pole.

The magnetic lines of force of the earth align the needle and pull or DIP one end of the needle below the horizontal. The DIP ANGLE varies from 0 degrees at the equator, to 90 degrees at the magnetic pole. To keep the needle horizontal, and to balance the effect of dip, the South end of the needle is WEIGHTED.

As you TURN the compass box, the needle continues to point toward the MAGNETIC NORTH. The reading you make is dependent upon the POSITION of the underlying, graduated circle card.

The MAGNETIC DEVIATION, variation or declination is the horizontal angle between the magnetic meridian and the geographical meridian.

There are three basic requirements in determining an angle. They are:
1. Your reference starting line.
2. Your direction of turning.
3. The value of the angle in degrees, minutes and seconds.

Computing BEARINGS and AZIMUTH are based on these 3 requirements.

The reference STARTING LINE or reference meridian is either the true geographic meridian or the magnetic meridian. The true meridian for any point on the earth's surface is a great circle passing through that point and the North and South GEOGRAPHIC poles.

True bearings are measured from the LOCAL GEOGRAPHICAL MERIDIAN. The bearing, is the angle measuring LESS that 90 degrees, as measured from a North or South reference line toward the East or West. The quadrant (a quarter of a circle) is shown by a letter N or S; the angle in degrees, followed by E or W.

Going North, measure CLOCKWISE from your meridian reference line. Going South, measure COUNTER-CLOCKWISE from your meridian reference line.

Assume you are moving North at a clockwise bearing of 40 degrees. You would write this as N 40° E. THIS IS YOUR FORWARD BEARING. On your return trip, your BACK BEARING is S 40° W. (Simply reverse direction with the same degree reading.) Forward bearings have the same numerical value as back bearings.

When you orient your compass correctly with your reference line meridian, for example, the true GEOGRAPHIC NORTH, AZIMUTH ANGLES are measured CLOCKWISE from the reference line.

Since a circle has 360 degrees in the full swing, your AZIMUTH DIRECTION requires only a numerical angle degree value measured from North only, or South only, depending on how you selected your reference line.

For example: If your BEARING is N 15° W, the DIRECTION using AZIMUTH from the North is 360 ^{o} less 15° or 345° as your path home.

Another example: S 68° E, as a BEARING, would read as 112° AZIMUTH path. (180° – 68° = 112°)

S 51° W BEARING would read 231° AZIMUTH path. (180° + 51° = 231°) (180° being half of a full circle, to include the quadrant you are moving in.)

MANY PROSPECTORS PREFER AZIMUTHS OVER BEARING FOR PATH LINE DIRECTIONS, BECAUSE THEY ARE EASIER TO FOLLOW.

POINTS TO REMEMBER WHEN DEPENDING ON YOUR COMPASS.

1. Don't confuse magnetic and true bearings.

2. Don't set declination off on the wrong side of North.

3. Don't mix bearing and azimuth readings.

4. Don't subtract from 360° 00' as though it is 359° 100'. Instead use 359° 60'. (60 seconds in a minute; 60 minutes in a degree.)

5. TO RELOCATE YOUR FIND write it down. Use bearing or azimuth angles to find your way back. (Along with landmark azimuths) Take frequent readings. Your notes can't be too detailed. (Someone else may need to use them.) A detailed sketch is your best friend.

6. DON'T FORGET PARALLAX. Sight your readings along the needle, not while looking from the side of the needle.

7. Don't read the wrong end of the needle.

ROCKPECKER TIP # 38 – It is possible that a local magnetic attraction can throw you off true-path. Highly mineralized zones with a lodestone effect can mess up your readings. That, in itself, may be an ore deposit location clue. After all, highly mineralized zones are the goal of your prospecting activities.

CHAPTER 17

GEOLOGIC & TOPOGRAPHIC MAPS

Public Lands are all lands administered by the federal government, no matter how acquired, which since 1785 are subject to the public land laws of the United States. Public Domain represents about 72% of the United States. It comprises thirty states, including Alaska, but excludes Texas and the original colonies.

The PUBLIC DOMAIN has been, or will be, subdivided into tracts forming a rectangle grid system. The purpose is to obtain sections one mile on each side. Any discrepancies are thrown into the sections bordering the North and West township boundary, to get as many regular-sized sections as possible.

37 INITIAL POINTS were established, (5 in Alaska) so that a principal meridian and base line passed through each initial point. From each initial point, a true N-S line called a principal meridian was run to the limits of the area to be covered. From the initial point, the E-W base line was extended, as a true parallel of latitude, to the limits of the area covered.

After the meridian and base lines were run, correction lines, known as STANDARD PARALLELS, are run as true parallels of latitude, 24 miles apart. Standard parallels are numbered consecutively N and S of the base line.

GUIDE MERIDIANS are run due N from the base line, E and W at 24 mile intervals. Any error in the 24 miles of the guide meridian is put in the northern-most half mile.

On United States WEST COAST maps, the principal meridians are shown on your maps as SBM (San Bernardino Meridian), MDM (Mt. Diablo Meridian), HM (Humbolt Meridian) and WM (Williamette Meridian) in Oregon.

In general, public lands are surveyed as QUADRANGLES, 24 miles on a side; then into 16 Townships, 6 miles on a side; then to 36 sections of Townships, 1 mile square. Subdivision of sections require the services of a local land surveyor.

PUBLIC LANDS are divided by N and S lines according to the TRUE MERIDIAN. The true meridian is crossed at right angles so as to form TOWNSHIPS 6 miles square. 6 x 6 means 36 square miles in a Township.

Townships are subdivided into SECTIONS of 640 acres each, or 1 square mile each, for a total of 36 sections per Township.

Sections are numbered beginning with section number 1 in the NORTHEAST CORNER, proceeding WEST, then alternating EAST with progressive numbers until 36 are completed. Therefore, section 36 would lie in the SOUTHEAST CORNER.

Because the earth is round, BASE LINES are run as circular curves on the map. The right hand top corner of the map has a QUADRANGLE NAME based upon the meridian line to identify the map.

The quadrangle is divided into RANGE AND TOWNSHIP LINES. Range lines are the true meridian through standard townships corners, parallel, and 6 miles on the base.

Range lines run N and S. N and S rows of townships are called Ranges. An individual township is identified by its number N or S of the base line, followed by the number E or W of the principal meridian.

TOWNSHIP 7 SOUTH, RANGE 19 EAST OF THE SAN BERNARDINO MERID-IAN is abbreviated as: T 7 S, R 19 E, SBM.

Townships, being technically square, are subdivided into SECTIONS, which are further subdivided into quarter-sections, and then quarter-quarter section corners on downward.

DESCRIPTION BY TOWNSHIP SECTION SUBDIVISIONS WOULD READ:
SE 1/4, NE 1/4, Sec. 14, Tp. 3 S, R 22 W , SBM. or:
SE 1/4 of SE 1/2 of NE 1/4 of SW 1/4, Sec 11, T 4 N, R 3 W, SBM.

Assume you are field exploring. You come upon a tag nailed to a tree. The tag is marked off in numbered squares. THE SECTION NUMBER, THROUGH WHICH THE NAIL IS DRIVEN, IS YOUR SPECIFIC SECTION LOCATION.

TOPOGRAPHIC SYMBOLS are used to show special map features such as streams, roads, mines, buildings. These are not explained on the map itself. You should order the SYMBOL EXPLANATION GUIDE for greater map reading efficiency.

TWO MORE POINTS ARE NOTEWORTHY:

At the bottom of the map you will find the deviation specified for the MAGNETIC NORTH DEVIATION for the CENTER of the map.

You will also find the MAP SCALE and CONTOUR LINE INTERVAL GIVEN IN FEET. You can project a location site by scaling the map.

If you were flying over the terrane shown on your map, you would see the graduations of CONTOUR as they appear in relief on your map. The arrangement and spacing of contour lines lets you see the hill, valley, or slope steepness. Interpretation of a topographic map takes both practice and imagination.

Notice that every fifth line appears darker with an elevation number shown. Look below the map scale for the contour scale. If the scale reads 50, that means each line is 50 feet from the one on each side of it. Therefore, the fifth darker contour line would be 250' lower or higher than the fifth darker line above or below it.

When you hike over the map area, you may meet a metal plate on a rock or concrete post. Note the elevation number. That is the precise elevation at that spot. It is called a BENCHMARK. NEVER DISTURB IT! YOUR MAP SHOWS AND LOCATES IT AS "BM"

Map making and reading is a precise art. Practice map interpretation until you are comfortable with it. It can be your best friend.

Several years ago I was talking to a Prospector by phone. He had twenty separate placer claims located over three counties. To help me at my end of the phone, I asked him to give me the name of the quadrangle map. He had the map in his hand, but did not know how to identify the quadrangle. (Name in upper right hand corner)

It proves that a successful claim filer does not need to know how to read a map. It just makes life easier !

When you need to record your claim, a correct land description of where the claim lies, is essential to secure your fortune.

CHAPTER 18

ROCK HEFT

CONFIRM WITH SPECIFIC GRAVITY

REVIEW OF ROCK TYPES

There are three basic groupings for rocks:

IGNEOUS ROCKS - FIRE FORMED with 2 classes:

1. INTRUSIVE - Molten magma cooled underground.

2. EXTRUSIVE - Molten magma cooled on the surface.

ROCKPECKER TIP # 39 - Intrusive igneous rocks are usually coarse grained due to the slow, underground cooling-off process. Extrusive rocks are fine grained due to a faster cooling in the air.

SEDIMENTARY ROCKS - Weathered, eroded sediments formed into strata layers to lithify under overlying sediment pressures. The 2 CLASSES are:

CLASTIC (fragmentary due to physical division)

CHEMICAL (due to the cementing that binds the pieces.)

ROCKPECKER TIP # 40 - Sedimentary rocks are generally stratified, fine-grained, composed of older rock fragments, sometimes with fossils, particles are cemented with iron, silicon, calcite or fine silt.

METAMORPHIC ROCKS - Alteration rock of igneous, sedimentary or metamorphic origin changed by temperature, pressure or other forces into a banded, striped or foliated new form.

As they are derived from previously existing rock classes, their appearance is variable. They are identified by the types of minerals they contain along with texture, and mode of origin. The 2 CLASSES are:

CONTACT METAMORPHIC - Country rock changed by the heat of a nearby intrusive.

REGIONAL METAMORPHIC - Country rock changed by heat and pressure from mountain building uplift or a cataclysmic event.

ROCKPECKER TIP # 41 - Rocks altered by heat border on igneous intrusions that changed them by contact heat transfer short of melting. (LOOK FOR THE INTRUSIVE) Regionally metamorphosed rocks were caused by heat and pressure from uplifting mountain ranges. (LOOK FOR MOUNTAIN ROOTS)

> > > > > > **IMPORTANT NOTE** < < < < < <

SPECIFIC GRAVITY MEASUREMENTS OF ROCKS ARE MEANINGLESS FOR IDENTIFICATION OF ROCK PURPOSES.

SPECIFIC GRAVITY MEASUREMENTS OF ROCKS FOR THE PURPOSE OF DETERMINING IF THERE IS AN ADDITIONAL METAL CONTENT DISSEMINATED IN THE ROCK CAN SAVE THE NEED FOR CHEMICAL ELEMENT TESTING.

IN A HYPOTHETICAL CASE, YOU JUST PICKED UP AN INTERESTING ROCK SPECIMEN. IT HAS PLENTY OF NEARBY BROTHERS. FROM THE HEFT OF THE ROCK, YOU BELIEVE YOU MAY HAVE SOMETHING OF VALUE. WHAT CAN YOU DO ?

YOUR FIRST STEP IS TO IDENTIFY THE CLASS OF ROCK YOU HOLD.
YOUR SECOND STEP IS TO MAKE A SPECIFIC GRAVITY TEST.
YOUR THIRD STEP IS TO CHEMICAL TEST ANY COLORED ROCKS THAT

SPECIFIC GRAVITY TESTED HIGHER THAN THE UPPER RANGE OF THESE ROCKS.

IGNEOUS INTRUSIVE ROCKS

SPECIFIC GRAVITY RANGE		ROCK NAME
2.6 - 2.7		GRANITE
2.7 - 2.9		SYENITE
2.8 - 3.0		DIORITE
2.8 - 3.1		GABBRO

IGNEOUS EXTRUSIVE ROCKS

2.5 - 2.8		PORPHYRY
2.8 - 3.3		BASALT
2.3 - 2.6		OBSIDIAN

SEDIMENTARY ROCKS

Too variable		BRECCIA
Too variable		CONGLOMERATE
2.0 - 2.7		SANDSTONE
2.7 - 2.8		CLAYSTONE
2.6 - 2.8		MARL
2.8 - 2.9		FIRECLAY
2.6 - 2.8		TRAVERTINE
2.4 - 2.5		BAUXITE

METAMORPHIC ROCKS

2.6 - 2.8		GNEISS
2.6 - 3.2		MICA SCHIST
2.7 - 2.8		PHYLLITE
2.6 - 2.8		MARBLE
2.7 - 2.8	(Greenstone)	CHLORITE SCHIST
2.7 - 2.8		AMPHIBOLITE
2.6 - 2.7		SERPENTINE

CHAPTER 19

ROCKS ASSOCIATED WITH METALS

INTRUSIVE ROCKS WORTH KNOWING

These are igneous rocks that never left the crust and have hardened beneath the surface. Heat exposure produced a texture of large grains of various minerals. Weathering has exposed them.

GRAIN COLORS - {TEXTURE}	ASSOCIATED METAL	INTRUSIVE ROCK
Pinkish to dark red with {blue specks.}	Rare Earth	ALKALINE FELDSPAR GRANITE
Granite gray with {mixed-size black inclusions.}	Copper & pyrites	TONALITE
Pale gray, pinks or light purple.	Important rare metals	SYENITE
Dark gray, greenish or reddish {small grains.}	Mineral deposits	MONZONITE
Green-gray, light gray, or gray with {coarse grains.}	Copper	OPHLIOLITIC GABBRO
Gray with green, brown or violet cast.	Chromium, nickel, cobalt, platinum	OLIVINE GABBRO
Very dark gray, nearly black. {Medium grains.}	Nickel-bearing pyrrhotite	NORITE
Light gray, pinkish or greenish.{Mixed grain size}	Rubidium, cesium, thorium, uranium.	FELDSPATHOID SYENITE
Light green, {fine grain size.}	Chromium & platinum	DUNITE
Brownish to yellowish green, overall darkish.	Nickel, chromium	LHERZOLITE
Light to dark green, overall dark. {Green pathways}	Nickel-pyrrhotite, platinum.	PERIDOTITE
Light grayish or yellow.	Rare elements	CARBONATITE

EXTRUSIVE OR VOLCANIC ROCKS

Extrusive or volcanic rocks crystallize on the earth's surface. Fast cooling produces a very fine grain size or none at all.

GRAIN COLOR - {TEXTURE}	ASSOCIATED METALS	EXTRUSIVE ROCK
Light gray, pink, violet, black, red. {Felsite groundmass}	Chalcopyrite, copper	QUARTZ PORPHYRY
Blackish brown or greenish. {Glassy groundmass.}	Porphyry copper ores	ANDESITE
Gray or pale brown. {Porphyritic texture.}	Rare earths, niobite, columbite-tantalite	MELILITITE

HYPABYSSAL ROCKS

Hypabyssal rocks crystallize at medium depths as in sills and laccoliths.

GRAIN COLOR - {TEXTURE}	ASSOCIATED METALS	HYPABYSSAL ROCKS
Dark green to blackish; {sometimes light brown spots.}	Copper	DIABASE
Intergrowth of large, {very light crystals.}	Beryl, muscovite	PEGMATITE

PRELUDE TO IGNEOUS AND METAMORPHIC ROCKS.

When you consider that 75% of the earth's crust lies under the seas, how much land surface would still be above water if polar ice and mountain glaciers were to melt? The point is: The ocean depths conceal terranes similar to those continental contours seen above water. That is: Solid rock of similar compositions evidenced by well and mine shafts. We find: Volcanic rock, rising from whatever depth, is virtually the same.

All *igneous* rocks were once molten. The cooling rate determines the final rock texture from basalt to granite. Periods of deep burial can change the rock texture and composition to further reform, through heat and pressure, a rock group known as *metamorphic* rock.

All rocks belong to three major groups: *Igneous, metamorphic and sedimentary.* Changes within the group give birth to other rocks. For example, the erosion of granite produces sand and clay as an end product. The clay, under compression, turns to *shale,* which changes to *slate* under heat and pressure. (or pressure alone) Slate, under different degrees of metamorphism, can turn to *schist, phyllite, or gneiss.* (merely descriptive terms.)

We can directly experience molten rock blowing or oozing from an active volcano. This molten rock is termed *magma.* If the magma blows out of crustal openings, we call the product *extrusive rock.* Often it will not surface. Magma running along cracks in the crust harden as *dikes.* Where it pushes between layers of sedimentary rock as a sheet, we call it a *sill*; where it bulges like a mushroom, we term it a *laccolith.* These under-the-crust magma depositions are called *intrusive rocks.*

IT IS WITHIN THESE THREE ROCK GROUPS THAT THE ECONOMIC MINERALS WE SEEK ARE HIDDEN.

CHAPTER 20

IGNEOUS - FIRE FORMED ROCKS

The discovery of the Mid-Ocean Ridge and of seafloor spreading due to basalt rising at the ridge, led to an earth-science theory revolution called plate tectonics. The word "tectonic" comes from the carpenter in the Iliad named Tekton. In geology, it means "construction", as in the building of the earth's structures. For our purposes, it throws light on where metals come from.

Seismic zones had been discovered that outlined large slabs of crust forming the outer skin of the earth. The rigid lithosphere is made up of 5 major slabs, called plates, along with many smaller ones. Comparatively, the earth's crust was as thick as an eggshell is to the egg. These plates float on a plastic-like molten rock known as the asthenosphere. Most of the plates move away from the mid-ocean ridge. Where they collide, the seafloor plate ducks under the edge of the continental plate at the ocean trench. This pressure buckles and cracks the continental upper plate raising it at the edges to form mountains ranges. The grinding friction of the subducting plate melts the overlying continental plate, and is itself melted as it plunges into the molten rock of the asthenosphere. The molten rock rises as magma into the cracked crust to fuel volcanoes in the newly formed mountain range.

Continental crust is mostly granite composed of lightweight quartz and feldspar. The Oceanic crust is iron-rich basalt, a dense, heavy rock. The continental crust being lighter, is more buoyant. That's why it floats on top as the heavier seafloor plate subducts the less dense continent.

INTRUSIVE IGNEOUS ROCKS - MAGMATIC ORIGINS:

The molten rock that penetrated the crustal passages at the crackage points is called magma. As the magma cools, metal-rich layers formed by progressively crystallized minerals settle out of the melt as a kind of rising slurry. The remaining melt is granite or some of its cousins of other names.

ROCKPECKER TIP # 42 - Recall that the subducting plate was the oceanic plate that melted to form the metal-rich magma. The dense, heavy, iron-rich basalt rock is known as ferromagnesium rock. It forms the darkest granites with funny names such as: *granodiorite and tonalite* as found in the California Sierras; *syenite* of the New York Adirondacks; *monzonite* in the Henry Mountains of Utah; *diorite* found in Minnesota; coal-black *norite* and violet-gray *gabbro* of San Diego, California; *anorthsite, essexite, pyroxenite, dunite, peridotite*, of the Stillwater Complex in Montana; and *diabase* as in the New York Hudson River Pallisades. These black granites carry the richest mineral concentrations. A very dark granite is worthy of close inspection.

This is the *MAFIC* rock class. When a mafic rock is on the blackish-overall side, it is know as *ULTRAMAFIC* rock. Mafic is short for MA (magnesium); F for Ferro (iron). Ultramafic, then would be a very dark granite such as pyroxenite, dunite, peridotite.

ROCKPECKER TIP # 43 - Some Mafic rocks associations have been: Tonalite with gold, copper and pyrites; syenite with Rare Earth Metals; monzonite with gold; gabbro with copper, cobalt, nickel, chromium, iron, platinum; norite with nickel-bearing pyrrhotite; essexite with rare earth minerals; dunite with platinum and chromium; peridotite with nickel-bearing pyrrhotite, chromite, platinum and rare earth minerals.

ROCKPECKER TIP # 44 - Don't panic over these rock names! You need to be an analytical chemist to distinguish them. See later for an easier way!

Igneous rocks are formed from molten magma rising from the mantle of the earth. The composition of the magma determines the mineral content of the rock. An igneous rock, by definition, is a fire-formed rock buried in the crust or blown to the surface. It makes no difference that the magma is buried deep in the crust or at rest on the surface of the earth.

IGNEOUS ROCK DIVISIONS:

EXTRUSIVE IGNEOUS ROCKS - Whenever an igneous rock is blown onto the surface of the earth by a volcano or bubbles to the surface through a crustal opening, it is defined as an *extrusive* igneous rock. It is extrusive because it exited the crust to reach the surface.

INTRUSIVE IGNEOUS ROCKS - If magma is buried in the crust to solidify as a sill, dike, stock, laccolith or batholith, it is an igneous rock. Because it is within the crust, it is by definition *intrusive* igneous rock.

What then would distinguish an intrusive from an extrusive igneous rock? The answer lies in volcanism and the cooling rate of the magma. When the magma extrudes the surface, it is quickly cooled. Much like dipping a red-heat poker in water. The metal grains in the poker expand when heated; sudden cooling contracts the grains. There is no longer any grain growth. The structure of the poker metal remains fixed. Therefore, because of sudden cooling, expect to see a tightly packed grain structure with microscopic crystal grains in an extrusive igneous rock. Also expect to see a mineral content of slightly differing mix, because the oxygen in the atmosphere will have burned out some of the volatiles.

INTRUSIVE IGNEOUS ROCKS - MAGMATIC ORIGINS:

Intrusive igneous rocks are often referred to as *Plutonic*, after the classical mythological name for Hades or Hell.

Intrusive igneous rocks are further divided into depth of burial. Those that form dikes, sills, or laccoliths are called *hypabyssal*. That is, they crystallize at medium depths at an atmospheric pressure sufficient to prevent too great a release of magmatic gases as they were injected into sedimentary rock layers or cracks in the crust. Those intrusive igneous rocks that were buried at great depths crystallize slowly under atmospheric pressures too great to allow the escape of magmatic gases.

ROCKPECKER TIP # 45 - Note particularly, the "release of magmatic gases" in the prior paragraph. This important point will become clearer as we discuss metal deposition in later chapters.

The slow cooling effect on a rock over a period of many thousand years causes the minerals grains to grow and become coarse. This grain distinction is referred to as *texture*, meaning grain size, shape, and internal interlocking grain arrangement. The intrusive rock texture will show definite individual grains. Think of granite.

Granite is the commonest of the coarse-grained igneous rocks. We know it is igneous and intrusive by the way it was squeezed into spaces between other rocks while in its molten stage. Granite is composed of quartz, orthoclase feldspar and a small amount of dark mineral, usually biotite mica.

There are three major feldspar groups found in granite. They are distinguished by their potassium, sodium-calcium, barium content. Orthoclase is K feldspar. (K is the chemical symbol for potassium) Albite feldspar is called soda-granite when sodium is present. Anorthite is the calcium phase, that when combined with quartz, is called quartz monzonite. The sodium-calcium feldspars are also called plagioclase feldspars. These distinctions are given here, to illustrate how a simple rock, can be made to look complicated to a layman. Whether the grains of feldspar are pink, white, gray, black or yellow-brown, granite by any other name, is still a generic granite.

Diorite is a coarse grained dark-colored rock with quartz nearly absent. It is like granite with the feldspar being chiefly plagioclase. It grades into granodiorite, tonalite and monzonite. If the plagioclase is calcium, the rock is gabbro.

Gabbro grades into norite, anorthite and diabase.

Peridotite grades into pyroxenite, dunite and kimberlite. The dark minerals in peridotite alter to serpentine. Serpentinite is an extensively altered peridotite.

Crystalline rocks are formed mainly through three basic physical-chemical processes: crystallization from the hot molten state; precipitation from solution; recrystallization in the solid state. Each of these processes result in a different end product rock. More on this a little later.

In a commercial sense, all intrusive rocks can be lumped into the term *granite*. They can be easily separated in your mind by thinking "colored" granite for all intrusive rock, then deducting the "black granites". The black granites will be gabbro - (DARK GRAY TO BLACK), diabase -(GREEN TO BLACKISH), anorthosite - (DARK GRAY) and pyroxenite - (DARK GREEN TO BLACK). You can usually pick out the anorthosite by its iridescent sheen caused by a bluish plagioclase called labradorite.

ROCKPECKER TIP # 46 - MENTALLY CLASSIFY GRANITES AS COLORED OR SHADES OF BLACK. THE BLACK GRANITES ARE GABBRO, DIABASE, ANORTHOSITE AND PERIDOTITE-PYROXENITE. THE BLACKS SHOULD BE YOUR TARGET ROCKS.

EXTRUSIVE IGNEOUS ROCKS - VOLCANIC ORIGINS:

Extrusive rocks were hurled from the mouth of a volcano or trickled from gap openings in the volcano's side or from the adjacent crust. By being thrust into this cooling environment, the crystal grains had little time to form. Therefore, we would expect to identify an extrusive rock as one with no discernible visible grains or tiny, close-packed crystals visible only with a magnifying glass.

Rhyolite is a light-colored rock equivalent to granite. It can sometimes show a flow pattern as in banding or flow streaks, indicating that the material moved while still molten. Rhyolite grades into *obsidian*, a natural glass that cooled so quickly that crystals had no chance to form. Where gas bubbles are contained in the obsidian, *pumice* is formed. Those built-in bubble pore spaces allow pumice to float on water. A variety of obsidian with a brown black cast is called pitchstone; one with a greenish-violet cast is known as rainbow obsidian. Apache tears are the result of incomplete obsidian weathering. They look like smoky colored, translucent pebbles.

Basalt is a fine-grained, dark-colored volcanic rock that we know as lava. It can be black, greenish, gray, or brown in color. When it weathers, it has a tan skin. Basalt grades into *scoria* when it looks like a porous, cinder rock. Almond-shaped gas pockets in basalt are called amygdules from the Greek word for almonds. Check the filled pore spaces for mineral values.

A mafic lava crystallizes into dark-colored basalt. It contains MAgnesium and Ferrite (iron) to make up the name *mafic.*

A felsic lava crystallizes into light-colored rhyolite. It contains FELdspar and SIlicon to make up the name *felsic.*

Based on the lavas: *MAFIC ROCKS are DARK-COLORED.*
FELSIC ROCKS are LIGHT-COLORED.
INTERMEDIATE are the IN-BETWEENS.

There is a mixed igneous rock that contains different grain sizes called *porphyry.* Imagine magma beginning to cool deep within the earth. Its mineral grains are growing ever larger as the magma cools. By the time the magma is half cooled those grains are of mixed sizes. Then imagine an eruptive force moving those partially-grown grains mixed with rising fluid magma bursting onto the crust. The lava will have a fine-grained texture because it fast-cooled. The half-formed mineral grains will be included in a texture that shows a mix of large mineral grains in a fine-grained background. The fine material is called groundmass and surrounds the larger crystals, called phenocrysts. Think of porphyry as "raisin-almond bread".

MAGMA AND MAGMATIC FLUIDS

Magma is defined as a molten mobile rock, naturally occurring within the earth, and capable of intrusion and extrusion. It is a molten mush of liquid and crystals at very high temperatures. Most magmas are thought to be a mix of materials rather than a single uniform mass of a similar material. The magma is constantly changing due to chemical reactions within the composition. When the mush crystallizes and solidifies, an igneous rock is formed. During the solidification process, the liquids in the mush separate into liquid concentrations that depend on pressure and temperature changes to permit a particular fraction of the liquid to leak off. Geologists call this fractional differentiation.

Where the mushy liquid is squeezed from the melt into the surrounding rock fractures, the process is called *magmatic injection.* An ore deposit would then be known as a magmatic injection deposit. Where the fluid fraction is squeezed away from the crystalline mush by external stresses, as by a tectonic event, (like grapes in a wine press) the process is termed *filter pressing.* Where the liquid fractions solidify directly as a true melt, they act as molten rock to become a mineral-rich *ore magma.* Here, the liquid fraction cannot escape the magma and becomes concentrated with the dark granite as in the gabbro phase.

Ore magmas, filter pressing and magmatic injection are concepts invented to explain deposits of nearly pure magnetite and copper found in a few hypabyssal intrusives and even fewer lavas.

During differentiation, the mafic parts of the magma are enriched in chromium, nickel, platinum and other elements; the felsic parts of the magma carry concentrations of tin, zirconium, thorium and other elements. Titanium and iron tends to concentrate throughout the entire magma range.

PEGMATITES

An interesting intrusive, igneous rock is known as *pegmatite.* It is theorized that the volatile lighter fraction rises from the magma (like oil on water) to collect near the roof of the magma chamber. When you consider that the magma can theoretically include every known element, it becomes easy to posit that the lighter fractions near the roof include exotic mineral combinations found only in pegmatites.

Pegmatite is basically a coarse-grained granite composed of quartz, K feldspar and mica. Usually the grains are twice normal granite size, but can be in the ton-grain size in a few very valuable intrusives. A special type of pegmatite has the feldspar and quartz grains intergrown to resemble alphabet letters or graphics. Not surprisingly, this is called graphic granite or alphabet granite. Pegmatites are usually found in dikes and veins. The entire rock body is called pegmatite.

As pegmatites are found, the simple coarse grain pegmatite is the normal find. Some define *coarse-grained* as grain crystals four inches in diameter or larger. Pegmatite, as a natural storehouse of minerals, is the most interesting rock group known. The rare mineral one is the exceptional find. Where erosion has disclosed a surface indication, pegmatites can run hundreds of feet deep, or pinch out in a few feet. Most occur in metamorphic rock such as schist and gneiss. Their shape and size is determined by the structure of the country rock in which they intrude.

During and after **WW** II pegmatites were studied as ore depositories under the Strategic Mineral Program. It was found that the rare-mineral pegmatite core is massive quartz; that the pegmatite had zones of fundamental structure units with fracture fillings and replacement bodies imposed on the zones; that crystallization occurs in successive stages from the fissure wall inward toward the center. The last finding has a logical cooling cycle format. The outside border of the pegmatite touching the colder fracture rocks would cause cooling to work inward. This gave rise to the idea that the pegmatite was marinating in its own fluids to grow its unique minerals. Understanding zoning in pegmatites can decrease slightly the inherent risk in mining them. Traditionally, the pegmatite mine is a glory hole, manually mined with valuable minerals selected and concentrated by hand. Dead work is a fact of life, for other methods can break or fracture the unique, valuable gem crystals.

ROCKPECKER TIP # 47 – Minerals found in pegmatites are: molybdenite, a 60% ore of molybdenum; quartz crystals of the smoky, ruby and rock variety; amblygonite with up to 8% lithium; fluorite, corundum, tantalite, brookite, chrysoberyl, garnet, topaz, titanite, spessartite, ilmenite, axinite, tourmaline, muscovite, lepidolite, K feldspar crystals, orthoclase, albite, spodumene, monazite, columbite, microcline, beryl and cassiterite.

As you can already guess, the granite rock specialists, called petrologists, look at the differentiation distinctions to make up a rock-type called "*granitoids*". The reason for these rock type distinctions is based on the differentiation source material. Generally, there are four types of granitoids designated by a type-lead-letter. The conventional granite is "I"-type granite for Igneous; "S" type granite for Sedimentary; "A"-type granite for Alkalic; "M"-type granite for Mantle derived granite. This can form a valuable "ore mineral occurrence" framework. Gold, silver, lead, copper, molybdenum, zinc are known to prefer I-Type granites; tin, tungsten, beryllium, uranium and lithium occur preferentially with S-Type granites.

MAGMATIC SEGREGATION DEPOSITS

This term is now applied to all ore deposits that are the direct result of crystallized magma products. Usually they are deep-seated intrusive bodies, but they have been found in hypabyssal sills and even in lava flows. A magmatic segregation deposit may be the entire rock mass or a layer in the igneous mass. The ore minerals may be early or late fractions concentrated by gravity settling or filter pressing. They remain in place, or are injected into country rocks. It is generally understood that mobile ore fluids, as overages, are released as the igneous rock begins to solidify. The magmatic stage of ore deposition then passes into the hydrothermal phase of ore deposition.

****

THE FOLLOWING IGNEOUS ROCK CHART MAY HELP YOU.

COMMON IGNEOUS ROCKS - CLASSIFICATION & IDENTIFICATION

ROCK NAME	ROCK COLOR	ROCK GROUP	CHIEF MINERALS	TEXTURE	INTRUSIVE OR EXTRUSIVE
RHYOLITE	LIGHT	FELSIC	QUARTZ, MICA ORTHOCLASE	FINE GRAINED	EXTRUSIVE
ANDESITE	DARK	INTERMEDIATE	PLAGIOCLASE AMPHIBOLE, PYROXENE	FINE GRAINED	EXTRUSIVE
BASALT	DARK	MAFIC	PYROXENE, OLIVINE CA. PLAGIOCLASE	FINE GRAINED	EXTRUSIVE
OBSIDIAN	DARK	FELSIC	QUARTZ, MICA, K & Na PLAGIOCLASE	GLASSY	EXTRUSIVE
PUMICE	LIGHT	FELSIC	QUARTZ, MICA K FELDSPAR, AMPHIBOLE	GLASS FROTH	EXTRUSIVE
TUFF	VARIOUS	VARIOUS	SAME AS ORIGINAL ROCK	VARIOUS	EXTRUSIVE PYROCLASTIC ASH
SCORIA	DARK TO INTERMEDIATE	MAFIC	AMPHIBOLE, PYROXENE, Ca. PLAGIOCLASE, OLIVINE	VESICULAR	EXTRUSIVE
GRANITE	LIGHT	FELSIC	MICA, QUARTZ K & Na. FELDSPAR	COARSE GRAINS	INTRUSIVE
SYENITE	LIGHT	FELSIC	MICA, HORNBLENDE K & Ca. FELDSPAR	COARSE GRAINS	INTRUSIVE
DIORITE	DARK	INTERMEDIATE	BIOTITE, AMPHIBOLE PLAGIOCLASE, PYROXENE	COARSE GRAINS	INTRUSIVE
GABBRO	DARK	MAFIC	OLIVINE, AMPHIBOLE, Ca. PLAGIOCLASE, PYROXENE	COARSE GRAINS	INTRUSIVE
PERIDOTITE	DARK	MAFIC	OLIVINE, PYROXENE Ca. PLAGIOCLASE	COARSE GRAINS	INTRUSIVE
PEGMATITE	LIGHT	FELSIC	QUARTZ, MICA K & Na FELDSPAR	VERY COARSE GRAINS	INTRUSIVE

CHAPTER 21

QUARTZ AND ITS VARIATIONS

A Rockpecker thinks of quartz as a repository for precious metals. For that reason, he is constantly in search of bull-quartz, which is intimately associated with gold. Yet other prospectors seek variations of quartz for its gem value.

Quartz is the most common form of silica which is SiO_2. Silica is a simple oxide which can crystallize in nine distinct forms. It belongs to the Silica mineral group. There are three broad quartz forms: Bull quartz is the *coarse crystalline* form; fine grained quartz, the gem material, is called *microcrystalline or cryptocrystalline* where the grains are microscopic. The high-temperature form above 573° C is *tridymite*. At 1470°, tridymite breaks down and reforms as *cristobalite*.

QUARTZ = SILICON DIOXIDE

[Color - white, colorless, purple, violet, blue, pink, yellow, red, brown, gray, black] (Many varieties of quartz are based on color)
[Streak - white] [Hardness - 7.0]
[Specific Gravity 2.65]
[Luster - vitreous] (waxy for chalcedony)
[Cleavage - conchoidal] (usual lack of cleavage)

LOOKS LIKE: Calcite, which has a hardness of 3.0.
Feldspar, which cleaves on 2 good planes.

VARIETIES OF QUARTZ - COARSE CRYSTALLINE:

1. Milky quartz - liquid flaws scatter light.

2. Rock crystal - colorless and transparent.

3. Smoky quartz - from pale smoky brown to black. Natural smoky quartz becomes colorless when heated above 225° C, but may assume a yellowish or yellow-brown color before becoming colorless. Most of the gem material sold as *citrine* is really heat-treated smoky or amethyst quartz.

4. Amethyst quartz - ranges in color from violet to red-violet, to purple. It can be decolorized by heating to 290° C. Heating to 500 - 600° C, often produces a yellow-brown color that is passed-off as citrine. It can be returned to purple color by X-ray irradiation.

5. Blue Quartz - owes its color to tiny needles of tourmaline.

6. Citrine Quartz - this is the yellow to yellow-brown variety. Yellow quartz is often sold as *topaz*. This is misleading. Don't confuse it with the mineral topaz.

7. Tiger-eye - yellow to yellow-brown retaining a fibrous crocidolite structure.

8. Rose quartz - color ranges from pale pink to deep rose. Commonly contains needle-like inclusions or *rutile* that scatter the light to give a star effect of *asterated rose quartz.*

9. Cat's eye quartz - color ranges from green, brownish or yellowish. May resemble chrysoberyl cat's eye, but not as valuable. Wavy light bands caused by asbestos fibers as inclusions.

10. Aventurine quartz - color green, reddish to brownish-yellow. Light reflected from inclusions of mica flakes.

MICROCRYSTALLINE — CHALCEDONY:

[Color - white, gray, blue, green-gray, yellow, brown, deep red, green.]

VARIETIES OF CHALCEDONY:

1. Agate - Opaque, a concentric banded variety which includes *eye agate.*

2. Onyx - Opaque, alternating light and dark banded layers.

3. Sard - Opaque, brown to yellow or red-brown.

4. Carnelian - Translucent, red to orange-red.

5. Chrysoprase - Translucent, apple-green.

6. Jasper - Opaque, dark red, red-brown, yellow-brown.

7. Prase - Translucent, dull green.

8. Plasma - Opaque, dull green.

9. Bloodstone - Opaque, green with red jasper spotting.

10. Chert - Opaque, dull black, white, gray or yellowish.

11. Flint - Opaque, dark gray to black.

12. Moss agate - Opaque, pine tree branching, white with black, brown, or red-brown treelimbs.

13. Silicified wood - Red or brown jasper or opal replace wood fibers.

OPAL — HYDRATED SILICA

[Color - Milky to colorless, bluish-white, black. Tints of red, yellow, orange, brown, blue, green]
[Hardness - 5.5 - 6.5]
[Specific Gravity - 2.0 - 2.2]
[Luster - Greasy]
[Crystal System - None. Always in amorphous masses]

LOOKS LIKE: Chalcedony which is harder with dull waxy luster.

VARIETIES - ALMOST ALWAYS OPAQUE OR TRANSPARENT.

1. Fire opal - precious opal with red-orange body color.

2. Hyalite - colorless but can fluoresce a brilliant green.

3. Hydrophane - opaque white with air inclusions. May be iridescent. Dipped in water becomes transparent.

4. Wood opal - opal petrified wood.

5. Geyserite - opal deposited around hot springs.

TOPAZ — THE MINERAL

[Chemical name - hydrous aluminum silicate]
[Color - white, colorless, light blue, green, yellow, pink, yellow-brown, wine-yellow, pinkish-brown, wine-red, violet]
(Color can fade from exposure to heat or sunlight.)
[Hardness - 8.0]
[Specific Gravity - 3.49 -3.57]

LOOKS LIKE: Quartz - which has a specific gravity of 2.65.
Beryl - which has a specific gravity of 2.66 - 2.83.

JUST BE AWARE - Heat-treated smoky quartz turns yellow. It is then called *Smoky Topaz, Citrine Topaz, or False Topaz.* Similarly, yellow *corundum* is sometimes called *Oriental Topaz. Yellow andradite garnet* may be called *Topazolite.*

QUESTION - IS IT QUARTZ, OPAL, CHALCEDONY OR TOPAZ ?

CHAPTER 22

METAMORPHIC ROCKS

So far we have examined two totally different kinds of rocks; igneous, fire-formed, melted rocks produced within the crust, and sedimentary rocks formed on the surface of the crust from chemically-cemented, fragmented materials.

But, when erosion and uplift exposes igneous rocks at the surface, we notice a third class of rocks at the contact point where the igneous rocks cooled. This third class marks the limit of the once-molten mass. We can also tell from the igneous matching-boundary contacts that these rocks were once in deep burial too.

Like igneous rocks which are crystalline and essentially homogeneous, these new rocks are layered like sedimentary rocks. Instead of being cemented together, the metamorphic rocks are crystalline; with intergrown crystals, that are not as well shaped as in igneous rocks. The crystals are irregular with a ragged texture.

The layering points to a sedimentary origin; the matching igneous boundary of these rocks indicate deep burial; the ragged crystal texture, with layering, points to non-melting. What has happened?

The answer lies in a sedimentary rock that was deeply buried, but did not melt. The usual cemented-fragment structure was erased with the birth of the crystalline rock. The deep burial caused a solid-state change, called recrystallization, which can take place under conditions of heat and pressure. This third class of rocks is called *metamorphic rock.* For purposes of future identification, when a rock description has the prefix *meta*, as in metasilicate, you know it is a metamorphic rock.

Metamorphic rocks are characteristically found around and

imbedded in a large igneous intrusion such as a sill, stock or laccolith. This contact point nearest the magma, is the most intense zone within the country rock. It is known as the *contact aureole*, and can reach as little as an inch into the invaded host rock, or be as much as a mile in thickness.

ROCKPECKER TIP # 48 - Where the contact intrusive enters a limestone formation, large amounts of iron and magnesium are introduced into the contact aureole to form lime-bearing silicates. Such a body is called a *skarn*. Many important ore deposits are associated with skarns in altered limestones.

ROCKPECKER TIP # 49 - Limestone and dolomite contact zones show the characteristic minerals of spinel, graphite, wollastonite, andradite, diopside, grossular, tremolite and vesuvianite. (idocrase)

ROCKPECKER TIP # 50 - If you find spotted rocks with a slaty cleavage, (easily broken into thin sheets) you know you are at the outer edge of the metamorphic aureole.

While there are four types of metamorphism, only two are usually mentioned. They are *regional metamorphism and contact metamorphism.* Regional metamorphism takes place over large rock masses with little change in mass composition. (The new texture and any new minerals form from the original rock contents.) It is not directly associated with an igneous intrusion. Regional metamorphism is classed as low- middle- or high-grade depending upon how much rock alteration has taken place. Certain minerals are indicators of each grade. For example: Temperature and pressure are stability factors in every mineral. If we consider mineral A, stable at 400 C, and mineral B, stable at 900 C, then thrust both A and B in a temperature-pressure environment outside of their stability range to 600 C, both A and B will attempt to adapt. A will alter to C; B will alter to D. Minerals such as kyanite, biotite, staurolite, garnet, epidote, andalusite are indicators of specific regional metamorphic zones of temperature-pressure.

ROCKPECKER TIP # 51 - Examine altered country rocks in the vicinity of regional metamorphism. They can tell you the extent of the metamorphic grade present at that location.

The second type of metamorphism places the sedimentary rocks close enough to the igneous intrusion to cause special effects to be produced in the country rocks by the magma. Magmatic heat and mineral stream fluids emanating from the intrusive induce the changes. This type is called *contact metamorphism.*

If you think of the igneous intrusion as deeply buried, or at least as hypabyssal, acting as a giant, sweating radiator, you can guess, that contact metamorphism will grade into zones of regional metamorphism, as the distance from the intrusive increases. This increasing zonal distance accounts for diminishing country rock recrystallization.

How can you tell a rock that has been metamorphosed?

While the affected rock can be one that has already undergone metamorphism in an earlier episode, a sedimentary rock such as shale will turn to slate; sandstone to quartzite. Sandstone conglomerate will turn to meta-conglomerate with a look of blood and tongue lunch meat. The gravels will look stretched into *eye shapes,* within a crystalline matrix intergrown with mica and quartz. Limestone will recrystallize into larger grains to become a *marble.* Impure limestone will produce a marble with delicate veining formed by the impurities.

Schist and gneiss are descriptive words to picture the degree of banding or layering in metamorphic rock called *foliation.* Shearing under compression causes rock foliation which gives the rock easy splitability at the banding-line.

In normal granitic rock the mineral grains are uniformly distributed in a random, intergrown fashion. In metamorphic gneissic rock, the grains are crushed and cracked to medium fine grains that produce a streaky texture. As grain-crushing advances, a fine foliation appears. The rock grains have been ground to a pasty mass with most particles too small to visually identify. This grain-crushing has been assigned to a the third type of rock change called *cataclastic metamorphism*. *Gneiss,* then is a long, platy, grained rock of alternating light and dark bands formed from granite under high-grade metamorphism.

Schist can be derived from any rock, and therefore is the most abundant. It is wavy-layered, banded in recrystallized minerals such as mica, talc, chlorite. It also is foliated, so that in tends to break between layers. Schists are named according to the minerals they contain. Greenschist from its green chlorite color; quartz schist from impure sandstone; calc-schist from impure limestone.

The slate cycle begins with weathered feldspar going into clay. Clay to shale; shale to slate in low grade: slate to phyllite in medium grade; thence phyllite to schist in high-grade metamorphism.

Technically, if you prefix "meta" in front of a rock name, you can call slate, "meta-shale"; quartzite, "meta-sandstone"; Marble, "meta-limestone".

ROCKPECKER TIP # 52 – METAMORPHIC ROCK CONCLUSIONS

However hydrothermal solutions form, they tend to create four general types of ore deposits: 1. *CONTACT METAMORPHIC DEPOSITS,* 2. *HYDROTHERMAL VEINS,* 3. *DISSEMINATED DEPOSITS,* 4. *HOT-SPRING DEPOSITS.*

Contact metamorphism can create ores of copper, silver, zinc, lead, and other metals in country rock. The country rock may be completely or partially removed and replaced by ore. This is particularly true of limestone beds, which easily react with hydrothermal solutions. Most hydrothermal ores are metallic sulfides, often mixed with bull quartz. (milky quartz) These ore bodies can be quite rich and very extensive.

Disseminated gold is mined in open pits and are often referred to as Carlin types, for the Carlin Nevada deposits. The largest gold mine in the United States is an underground mine, where finely disseminated gold is extracted from *FOLDED METAMORPHIC ROCK.*

METAMORPHIC ROCK IDENTIFICATIONS

ROCK NAME	METAMOPHISM TYPE	TEXTURE TYPE	GRAIN SIZE	CHEMICAL COMPOSITION	COLOR APPEARANCE	PARENT ROCK META GRADE
GNEISS	Regional	alternating light/dark bands	long platy	biotite, quartz,feldspar	black/white gray bands	granite hi-grade Meta
SCHIST	regional	parallel bands elongated	coarse to fine mix	quartz,feldspar chlorite,mica	shiny raisiney	shale hi-grade Meta
PHYLLITE	regional	slaty crinkled	aphanitic can't see	quartz, fledspar mica	silky sparkling	shale medium grade M.
SLATE	regional	slaty	dense smooth	quartz, feldspar mica	variable	shale low grade M.
METACON-GLOMERATE	Cataclastic regional	deformed pebbles	NF variable in fine matrix	quartz,feldspar clay	pressed pebbles	granite,clay sand grade-L to M
QUARTZITE	Contact	welded sand grains	non-foliated variable grs.	quartz	various sugary	quartz sandstone grade-LtoH
HORNFELS	contact & regional	fine grained dense	NF aphanitic	quartz,feldspar	dark Gr., Bk.	shale basalt grade-MtoH
MARBLE	contact & regional	crystaline coarse grs.	NF	calcite dolomite	various	limestone dolomite grade-LtoH
ANTHRACITE COAL	Regional	dense	NF	carbon	shiny black	bituminous coal peat grade -LOW

CHAPTER 23

MINERAL EMPLACEMENTS

WE WILL NOW FOCUS ON ORE BEARING FLUIDS AND THEIR MIGRATION TO FAVORABLE EMPLACEMENT SITES AS ECONOMIC MINERAL DEPOSITS. THINK IN TERMS OF GUTTERS OF GOLD.

MAGMATIC SEGREGATION DEPOSITS - Magmas are an important source of many mineral deposits. Crystal separation into oxides, sulfides, and chlorides combine with metals to form common ore deposits. Native metals can be deposited within an intrusive, or be injected into country rock as phases of the separation process. Direct crystallization products are referred to as *magmatic segregation deposits.*

In general, igneous rocks have only a few main constituents. Among these are the feldspars, pyroxenes, amphiboles, micas, together with quartz and minor amounts of accessory and accidental silicate minerals. Silicates,(as we have seen in the mineral class chapter) are distinguished from other inorganic substances by a type of atomic structure represented by a tetrahedron network of oxygen atoms linked with atoms of silicon. The spaces between the regular groups of oxygen atoms are occupied by metal atoms. In such a network there are no molecules. Except for oxygen, any atom in the structure may be replaced by any other atom of similar atomic radius. Most inorganic compounds are ionic, so the term ionic radius is preferred. It is important to remember that an ionic radius is not a fixed quantity. Its ionic size varies with its electrical charge. A negative ion, being an atom that has picked up one or more electrons, is larger than a positive ion, which is an atom that has lost one or more electrons. An element can have as many ionic radii as it has valences. *Valence* determines the number of atoms that will unite chemically.

The *sharing* of electrons between two or more atoms is called a covalent bond. The *attraction* that an atom has for the electrons it shares with another atom is called *electronegativity.* Compared to nonmetals, which have a high electronegativity, metallic elements have a low electronegativity. This means that atoms of metals tend to lose electrons more readily than atoms of nonmetals. Since opposites attract, the positive-negative attraction is what holds atoms together. When one element gives up electrons to another element, this is called *ionic bonding*. The process of removing an electron from a neutral atom is called *ionization.* When two or more atoms combine to form a molecule, we say that they are joined together by a chemical bond. How atoms combine, solves the mystery of how just over 100 elements can combine to form more than three million known compounds.

As you can imagine, the silicon tetrahedron can accommodate aluminum, chromium, vanadium and ferric iron, as well as magnesium, ferrous iron, manganese and nickel in its structure, without causing serious distortion of the structure. In other words, two minerals of *different* chemical composition can crystallize into the same atomic crystal structure. Two minerals with the *same* chemical composition can have different atomic crystal lattices. For example, $CaCO_3$, calcium carbonate, is the chemical formula for both calcite and aragonite. The crystal lattice for calcite is hexagonal, with crystal faces that meet at 120 degrees; while that of aragonite is orthorhombic with brick-shaped crystals. These are called *polymorphs.*

Certain ore minerals are characteristic of specific igneous rocks. Ores commonly found with mafic rocks (black granite) include chromite, ilmenite, apatite. By broad genetic class, chromium, tantalum, thorium, platinum group metals, asbestos, nickel, titanium, vanadium have mafic connections. Again, by broad genetic class, felsic rock (colored granite) associations are silver, tungsten, copper, molybdenum as accessory minerals. Siliceous rock accessory minerals are zircon, monazite, uraninite, cassiterite. Many of the precious metal deposits fall in the felsic category.

HYDROTHERMAL ALTERATION - Mineral streams, defined as metallized liquids, gases, or both in combination, can deposit on, and alter, the enclosing country rock. This can occur both at the point of intrusive contact, or at a distance from the magma intrusion. This process is referred to as *hydrothermal alteration.* The hot mineralized solutions travel and trespass the available country rock passages, to paint a little of themselves on the passageway walls. In doing so, the wallrock is altered by enrichment of metal additions, or mineral-depleted by the leaching action of the mobile fluids. Metal additions to the enclosing rocks are termed *epigenic deposits.* That is, "later formed" than the rocks which contain them.

By broad genetic class, manganese, cobalt, antimony, asbestos, fluorine, gold, bismuth, silver, tungsten, cadmium, beryllium, zinc, selenium, mercury, lead, barite, sulfur are minerals of hydrothermal origin.

REPLACEMENT ORE DEPOSITS - As the mineral streams pass through rocks of carbonate beds, such as limestone, a slightly acid solution will dissolve the carbonate and then replace the dissolved minerals with some of its metallic load. The replacement is usually on a one for one basis. That is, no gain or loss in the sense of an unfilled space. The last statement is not precise enough, in that it does not consider a continuing acid solution, now barren of any metal to offer replacement. These are termed *replacement deposits.* Sometimes you will hear the word "metasomatism". This is the type of deposition that replaces sap in tree growth-rings with silica solutions to petrify wood. Metasomatic replacement is the most important process in later-formed mineral deposits. It is thought that the substitution-replacement exchange is like a "cut and fill" nearly simultaneous process. Theoretically, given the proper conditions, any mineral can replace any other mineral. Think of this as a new mineral growing within the body of an older, different, mineral deposit.

CAVITY FILLING ORE DEPOSITS - Where metallics simply precipitate to fill rock openings, this process is called *cavity filling.* More mineral deposits are formed by the filling of cavities than by any other process. The most important filled cavity is known as a *fissure vein.*

FISSURE ORE DEPOSITS - These are table-like cracks in the country rock. They can run continuously for a great distance and depth. Tectonic events can cause a fissure which is called a *fault.* However, all fissures need not be faults. They are solution pathways for metal or minerals. When the crack is filled, it forms a fissure vein.

SADDLE REEF ORE DEPOSITS – Folding of sedimentary strata results in a bent roller coaster-like structure. Where the anticline (upward rising top portion) of the fold is tightly bent, the crest develops multiple joint-like cracks which provide avenues for cavity filling. Unless the ore contains gold or other high-value mineral, forget it. There will not be enough economic ore to work profitably.

FOLD CRACKS ORE DEPOSITS – As in saddle reefs, folding of brittle sedimentary rocks on the syncline (downward dropping top portion) can cause a series of disconnected tension cracks. These fractures are wedge-shaped but run along the axis of a fold for a good distance. Anticlinal gold and synclinal copper has been found in these cracks which are known as *pitches and flats.*

ROCKPECKER TIP # 53 – Where the crest of the anticline is overlain with shale or other impermeable rock, expect the shale to have acted as a cap to contain and concentrate the upwardly mobile volatiles in the crest-saddle. Look for gold and pyrites.

SHEAR-ZONE ORE DEPOSITS – The stresses that cause faults tear the adjacent wall rocks into multiple closely spaced ruptures. Generally these are thin sheet-like deep-seated fractures. They may be very thin and narrow, but with large flat overall areas, very favorable guideways for cavity filling and replacement deposits.

STOCKWORK ORE DEPOSITS – These have yielded ores of tin, gold, silver, copper, cobalt, lead, zinc, molybdenum, mercury and asbestos. If you look at the individual hairs of a shaving brush, imagine each hair as being a veinlet. These are called *vein swarms.* While there will be many veinlets, there will also be so much gangue that the ore will have a low tenor, for the entire mass must be mined. It is thought that the hair-like veinlet-cracks formed when the upper igneous area quench-cooled.

LADDER VEIN ORE DEPOSITS - Gold quartz, molybdenite and copper ladder veins have been found in dikes with mineral-filled ladder rungs. It is thought that these cracks, rising upward in the dike, are cooling contraction joints, or based on fissures running into the wall rock, were caused by the movement of weak wall rocks. This type ore may be economically mined if the mineral-filled rungs are closely spaced.

BRECCIA PIPE ORE DEPOSITS - When during a tectonic event, a fault line abrades to fill the fault with pieces of angular wall rock debris, an easy solution pathway for deposition is formed. These deposits are sometimes called chimney breccia deposits.

CAVE ORE DEPOSITS - Usually found in limestone at shallow depth, where a stream has solution-dissolved a rock cave. The roof rock and other debris drops to the floor of the cave with later mineralized solutions filling and concentrating a floor deposit. Replacement has also been credited with mineral deposition. Zinc, lead, silver, vanadium and copper have been found as cave fillings.

SANDSTONE PORE-SPACE FILLINGS - Pore spaces in rocks may contain ores of copper, silver, lead, vanadium, uranium and mercury. These ores are mostly low grade, but deposits are huge. *Red bed ores* are copper ores which have cemented the quartz grains in sandstone.

LATERAL SECRETION ORE DEPOSITS - Where the country rock is leached of its metals by barren hot solutions that carry, concentrate, and deposit those metals to some other favorable site, we have the theory of *lateral secretion.* Another leg of this theory has the hot solutions bursting through the cap of the intrusive to distribute the metallic solution in the cracks over the top area of the intrusive. (Think of a shower bath.)

The concept that an orebody can be chemically scavenged and depleted of its metals, then transported over distance, to be redeposited in fractures as vein deposits has been around since the 16th century. Many geologists assign a low level of importance to this idea. They firmly believe that later-formed metallic ore deposits have their roots in igneous activity. There is some slight evidence that *thermal metamorphism* plays a part in *solution remobilization.* Quartz-carbonate-gold vein systems; quartz-ankerite-gold veins with talc-carbonate alteration in mafic volcanic rocks, seem to point to lateral secretion with metamorphic remobilization. This is for the reason that ankerite is a carbonate, which is of chemical sedimentary origin. By analogy, when you think of heap leaching, chemical leaching as a natural counterpart, sets lateral secretion a short step away.

METAMORPHIC ORE DEPOSITS - Where a deep-seated intrusive takes thousands of years to cool, the pressure, temperature and emanating hot solutions can change the form and texture of an ore deposit. Recombination of minerals or recrystalization to adapt to the new environment, can form a new and different mineral. This geologic process is called *metamorphism.* Think of a caterpillar turning into a butterfly.

SECONDARY ENRICHMENT ORE DEPOSITS - After a lean-ore mineral deposit forms, secondary processes involving weathering, erosion and enrichment can upgrade and sweeten an ore into an economic deposit. This process is termed *supergene enrichment or secondary enrichment.* In a later chapter we will examine a gossan.

PLACER ORE DEPOSITS - Weathering along with the transportation, sorting and deposition action of water can form deposits of resistant minerals called *placers.* We will examine a tertiary placer later.

EVAPORATION ORE DEPOSITS - Chemical weathering of soluble substances removed to favorable deposition sites form valuable clay deposits. Where the dissolved mineral solutions settle in pond-like basins, evaporation over time, forms gypsum, bauxite and borax residue deposits. These are the *evaporites* that provide valuable industrial minerals.

FUMAROLIC ORE DEPOSITS - Where a volcano was the source of OCEAN FLOOR stratabound layers of tuff, then later-on impregnated by mineral rich smokers, the metals thusly deposited are held in the tuff strata. As the seas receded or tectonic events raised the ocean floor to continental levels, the sulfur-rich metallized tuff ore deposit became accessible. It has been found by measuring from the volcanic center outward, that as the iron, copper sulfide strata content decreases, the zinc, lead, silver, manganese, barium and fluorine content increases.

PEGMATITE ORE DEPOSITS - A pegmatite is an intrusive, igneous rock which is basically a coarse-grained granite. It is composed of quartz, feldspar and usually mica. The economic minerals are: beryl, pollucite, bertrandite, corundum, gemstones, spodumene, lepidolite, amblygonite, petalite, tantalite, columbite, wolframite, ferberite, scheelite, molybdenite and uraninite. The chief feldspar is microcline, which often is intergrown with quartz in patterns that resembles letters of the alphabet or semitic letters. This special type of pegmatite is called *graphic granite or alphabet granite.*

Pegmatite rock is a natural storehouse of minerals. It is probably the most interesting rock group known. Most pegmatite is merely coarse giant granite. This is the normal find. The rare-mineral pegmatite is the exceptional find, for its core is massive quartz with outward zones in successive stages of unique beryl, monazite, columbite, lepidolite and topaz crystallization. Spodumene in log-shaped, crystals 20 feet long and five feet in diameter have been found in the Black Hills, Maine and New Hampshire. Giant crystal growth is a pegmatite characteristic.

Pegmatites form as dikes, sills and lenses that can run very deep or pinch out in a few feet. Faults control cross-cutting relationships, branches and abrupt changes in direction. They may be concordant or discordant bodies. Many occur in metamorphic rock, particularly with schist and gneiss.

Pegmatites are believed to be injections of late-stage magma from a parent batholith; that crystallization proceeds from the fissure wall inward in successive stages; that fractures within the partially frozen pegmatite allow, hot gases and unique mineral-forming solutions that contain the rare elements, fluorine, lithium, beryllium to enter, so as to stew in its own juice.

BUT CONSIDER THE WORDS OF A MESA GRANDE PEGMATITE MINE OWNER WHO WORKED THESE MINES FROM 1901 TO 1918: The pocket was fifty feet long and not over twelve feet wide, it was very productive of morganites and stibiotantalite. At the north end of the pocket, the pay streak pinched down and scattered. I continued down another fifteen feet when I struck a vein about twenty-four inches thick. A chunk of the hanging wall, about four by three feet and 12 inches thick had tipped out. It lay with the top end touching the footwall. The lower end still retaining its footing on the hanging wall. The slab lay just as it had when it was part of the hard rock hanging wall. I could have fitted it back into place. There I saw a fault fissure pitched up and down lengthwise with the slab. This also coincided with the mineral streams *which came up from below, and in the same direction.* I have found several such examples as this slab, but none so perfectly preserved. If the slab had fallen over when the fault pitched at the angle and direction as when I found it, the slab would have fallen into the fissure. This leads me to believe that the slab had fallen *after* the pegmatite was emplaced and *before* the vein was filled.

This writer also concludes that the pegmatite was injected first; that the pockets were formed by barren fluids or gases that escaped to leave a hollow void; that later faulting cracked the pegmatite to allow rare, mineralized, hydrothermal fluids to enter and grow their crystal tourmalines and related gemstone minerals; that the source of the exotic fluid was either of magmatic origin or from thermal metamorphic lateral secretion fluids; that pegmatites grow their minerals in an open fluid system.

ASIDE: THIS SAME GEM MINER IS QUOTED:
"I should have taken pictures of many things I have seen in the mines. There were wonderful specimens destroyed in getting tourmalines out of them. I was guilty of this waste, along with others. But we were after tourmalines, not specimens. Specimens wouldn't bring the money the tourmaline crystals would when cut, and cutting was my job."

PRELUDE TO SEDIMENTARY ROCKS AND STRATA LAYER INTERPRETATION.

As older sediments are pushed down by the overlying weight of younger deposits, they gradually compress into solid rock. Gravel beds become *conglomerate*; sand grains turn into *sandstone;* clay muds compress to *shale;* dissolved calcium carbonate forms *limestone.* A limestone layer found above sea level suggests that it once formed the bed of a sea; coal seams indicate an ancient swamp; layers of salt or gypsum point to a dried-up salt water body.

Geological forces can uplift the sedimentary beds, tilt them at angles to produce *unconformities*, then hide them in younger sediments. Where your economic mineral rests within a terminated strata layer, deciphering the sequence of events can help you relocate your mineral find through strata interpretation.

It is the opinion of the author, that valuable economic minerals disseminated in sedimentary rocks will far exceed all previous ore-value finds. It is for this reason that chapters on sedimentary rocks and their formation processes are detailed.

CHAPTER 24

SEDIMENTARY ROCKS

THE MAKING OF A PLACER

THE BREAKDOWN OF ROCKS INTO SAND & REBIRTH AS NEW ROCKS.

Sedimentary rocks cover roughly 75% of the earth's surface. If you were to take a grab-sample of loose sediment from the surface of the earth, you would hold 8 of the most abundant elements found in the earth's crust. The quartz sand grains provide (1) silicon and (2) oxygen; feldspar grains, the (3) calcium, (4) potassium, (5) sodium; biotite, the (6) aluminum, (7) magnesium and (8) iron. These 8 elements are found in 98% of all rocks. Your hand-held sediment sample may once have been part of a granitic magma intrusion fire-formed and deposited deep within the bowels of the earth's crust.

All sedimentary rocks start their life as bits and pieces of other rocks. In their former life they may have been fire-formed rocks, heat baked rocks or other sedimentary rocks. These comprise the three major rock groups: igneous, metamorphic, and sedimentary. It follows that the destruction of these rock groups into sediments, can release their contents to form new, valuable, economic mineral concentrations in a younger, later formed sedimentary rock.

ROCKPECKER TIP # 54 – Be alert and search for surface ores exposed to weathering and found in surface environments known as the zone of weathering. Low grade ore, which in itself is not economic, may be sweetened or enriched by the weathering process. The ores may then form in the zone of weathering. The most important are ores of iron, nickel, and aluminum. Two examples are:

1. Where the host rock and gangue are destroyed to locate

generally in place as "lag" deposits. Barite, fluorite, galena, gold, ilmenite and phosphate are possible mineral finds.

2. Weathered rock on destruction slopes known as *eluvial deposits*. These deposits are still nearly in place, but erosion has removed the weathered material. Slight transportation has moved the mineral ores down-slope to deposit in a feather-like zone of accumulation. Free gold, beryl, cassiterite, chromite, diamonds, platinum, sheelite and wolframite have been found in this type of weathered environment.

It can be helpful to explore the destructive processes of: *weathering, erosion, transportation and deposition*. Each of these contribute in a specific manner to the production of a younger placer deposit and of a new sedimentary rock.

Weathering is a physical or chemical action on a rock at or near the surface, that when given enough time, disintegrates the rock surface. Weathering is a general term for all the ways in which a rock may be broken down. It is the interaction of crustal matter and atmospheric conditions that can effect rocks down to the water table.

LET'S PRESUPPOSE A GRANITE FORMATION WITH A QUARTZ/GOLD VEIN AND FOLLOW THE GOLD TO EACH LIKELY PLACER DEPOSIT.

In the case of physical weathering (also called mechanical weathering) the rock simply breaks into smaller pieces with no change in the original rock composition. Look at it as a simple division process.

On the other hand, chemical weathering decomposes the rock along the cemented mineral grain boundaries. As the mineral grain contact points weaken, the tougher mineral grains form a crumbly grain mass. Weathering is a long time-frame process. Essentially, a crack opens to permit chemical or physical attack to the rock surface. An important point to remember is that chemical weathering can produce surface alteration on an outcrop to form new chemical compounds. It can also leach-out or concentrate surficial minerals. AT THIS POINT OUR QUARTZ AND GRANITE

ARE BEGINNING TO DETERIORATE.

ROCKPECKER TIP # 55 – Chip out all the weathered material before taking an assay sample. Chemical weathering can distort the assay results to show a rich or poor sample, while the underlying rock may contain economic or sub-marginal values. Be sure your sample is taken from a fresh and unweathered outcrop to avoid false results.

Overall, weathering has beneficial effects in that it makes the rock pieces available for removal so that the next cycle process can operate. Consider too, these effects attributed to the weathering process: The furnishing of the raw material for soil formation; creation of landform slopes; alteration to new mineral forms; concentration of metals into economic deposits. Newly formed soil, combined with organic debris, releases nutrients for the growth of plant life and other living organisms.

ROCKPECKER TIP # 56 – The raw material produced by weathering, not only forms future economic mineral concentrations of copper, aluminum, iron, gold and uranium, but also points the way to the lode-source of some of these metals. THAT IS, THE SITE OF THE ORIGINAL GOLD/QUARTZ VEIN.

Last, but not least, weathering is the beginning of the production cycle of raw materials that will form future sedimentary rocks.

There are various types of rock break-up caused by weathering. If we examine *granular disintegration*, we can see shale breaking into clay particles. Where a granite boulder surface is flaking off into rounded onion-like layers, the spalling is called *exfoliation*. Metamorphosed shale, called slate, breaks into parallel, horizontal layers at the weak bonds to give a *cleavage splitting*.

The rate of weathering can vary as to depth and type depending on rock composition and environment. For example: The rate will be slower under a vegatative cover such as trees, and speedier in climates with hot and humid conditions. Then again, the rock mineral grain bonding and grain interlocking arrangement, as described by MOHS hardness scale, together with chemical composition, can accelerate or slow the rate of decomposition. A granite type that is resistant to physical weathering because of its hard quartz content, will decompose more readily under chemical weathering where the feldspar content is attacked by moist air vapors combined with plant CO^2 emissions to form carbonic acid. In the case of feldspar, the weakest links are the elements potassium, sodium or calcium, which the acid dissolves into solution to alter the feldspar to hydrous aluminum silicate. We know that mineral as clay. OUR GOLD/QUARTZ VEIN IS BEING RELEASED FROM ITS GRANITE MATRIX.

The surface position of a rock lying on a slope will weather faster than one on a plain where only shallow weathering can occur. Mechanical or physical weathering at the higher mountain elevations invoke a freeze-thaw-melt cycle through frost action. This can be frost wedging, frost heaving or frost shattering.

Moisture freezing and expanding in a rock fissure exerts lateral pressure on the crack to slice it like a loaf of bread. This type of weathering is called *frost wedging.*

When the soil beneath a rock gets wet and runs through the freeze-thaw-melt cycle, expansion of the soil as a bulging pressure upward, or vertical pressure, raises the rock. This heaving causes rock stress cracks and fissures to appear, Weathering will operate as *frost heaving.*

A rock containing water within its pore space can be fractured by alternate freezing, expansion and thawing of that water. Rock deterioration by this method is known as *frost shattering.*

Weathering breaks down rocks whether stationary or moving. For example, a landslide releases unweathered rock into a stream. During the rock's journey in the watercourse, it grinds and abrades on other rocks. This friction rubbing also is considered physical weathering. THE DURABLE GOLD IS RUBBED SMOOTH DURING TRANSIT.

ROCKPECKER TIP # 57 – Where transport and deposition operate to bring the sediments to rest, alluvial or marine placers can be formed. Look for free gold, cassiterite, chromite, diamonds, magnetite, columbite, monazite, ilmenite, garnet, sheelite, wolframite, platinum and rutile minerals as surface ores.

The weight of overlying mass or overburden places downward pressure on the underlying rock structures. Where the ground cover is removed, or when a glacier melts, the downward pressure is relieved and the rock expands. In both these cases the slow or fast release will cause cracks to appear as joints to permit entry of weathering agents. This phenomenon is a common occurrence in rock quarries when large slabs are cut from the face of the work. The expansion and jointing is heard as a distinct snapping sound, much like the breaking of a rubber band.

Root penetration into existing cracks expand and widen the opening as growth proceeds; burrowing animals that remove underlying rock support; desert area salt condensation, deposition and expansion of salt crystals; forest fire heat cracks, all further the work of weathering. In all these cases tiny hairline cracks are enlarged to allow weathering to disintegrate and crumble the rock into particle sediments that allow the second phase know as erosion to operate.

ROCKPECKER TIP # 58 – Burrowing animal diggings can bring to light some of the underlying minerals. It can be profitable to examine this loose material for evidence of mineralization.

Erosion is the weathering stage where the loosened particles are removed from the rock. In this phase, an agent such as a stream or glacier, picks up the particles to remove them to a new location. In the case of an eroding agent such as a light rain, removal may consist of merely droppings to the base of the rock to await further transport by other means. We tend to think of sediments much as dust particles settling on furniture, but we need to expand our thinking so that sediments include, not only loose grains, but boulders, pebbles and other products of weathering. If a solid, loose particle originated from weathering and erosion of pre-existing rock, or chemical precipitation from a solution, or secretion from a living organism, that piece of matter is considered sediment. For the sake of technical minds, sediment is sorted by decreasing size into *gravels* that are rounded boulders, cobbles or pebbles; sand grains, silt and the finest size called clay.

ROCKPECKER TIP # 59 – Consider whether weathering has leached the rock of impurities and the residue has been enriched to ore grades. Examples are bauxite, iron, nickel, cobalt, chromium and manganese.

ROCKPECKER TIP # 60 – Consider whether chemical weathering has formed new minerals from dissolved deposits. Particularly, search for enrichment zones in sulfide orebodies. Examples are copper, nickel, silver, zinc, lead, mercury and cadmium.

ROCKPECKER TIP # 61 – Attempting to associate oxidized minerals to their parent-source can lead to much ambiguity. The leaching can react with a myriad of minerals to reach diverse and uncertain results. Attention to small details in evaluating a leached outcrop is time well spent. The alternative is back-breaking digging or expensive depth work.

Transportation and sorting work hand in hand. Transport moves the sediment along to round off the sharp edges of rock fragments through grinding and abrasion. Stream transportation provides the most rapid rounding; from a sheer speed standpoint, it is the most important. Sorting works to arrange sediments according to size. It stands to reason that a streamflow of specific velocity will drop its heaviest burdens first, followed by diminishing sediment sizes as it runs its course. It also follows that the energy of the stream is directly related to stream slope.

Once the sediment comes to rest because the transporting agency can no longer carry the burden, **deposition** occurs. The location of deposition can be a bend in the stream, a lake bottom, a desert valley, or even a sand dune. This is called the *environment of deposition*, which brings us to the word *facies*.

In a sedimentary rock, the origin that formed its environment is called its facies. For example, sandstone in a river facies, coal in a swamp facies. At the point of final deposition, compaction of mineral grains takes place. The voids between contacting grain boundaries fill with calcite, iron, silt or silica in solution. This provides the cement that seals the sediment into a cohesive rock unit. Geologists are fond of short-cut talking so they give names to processes that work as a group to form a rock. Loose sediment squeezed and compacted, then cemented into a sedimentary rock is known by the general term *lithification.* This term also includes solutions that crystallize directly into rock, and the consolidation of plant and animal remains, into limestone and coal. These are different types of rock called *clastic*, if from sediment; *chemical*, if from direct solution; and *organic*, if from the remains of living organisms.

Clastic rocks are divided into divisions known as *conglomerate, breccia, sandstone and variations on shale*. Note that all of these are based on compacted, cemented sediments. If you pick up a rock with large rounded pebbles cemented together, it is conglomerate. If the rock has sharp, angular pieces, it is breccia. Breccia comes from the German word *brechen* meaning *to break*. The fractured nature of breccia tells us it lithified close to where it was found, whereas the rounded pebbles in conglomerate abraded over a longer travel course.

Sandstone can vary by size of grain, degree of roundness, or mineral content, however, it all forms in a water environment that is needed to sort and grade the sediments. Quartz sandstone and arkose have in common a mineral precipitate cement, whereas graywacke has a cement of fine-grained silt and clay. If the sandstone has 95% or more quartz grains, it is *quartz sandstone*; if it has a mix of over 25% feldspar, it is *arkose*; if less than 25% feldspar, it is *feldspathic sandstone*; if it is a mix of over 25% feldspar and sand-size grains of rock, it is *graywacke*. Quartz sand grains make up the difference in arkose and graywacke composition.

Shale, practically all of which is marine, is compaction of silt or clay, wetted to form mud. Marine environments such as lake bottoms, river deltas, and ocean floor contain silt and clay, that when compacted by the weight of overlying material, removes the large water content of the mud, that after cementation, becomes shale. Shale is fine grained, smooth, laminated and splits easily. Mudstone looks and feels like shale, but lacks the laminations for easy splitting. Siltstone consists mostly of silt, which has a coarser grain than clay; It also lacks the laminations. These composition distinctions are purely geological. Shale is the commonest of sedimentary rocks. In commercial use it competes with the clays. It is a raw material for portland cement manufacture that must include lime, silica, alumina and iron oxides. Gypsum is added to regulate setting time. Impure limestone, as a natural cement, is available for less that 20% of the portland cement manufactured. High calcium limestone is used for the other 80%, but it is deficient in silica and alumina. Shale, being a hydrous aluminum silicate, is added to complete the cement formula. Portland cement cannot contain more than 5% magnesia, nor iron in the form of iron sulfide that may cause staining. Dolomite, with a 5 to 10% magnesia content, cannot be substituted for limestone as the raw material. Since shale can be replaced by clay in the cement formula, the use of one or the other rests solely on an economic decision. That is, which is cheaper to ship to the cement plant.

This anecdote is COURTESY: DR. RALPH E. PRAY, MONROVIA, CA.

Imagine, if you will, a single goldpan impacting the giant of the domestic steel industry, U.S. STEEL. (Now USX.) Their Universal Atlas Cement Division operated a white cement plant at Northhampton, Pa. The cement produced in this facility had a distinctive gray color, when compared with the white cement of the four other competitors. Needless to say, large orders for high-rise steel building frames were not being followed with the expected white cement orders. The cement contracts were being lost to the competition. Top management was at a loss to come up with the real reason for the lost business. Years of inaction were interrupted by the initiative of a newly hired, young engineer. He obtained samples of the Northhampton clay and ran the usual tests for quality. Everything in the sample appeared normal, except for a tiny extra iron content.

Picture a laboratory with the latest equipment entered clandestinely by our engineer. Under his bulging trenchcoat, he carries a goldpan to secretly pan the fine clay. The bottom of his pan shows consistent results; a small amount of black sand, just enough to color the final product. Simple magnets placed on the inside of the clay slurry pipeline solved the color problem. The discharge from the iron separator was found black as the darkest licorice; the cement was white as snow. Dr. Pray never reveals the engineer's reward.

ROCKPECKER TIP # 62 - Pastel colored clays have been found among the wall rock alteration products in certain ore deposits. These lightly colored clays may be used as a guide to further ore deposits.

Before sediments turn to rock they can develop characteristic identification features. These are called *sedimentary structures* and may be a series of horizontal layers within the rock called *bedding*. When two layers of rock are separated by a thin flat deposit layer, the flat layer is called a *bedding plane*. Bedding planes are usually caused by a change in particle mix or grain size. Whenever a current begins to slow gradually, it will release its sediment in layers, dropping the largest grains on the bottom, then proceeding to sort the grains to the next smaller sizes with a silty top layer as the icing on the cake. This type of bedding is called a *graded bed.*

Sandstone will occasionally show a bedding within the rock that appears to cut it diagonally. This is called *cross-bedding*. It can be an important indicator of the flow at the time of deposition. As the stream begins to wane and dump its load of sediment, it does so at the top or brink of the deposited sediment slope. Then, being freed of its burden, the water races downward along a steepened angle. The sand and silt carried by the water as it ebbs, writes a nearly diagonal line in the sand. It usually appears as a darkened, diagonal, shadow line and contrasts with the rest of the sandstone.

ROCKPECKER TIP # 63 – You have just located a cross-bedded sandstone formation in your search for the source of the float you found earlier. The terrane is now nearly level. There is no apparent high ground. The ancient highland has flattened. Which way do you turn to follow the ancient water-course in an upstream direction? Remember the "top of the brink" rule, and turn toward the brink; the lower portion of the diagonal points to the ancient downstream course.

Sedimentary rock history decoding is based on four simple principles. We may seem to belabor the obvious in reciting these laws. Even in their simplicity they must be spelled out, because they ground the study of strata analysis.

Principle #1. In determining the strata age, (in the sense of younger versus older) the lower stratum is the oldest. It had to be there before the strata laying atop it could be laid down. Geologists call this the *law of superposition*. *Super* meaning higher up. (This law also applies to the laying down of multiple lava flows.)

Principle #2. Most sediments, falling to the bottom of a lake or sea, were originally laid down on a level plane, or perhaps only on a slight sloping plane. Therefore, when we see tilted strata, we know that the deforming cause was a later event. Geologists call this principle the *law of original horizontality*.

Principle #3. Sediments, at first, form in a continuous stratum until interrupted by a facies environment change. Where there is an abrupt strata end, we have cause to believe that an intervening event occurred after the strata formed.

Now we can get into the meat of our analysis. Was the strata terminated by a fault? Was it stream erosion? Was it cut by a dike? Whatever the cause, the strata had to be there first only to be followed by the later severing event. The deforming event, if a fault, may carry mineralization. Does the strata gap continue on the opposite side of the stream valley? If the strata beds are tilted, the higher side was later formed by that unknown intervenor. (Reason - The strata formed level and now runs to a tilt. The tilting was the later event.) This the *law of lateral continuity*. Such a stratum can be mapped for miles across the countryside. The true order in which the sediments were laid down is called a *stratigraphic sequence*. Sooner or later, the formation will end only to be replaced by rocks of a different environment called a *separate facies*.

Principle #4. This is the *law of cross-cutting relationships*. If the stratabed is cut by another rock stratum at an angle, such as by a dike or fault, the bed is older than the magmatic intrusive or the distorting event. Conversely, the dike or fault is younger than the rocks they cut across. If a lava flow intrudes between two stratabeds, it is younger than the strata. There is a point to be aware of when you see a lava flow intrusion between strata beds. That is, how do you know that the lava was not laid down with later sediments covering the lava to make it appear as an intrusion of lava? The intrusion will show evidence of slow cooling in the form of crystal grain growth on both the top and bottom of the intruded strata. Since the upper strata was deposited after the lava flow, heat would not show any tell-tale effects. The exception to finding larger crystals lies at the chilled edges of the lava. There, a faster cooling rate was promoted by the cooler country rock strata. The country rock edges also will show heat effects. If the intrusion between two beds is discovered to be a sill, then the area of mineralization that will be most profitable to examine will be at the bottom of the sill. This is because during a long cooling period, the plastic magma of the sill may have had an opportunity to settle out the heavier minerals.

These four principles define the theory of *younging* as to sedimentary rocks.

ROCKPECKER TIP # 64 - The history of strata formation, as interrupted or altered by visible or hidden intrusives, which raveled the fissures to deposit their mineral streams, is an important guide to ore deposits. *Younging* lets you reconstruct the sequence of geological events.

ROCKPECKER TIP # 65 - Be notebook precise when you measure the width of each stratum near a site of known mineralization. Select a unique stratum layer as your benchmark. Make a strata map! That strata may reappear with precise exactness many miles away, and with it a similar mineral continuation worthy of exploration. A unique stratum to select would be limestone, a marine environment sedimentary rock composed mostly of calcite. This is a crystalline rock formed by precipitation of clastic marine organism fragments. Usually it is recrystallized to erase the clastic texture and fossils.

ROCKPECKER TIP # 66 - Calcite, being a carbonate, fizzes vigorously when a dilute acid is applied. Vinegar can be used as this acid.

Solutions containing magnesium flowing through limestone partially replace some of the calcium in calcite to form the sedimentary rock dolomite. Here too, recrystalization erases original rock history. Dolomite is both a sedimentary rock and a mineral. Distinguishing between dolomite and limestone, *in a visual way*, is mostly a 50% right or wrong guess.

ROCKPECKER TIP # 67 - Dilute hydrochloric acid will cause a slow, slow fizzing. Though the fizz will still be slow, you can speed up the reaction by first powdering dolomite, then adding the acid.

At least 35 different rocks and minerals are of fundamental importance to industry. They are the *non-metallics*, a class apart from the metallics, fuel and water. They are the *industrial minerals*.

Ancient seas and lakes can still be imagined analogous to the Great Salt Lake and the Bonneville Salt Flats, but now are covered with million year plus sediments. These sedimentary deposits are known as *evaporites*. How long did it take for evaporation and deposition to form the 200 foot deep, 625 square mile *gypsum* beds of Nova Scotia? When heated, gypsum loses three-quarters of its chemically held water. It is then known as *calcined gypsum or plaster of Paris.* Nearly all is used in making gypsum plaster, lath and wallboard.

In closing, one more point about limestone may be of interest. When it contains significant *silica or chert*, in weathering the calcium carbonate can be leached out leaving a residue we know as *tripoli* or *rottenstone.* It has an industrial use as a buffing or polishing abrasive, and looks much like *diatomite.*

WITH THE BREAKDOWN OF ROCKS INTO SAND AND SMALLER PARTICLES, A NEW SEDIMENTARY ROCK IS BORN AS A RESULT OF NATURE'S PROCESSES. BUT IN THE INTERIM, A PLACER DEPOSIT WAS FORMED. OUR ORIGINAL GOLD/QUARTZ VEIN NOW LIES IN THE SETTLING SITE AS FREE GOLD AND GRAINS OF QUARTZ SAND.

THE CHART BELOW SHOWS A FEW OF THOSE SEDIMENTARY ROCKS, AND HOW NATURAL PROCESSES COMBINED TO FORM THEM.

ROCKPECKER TIP # 68 - It is the author's opinion that economic discoveries of valuable minerals, disseminated in a sedimentary environment, far exceeding prior hardrock recoveries, will come to light.

EXAMINE THE FOLLOWING CHART OF SEDIMENTARY ROCKS.

UNDERSTANDING THEM WILL MAKE YOUR SEARCH MORE PROFITABLE.

COMMON SEDIMENTARY ROCKS - CLASSIFICATION AND IDENTIFICATION

ROCK NAME	SOURCE TYPE MATERIALS	TEXTURE	MINERALS COMPOSITION	STRUCTURE FORMATION	PROCESS LITHIFICATION	TYPE ORIGIN	GRAIN CHARACTER HOW DEPOSIT
SHALE	SILT & CLAY	FINE GR.	FELDSPAR	HORIZONTAL	COMPACTION	CLASTIC	LAKE, OCEAN
			SILICON	BEDDING	CEMENTATION		RIVER DELTA
CHALK	ORGANIC	EARTHY	$CACO_3$	HORIZONTAL	COMPACTION	BIOCLASTIC	LAKE,OCEAN
		FINE GR		BEDDING	CEMENTATION		USED LIMESTO
DIATOMITE	ORGANIC	EARTHY	$CACO_3$	HORIZONTAL	COMPACTION	BIOCLASTIC	LAKE, OCEAN
		FINE GR		BEDDING	CEMENTATION		LAKE, OCEAN
CONGLOMERATE	INORGANIC	ROUNDED	SiO_2	HORIZONTAL	COMPACTION	CLASTIC	WATER ENVIR
		GRAVEL,SAND	QUARTZ	BEDDING	CEMENTATION		MIXED GR.
ARKOSE	INORGANIC	MIXED	QUARTZ	HORIZONTAL	COMPACTION	CLASTIC	WATER ENV.GRS.
		SIZE GRS	25%+FELDSPAR	BEDDING	CEMENTATION		COARSE ANG.
GRAYWACKE	INORGANIC	MIXED SAND.	MIXED SAND	HORIZONTAL	COMPACTION	CLASTIC	WATER ENVIR
		& SILT	+15% SILT	BEDDING	CEMENTATION		DENSE GRS.
COQUINA	ORGANIC	COARSE	MARINE	HORIZONTAL	COMPACTION	BIOCLASTIC	LAKE, SEA
		SHELL-LIKE	ORGANISMS	BEDDING	CEMENTATION		
COAL - BITUMINOUS		DENSE	PLANT	HORIZONTAL	ACCUMULATION	ORGANIC	SWAMP
	ORGANIC	CRYSTAL	MATERIALS	BEDDING	COMPACTION		FUEL
LIMESTONE	CHEMICAL	CLASTIC	$CaCO_3$	HORIZONTAL	RECRYSTALIZATION	ORGANIC	WATER ENV.
	PRECIPITATION			BEDDING	CEMENTATION		CEMENT MFG.
TRAVERTINE	CHEMICAL	DENSE	$CaCO_3$	HORIZONTAL	HOT WATER SPRING	CHEMICAL	HOT H_2O
	PRECIPITATION	POROUS	DRIPSTONE	TERRACES	EVAPORATION		WATER,DRIPS
CHERT	CHEMICAL	AMORPHOUS	SiO_2	WITHIN OTHER	SILICA	CHEMICAL	UNDERGROUND
	PRECIPITATION	DENSE		ROCKS	LUMPS		WATER PRECIP.

CHAPTER 25

STRATA INTERPRETATION

ROCKPECKER TIP # 69 - The history of strata formation, as altered or interrupted by visible or hidden intrusives which traveled the fissure to deposit their mineral streams, is an important guide to ore deposits. *YOUNGING LETS YOU RECONSTRUCT THE SEQUENCE OF GEOLOGICAL EVENTS.*

ROCKPECKER TIP # 70 - When you measure the strata width in order to map each strata layer near known mineralization, be notebook precise. That strata may reappear miles away, and with it a similar mineralization.

Sedimentary rock history deciphering is based on four simple principles. These simple, obvious rules are based on logic. They found the basis for strata analysis.

#1 - In determining strata age, (in the sense of younger versus older) the lower strata is the oldest. It had to be there before the strata laying atop it could be laid down. Obvious, yes! Geologists call this the "law of superposition". "Super" meaning "higher up". This law also applies to the laying down of multiple lava flows.

#2 - Most sediments falling to the bottom of a lake, pond, or sea were originally laid down on a level plane, or perhaps on a slight sloping plane. Therefore, when we see the strata tilted upward, we know that a deforming cause happened AFTER deposition of the strata. Geologists call this the "law of initial horizontality".

#3 - Sediments, at first, form in a continuous strata until interrupted by a facies environment change. Where there is an abrupt strata end, we have cause to believe that an intervening event occurred AFTER the strata was formed.

FACIES DEFINED - In a sedimentary rock, the ORIGIN that formed its environment. For example: A sandstone in a river facies; coal in a swamp facies.

ASK YOURSELF: Did the strata terminate because of dislocation by a fault? Was stream erosion the cause of the strata ending? Does the strata end because it was severed by a dike? Whatever the cause, the strata had to be there first, followed by the severing event as second or later. The severing event, if a fault, may carry mineralization. Does the strata-gap continue on the opposite side of the stream valley?

If the strata beds are tilted, the higher side was LATER-FORMED by that unknown intervener. (Reason: The strata formed level, but now runs to a tilt.) THE TILTING WAS THE LATER EVENT. Geologists call this the "law of lateral continuity". Such a strata can be mapped across the countryside.

THE TIME ORDER IN WHICH THE SEDIMENTS WERE LAID DOWN IS CALLED A *STRATIGRAPHIC SEQUENCE.* Sooner or later the formation will end to be replaced by rocks of a different environment called a *SEPARATE FACIES.*

#4 - If the strata bed is cut by another rock strata AT AN ANGLE such as a dike or fault, the bed is OLDER than the intrusive. Conversely, the dike or fault is YOUNGER than the rocks they cut across. Call this the "law of cross-cutting".

If a lava flow intruded between two beds, it is YOUNGER than the strata. But how can you know that the lava flow intruded between two beds? It could have been first laid down with later sediment deposited atop the lava. Answer: When the lava intrudes two beds, it will show chilled edges on the top and bottom of the lava, because of faster cooling from the colder country rock layers. The lava intrusion will show crystal grain growth in the main body. It took longer to cool in its country rock blanket.

THESE FOUR PRINCIPALS DEFINE THE THEORY OF *YOUNGING AS TO*

SEDIMENTARY ROCKS. BUT WHY BOTHER?

TECTONICS

Agricola and Steno pointed our way to ore deposit formations as mineral streams deposited in channelways. THESE ORE PASSAGEWAYS ARE FOLDS, FAULTS AND FISSURES.

Just how did these convenient aisles for mineral streams occur?

What cataclysmic events of what magnitude could cause the crustal layer of the earth to crack so to pave the way for mineral deposition into those cracks?

What mechanism caused and propelled the mineral streams to their destinations?

The earth today is hot. Its temperature at the center had been estimated at 3,000 degrees C. In 1989 that estimate was raised to 6,000 degrees C, caused by radioactive heat.

Below the thin skin of the earth's crust lies a thick mantle composed of melted matter rich in elements similar to the continental crust. In a dross-like manner, the mantle's lighter elements rise to the top of the melt. The denser, heavier elements drop deeper into the mix-melt. This depletes the upper mantle of the lighter elements, to enrich the dross, which becomes the crustal layer of the earth.

With nearly five billion years of recycled crust with dross, this melting pattern melts and renews the crustal surface. The dross-like minerals just need a shove upward to deposit their ore materials.

Thanks to earthquakes which crack the crust, and volcanoes that raise and force the plastic, enriched melt through the cracks, we have a healed crust rich in possible ore deposits.

CHAPTER 26

MINERALIZING SOLUTIONS

Basically, hydrothermal mineralizing solutions are hot water solutions. The water can originate as rain, surface run-off, or thermal springs. These waters are called *meteoric waters.* The water can arise from within rock pores below the crust of the earth in which case they are called *connate waters.*

Connate water is water trapped in sediments at the time they were deposited. They are actually *fossil waters.* These fluids are rich in brines, i.e.. large amounts of soluble salts.

The heat that raises the water temperature to hot, very hot, or to the vapor state we know as steam, can come from magmatic molten rocks, or metamorphic activity.

As in a brine solution, these hydrothermal fluids can carry trace amount of minerals. It has been suggested that metals can act as electrolytes in a solution. That is, like electroplating or bronzing baby shoes.

To give you an idea of the metal carrying power of hydrothermal solutions, geothermal power wells drilled in Imperial County, California, reached 5200 feet into hot brines. The water reached 370 degrees C. with residues of copper, silver, lithium, potassium, lead, antimony, arsenic, boron, beryllium, gold, gallium, and bismuth. Within three months the discharge pipe lining had accumulated a 20% copper, 2% silver, silica-like precipitate, estimated to weigh 5 tons. It was also concluded that metamorphism was the heat source; that these metals had been leached from those country rock sediments that were undergoing metamorphism.

Ore deposits formed from hydrothermal solutions are zoned and described as *high, medium, or low temperature solutions.*

Deposits nearest the heat source (magma) are high temperature solutions. Moving away from the intrusive, we find medium or intermediate temperatures, and the farthest the low temperature solutions. This is because the hot solutions gradually lose their heat to the rock structures through which they pass and deposit characteristic identifying minerals.

As the upward magmatic pressures reach channel obstructions or previous barrier ore deposits, the pent-up pressure will open new areas to relieve itself. As new channel exits burst forth, the pressurized solution deposits its load along with the drop in pressure. Particularly in the vapor state, at constant temperature but with increasing pressure, non-volatile substances can become soluble. Consequently, where the vein wall temperature is constant, a sudden decrease in pressure, as through a new pathway, a mineral stream content dump will take place at depth.

Hydrothermal solution deposits in rock pore space and cavity filling were once thought to be the only mode of mineral formation. Now it is thought that the majority of deposits fall in the cavity filling and filled fissure vein area.

ROCKPECKER TIP # 71 – If the cavity is a fissure with crusts of different minerals in a layer cake pattern, suspect more than one hydrothermal filling episode. Ribbon quartz and crustification may be abundant in the vein. *Cockade ore* can result from crusts that form around angular rocks or breccia. Cockade ore looks like rosettes or knots in a ribbon. Look for silver and silver sulfosalt minerals.

ROCKPECKER TIP # 72 – Crustification layers are avidly sought after by specimen collectors. Even if small, they are valuable.

ROCKPECKER TIP # 73 – The presence of a waxy green quartz has been reported from many silver epithermal deposits. This green quartz is known to the miners as a *pathfinder* material indicating that ore is nearby.

When a cavity fills from the walls toward the center, it is referred to as *comb* structure. Comb structures remind me of false teeth. If the cavity is not completely filled, it is termed a *vug*.

ROCKPECKER TIP # 74 - Where you find fissure vein filling, give the wall rock extra close scrutiny. Pore space filling may have enriched the rock.

ROCKPECKER TIP # 75 - Where you find *float* with a rock-like look, examine the pore spaces for values. Then continue to look for that particular rock structure in the upper elevations. In sandstone, look for copper, lead, silver, vanadium, and uranium.

ROCKPECKER TIP # 76 - Alteration fogs, auras or halos usually due to wallrock oxidation or dehydration are prime exploration targets. This takes us to the *iron hat* clue. (see gossan)

ROCKPECKER TIP # 77 - A valuable ore vein may be only paper-thin at the point of discovery in the outcrop. Dig! Dig! Dig!

ROCKPECKER TIP # 78 - Don't overlook the concept of *drag ore*. This is where an older ore vein has been cut by a younger vein. When the younger vein filled, it dragged old vein ore with it. This might be a tough greenstone cutting a more brittle rock such as sandstone.

ROCKPECKER TIP # 79 - Where you meet an ore vein with included breccia that terminates, your continuation-of-the-vein clue is to follow the breccia, even though breccia continuation is found miles away. This is especially true where tectonic forces of folding, faulting, and shearing have shattered the rocks. The later birth of the vein filling simply included the breccia.

ROCKPECKER TIP # 80 - In limestone areas where sinkholes are prominent, small caves can be completely filled with ore minerals of lead, zinc, copper, vanadium, silver-lead, fluorspar and barite. These are small, rich, deposits termed *solution caves* or *gash* fissures. Geologists refer to these limestone terranes as *karst topography*.

The movement of mineralized fluids produces zoning in ores.

ZONING & PARAGENESIS

Zoning in ore deposits is based upon size regardless of origin. It simply means that within a single ore body, whether of sedimentary, metamorphic, or magmatic origin, moving ore fluids produce changes in ore and gangue along the pathways or channels of deposition.

The fluid deposition may be parallel to a stream channel leading away from the ore body, or in zones at increasing distances from the ore body. Two to three miles distant would not be unusual. Field observations have recorded groups of minerals forming in predictable sequence outward from the source of ore body fluids. Since the character of ore-bearing fluids changes gradually in its upward travel, different minerals will be formed along different parts of a channel. Tectonic activity allows mineralizing surges into newly opened fissures.

Many reasons have been advanced for the theoretical ore deposition sequence, but as in all controversies, no consensus has been reached. Supergene enrichment, (which we will meet later) and idealized metamorphic zone mapping are other instances of zoning.

Paragenesis, then is the order or sequence of mineral stream deposition, *as in a snapshot of a specific moment in time.* This concept has been classed as:

ORE BODY ZONING - That is within the body itself, or in an ore shoot within it.

DISTRICT ZONING - A series of closely grouped mines in the same area.

REGIONAL ZONING - A large association of ore bodies, such as the Sierra Nevada's in the West, and the Piedmont region in the East.

Strangely, the Mother Lode District of California shows no

zoning either in individual ore bodies or in the district. Yet, from map plots, a regional zoning is apparent.

Zoning boils down to a *spatial distribution* of mineral deposition. Paragenesis refers to the *order of ore deposition in a time frame of from the first to the last ores to deposit.*

Extensive studies have taken paragenesis and sequenced the deposition of ore-bearing fluids as it migrates from its source. Remember, this is a single event sequence theory. Tectonic events are usually multiple and repeating over different time spans.

ROCKPECKER TIP # 81 – Where a known hydrothermal fluid has deposited an ore body, zoning encourages you to investigate, not only the immediate area in the vicinity of the deposit, but along the strike of the entire region. If the terrane lies in such a way that the veins can radiate from the intrusive fluid source, then the unwinding coil search may lead to discovery. This is especially valid if skarn is extensively developed around the intrusive.

ROCKPECKER TIP # 82 – We will see how supergene enriched sulfide zones in a depth dimension, will show as a stacking of associated minerals in a generally predictable order.

ROCKPECKER TIP # 83 – Supergene wallrock alteration can be a *lateral guide* to a blind ore deposit. Where weathering has bleached biotite black mica to a white fine-grained sericite, or changed the biotite chemically to chlorite, that alteration process may show from one wallrock type to another near the surface. The bleaching will show as a very narrow line, behind which may lie hidden veins that measure in centimeters. For this tip to operate you must be able to relate the bleaching to weathering of hidden ore or gangue materials.

CHAPTER 27

GUIDEPOST MINERAL ASSOCIATIONS

Certain mineral associations can tip you to the possible presence of metal and semi-metal ores. MINERALS CONTAINING SPECIFIC GROUPS OF ELEMENTS FREQUENTLY OCCUR TOGETHER. THE KNOWLEDGE OF SUCH ASSOCIATIONS IS AN IMPORTANT GUIDEPOST TO UNDERSTANDING MINERAL OCCURRENCE. This chapter is presented to acquaint you with the detailed specific information you can glean from the ***MINERAL CHARTS***. *Use the charts to your advantage.*

Some minerals can fool you because of similarities in appearance. That's why fool's gold is so named. Many metal ores are in combination with minerals that are near-guides to their presence. They are called ***MINERAL ASSOCIATIONS***. BE SURE TO DISTINGUISH THEM FROM ***INDICATOR MINERALS***. Indicators are *"here I am"* signals. Associated minerals are clues to look deeper for OTHER probables. Once you locate a mineral associate you still need to identify it. *Color, luster, hardness, specific gravity, and streak tests* can go a long way to eliminate possibles. Each of these physical tests will be detailed in later chapters. Use every crutch you have!

Finally, some simple *chemical tests* will aid in the identification or elimination of suspected elements.

MINERAL DESCRIPTIONS, THEIR LOOK-ALIKES AND ASSOCIATIONS.

The mineral and mineral chart number appear in brackets.

GOLD:

[Color - gold] [Streak - gold-yellow] [Hardness 2.5 - 3.0.] [Specific Gravity - 19.3] (pure gold)

LOOKS LIKE: [Pyrite - 207] Brittle with black streak.

[Chalcopyrite - 068] Black streak.
[Phlogopite - 196] Yellow mica, crushes easily.

ASSOCIATED WITH: [quartz - 213] [pyrite - 207] [pyrrhotite - 212] [arsenopyrite - 026] [sphalerite - 234] [galena - 116] [molybdenite - 173] [phonolite (rock).]

TEST FOR GOLD: Powder your mineral. Place on charcoal with sodium carbonate. With blowpipe, heat and fuse the unknown mineral. Gold compounds are reduced to a malleable, golden button.

TELLURIDES: [sylvanite - 244] = Gold silver telluride.
[calaverite -] = Gold telluride.

TEST FOR TELLURIUM: Powder your mineral, add hot concentrated sulfuric acid. A deep crimson color results.

SILVER:

[Color - silver] [Streak - silver-white]
[Hardness - 2.5 -3.0] [Specific Gravity - 10.5] (pure silver)

LOOKS LIKE: Platinum, is harder and heavier.
Bismuth, is softer with iridescent tarnish.

ASSOCIATED WITH: [calcite - 056] [barite - 036] [fluorite - 112] [copper - 083] [uraninite - 260] [rhodocrosite - 215] [siderite - 228] [cerrusite - 063]

TEST FOR SILVER: Powder your mineral. Mix equal parts of sodium bicarbonate and powdered charcoal and with blowpipe fuse the unknown mineral on charcoal block. Silver-white button, soft and malleable will remain bright when cooled.

SULFIDES: Form an important group of ore minerals. When sulfur combines with a metal, it is a sulfide. When a metal and a semimetal combine with sulfur, it is a double sulfide, called a *sulfasalt.* Arsenic, bismuth, and antimony are common semimetals.

SILVER SULFIDES:

[argentite - 024] = Silver sulfide.
[proustite - 203] = Silver arsenic sulfide.
[pryargyrite - 206] = Silver antimony sulfide.
[polybasite - 201] = Silver copper antimony sulfide.
[stephanite - 238] = Silver antimony sulfide.

TEST FOR SULFUR: Powder your unknown mineral. Place it on a charcoal block and fuse it. Sulfur dioxide fumes will smell like "rotten egg water".

TEST FOR ARSENIC: Powder and heat your unknown mineral on a charcoal block. The fumes will give a garlic odor and a white coating will appear on the charcoal.

TEST FOR ANTIMONY: Powder and heat your unknown mineral on a charcoal block in an oxidizing flame. Dense, white, odorless fumes will rise along with a white coating on the charcoal.

COPPER

[Color - copper] [Streak - shiny copper]
[Hardness - 2.5 - 3.0] [Specific Gravity - 8.95]

LOOKS LIKE: Niccolite, has brown-black streak.

ASSOCIATED WITH: [malachite - 162] [cuprite - 092]
[azurite - 034] [chalcopyrite - 068] [gold - 124]
[silver - 230] [calcite - 056]

TEST FOR COPPER: Powder your mineral and roast it with your blowpipe to remove sulfur, arsenic, and antimony. Heat residue with an equal mixture of powdered charcoal and sodium bicarbonate. Copper-red malleable button will turn black upon cooling.

PLATINUM

[Color - silver-gray] [Streak - shiny gray]
[Hardness - 4.0 - 4.5] [Specific Gravity - 14.0 - 19.0]

LOOKS LIKE - Silver - hardness of 2.5 - 3.0 with shiny silver streak. Perhaps gray-black tarnish.

ASSOCIATED WITH: [olivine, pyroxene rocks] [magnetite - 161]
[chromite - 071]

TEST FOR PLATINUM: Infusible in blowpipe flame; may be magnetic. Dissolves in aqua regia, but not other acids.

BISMUTH

[Color - pale reddish-silver] [Streak - shiny silver]
[Hardness - 2.0 - 2.5] [Specific Gravity - 9.7 - 9.8]

LOOKS LIKE - Silver - with specific gravity of 10.5.

ASSOCIATED WITH: Ores of: [silver] [nickel] [cobalt]
[tin] [tungsten]

TEST FOR BISMUTH: Powder your mineral, mix it with sodium carbonate and fuse it on charcoal. It will be reduced to a brittle metallic button with a yellow coating near the COLD fused mass.

PYRITE = Iron Sulfide

[Color - brass-yellow] [Streak - green-black]
[Hardness - 6.0 - 6.5] [Specific Gravity - 5.01]

LOOKS LIKE: [chalcopyrite - 068] [marcasite - 164]
[gold - 124] [pyrrhotite - 212] [pendlandite - 191]

ASSOCIATED WITH: [chalcopyrite - 068] [sphalerite - 234]
[galena - 116] [found alone]

ALTERS TO: [limonite - 158] [goethite - 123]

PYRRHOTITE = Iron sulfide

[Color - bronze-brown] [Streak - black]
[Hardness - 3.5 - 4.5] [Specific Gravity - 4.6 - 4.7]

LOOKS LIKE: [pendlandite - 191] [pyrite - 207]

ASSOCIATED WITH: [pendlandite - 191] [chalcopyrite - 068]
[pyrite - 207] [magnetite - 161]

ALTERS TO: [limonite - 158] [goethite - 123]

GALENA = Lead sulfide

[Color - lead-gray] [Streak - lead-gray]
[Hardness 2.5] [Specific Gravity - 7.58]
[Other - cubic crystal]

LOOKS LIKE: [galena is distinctive]

ASSOCIATED WITH: [silver minerals] [pyrite - 207]
[sphalerite - 234] [chalcopyrite - 068] [calcite - 056]
[barite - 036] [fluorite - 112] [quartz - 213]

ALTERS TO: [cerussite - 063] [anglesite - 015]
[pyromorphite - 209] [mimetite - 172] [phosgenite - 197]

TEST FOR LEAD: Powder your mineral and heat on charcoal with your blowpipe. A metallic lead button and yellow coating will appear on the charcoal.

ROCKPECKER TIP # 84 - This is the most important ore of lead. With economic silver content, it is often mined as a silver ore. Found in hydrothermal veins and as replacement deposits in limestone and dolomite.

NICCOLITE = Nickel arsenide.

[Color - pale reddish copper] [Streak - brownish-black]
[Hardness - 5.0 - 5.5] [Specific Gravity - 7.78]

LOOKS LIKE: Copper - which is softer.

ASSOCIATED WITH: [silver - 230] [cobaltite - 078]
[skudderudite - 231] [pendlandite - 191] [pyrrhotite - 212]
[millerite - 171] [annabergite - 017] [erythrite - 108]
[chalcopyrite - 068] [arsenopyrite - 026]

TEST FOR NICKEL: Powder your mineral and dissolve it in nitric acid. Filter out brown material. To the remaining solution add dimethylglyoxime. The precipitate will be deep pink.

ARSENOPYRITE = Iron arsenic sulfide.

[Color - silver-gray] [Streak - black]
[Hardness - 5.5 - 6.0] [Specific Gravity - 6.07]

LOOKS LIKE: Marcasite - which is pale brass in color.
Cobaltite - hard to distinguish.
Skudderudite - hard to distinguish.
Glaucodot - hard to distinguish.

ROCKPECKER TIP # 85 - The arsenopyrite look-a-likes of cobaltite, skudderudite, and glaucodot should be assayed for cobalt and nickel content. You may have a cobalt or nickel ore. The mineral is the principal source of arsenic. Strike it with your hammer and take a sniff, you may sense the garlic odor. THIS IS A VALUABLE GUIDE MINERAL TO ORE DEPOSITS.

ASSOCIATED WITH: [gold - 124] [silver - 230]
[wolframite - 271] [scheelite - 223] [molybdenite - 173]
[cassiterite - 060] [galena - 116] [pyrite - 207]
[chalcopyrite - 068]

TEST: Same as for arsenic, garlic odor.

SPHALERITE = Zinc sulfide.

[Color - variable] [Streak - pale yellow-brown]
[Hardness - 3.5 - 4.0] [Specific Gravity - 3.9 - 4.1]

LOOKS LIKE: Siderite, when brown.
Galena, when black.
Cassiterite, which has a hardness of 6.0 -7.0.

ASSOCIATED WITH: [galena - 116] [pyrite - 207]
[chalcopyrite - 068] [marcasite - 164] [greenockite - 126]
[calcite - 056] [smithsonite - 232]

TEST: Same as for sulfur, rotten egg smell.

ROCKPECKER TIP # 86 - Sphalerite is the most important ore of zinc. It is the source of the co-product elements of cadmium, germanium, thallium, indium, and gallium. If you locate sphalerite, you normally also locate galena as its companion mineral.

CHALCOPYRITE = Copper iron sulfide.

[Color - gold] [Streak - greenish-black]
[Hardness - 3.5 - 4.0] [Specific Gravity - 3.9 - 4.1]

LOOKS LIKE: Pyrite, which has a hardness of 6.0 - 6.5.
Gold, which has a pale gold streak.

ASSOCIATED WITH: [pyrite - 207] [galena - 116]
[sphalerite - 234] [chalcocite - 067] [malachite - 162]
[covellite - 086] [quartz - 213] [calcite - 056]

TEST FOR IRON: Powder your unknown mineral and fuse it on charcoal. The metallic button will be magnetic. (But so will nickel and cobalt) To chemically test iron, you must do two tests, one for ferric iron and the other for ferrous iron. We'll just pass.

REALGAR = Arsenic sulfide.

[Color - reddish-orange] [Streak - orange-red]
[Hardness - 1.5 - 2.0] [Specific Gravity - 3.56]

LOOKS LIKE: Cinnabar, which has a specific gravity of 8.1.
Crocoite, which has a specific gravity of 6.1.

ASSOCIATED WITH: [orpiment - 186] [stibnite - 240]
[ores of gold, silver, and lead.]

TEST: Same as for arsenic, garlic odor when heated on charcoal.

NOTE: Realgar, when exposed to prolonged light, will decompose and disintegrate into an yellowish-orange powder.

ROCKPECKER TIP # 87 - If your prospecting centers on the precious metals, this chapter on sulfides will make life easier for you.

CHAPTER 28

OUTCROP INTERPRETATION

OUTCROPS are exposures of rock materials that can be examined *IN PLACE*. They tend to be located on steep slopes, ditches, landslide scars, and where streams have undercut the bank. Rock material essentially means "rock formations." We are looking for field evidence of hidden mineralization. Searching for visible guides that can indicate a mineral stream raised by the emplacement of an intrusive that blessed our search with a shallow deposit. It sits shallow now because the forces of weathering have exposed it to our inspection.

You must look carefully, and in detail, to study the outcrop. Ask yourself:

1. Is the rock body continuous, or does it butt up against a rock body with some verifiable difference? An obvious difference would be slate grading into shale. We are really looking for an interfering cause that sliced the rock formation, or a factor such as metamorphism that has changed some of the rock body.

a. Does the outcrop end against a fault or intruding contact? An intruding contact would be black granite penetrating colored granite. Faults are mineral stream pathways. Did the intruding contact find a path through the fault? More important to our purpose, did the intrusive carry mineral values?

b. Does the rock structure appear to have forced itself into the rock body in a way that is visually apparent? A younger dike cross-cutting strata layers. Could it be an older fault later filled with mineralization? Do we have multiple faulting ?

c. Does the outcrop vary in color or rock texture? Color and texture variations are clues to rock events that may carry mineralization. Contact metamorphism can alter the grain size through grain growth of the intruded country rock.

AS YOU CAN SEE, *LOOKING* IS THE MAJOR TASK OF THE PROSPECTOR.

Next ask:

2. Does the rock body slice across strata in sheet-like fashion or does it look like an inverted saucer stuck between strata? Is it a lens or a dike? Check the rock body at the point of difference; does it appear to have been heat changed where they meet?

3. Does the outcrop have a recognizable, distinctive shape? Can you visually follow the shape for any distance? Is it a sedimentary strata, filled fault or folded rock?

4. Can you see rock fragments within the outcrop that may have come from another source? Is it breccia, gouge or sediment? Does it look like it may have pushed through and carried with it other rocks on the path to emplacement? If a fault, did multiple faultings grind the wall rock into gouge?

5. Examine the weathered outcrop surface with your hand lens. Does it show pits where chemical erosion worked? Can you identify the less soluble grains or the rock itself? Are the less soluble grains metallic? Remove some of the weathering. Is it hiding, covering, or disguising the true rock?

NEXT CHECK FOR FOLDING:

6. Break off a sample and examine the newly fractured surface for grain size and shape.

a. Are the layers bent so as to indicate folding?

b. Does the grain orientation run differently than the grains in the abutting rock structures?

c. Does the grain orientation indicate structural deformation?

d. Are the mineral grains flattened as if pressed into plates?

e. Are the mineral grains stretched or squeezed?

A deformation, known as *shear*, can be indicated by flattened, pressed, or stretched mineral grains. Shear is an important clue to strains found in fault zones or near contact intrusions. If there is any mineralization, that's where it will be.

If your rock specimen contains a recognizable crystal, is there a *strain shadow*? If you find a shadow, the crystal may have rotated due to simple shear during the deformation. From the evidence in your specimen, can you visualize the direction of the shear forces?

ROCKPECKER TIP # 88 – To find the hingeline of a fold, check for lines, bands, or foliation in your rock specimen. Then seek outcrop surfaces perpendicular to those lines or bands. Next look for other line-sets that are parallel to those stretch-marks in your original rock specimen. Think of a caterpillar, the hingeline will be in the direction of the caterpillar's travel.

Next connect the imaginary hingeline to the fold lines, much like connecting the ribs of a umbrella to the hinge. This can give you the strike and dip of the overlying surface. You can now visualize the caterpillar moving by the corrugations of this fold. Severe tectonics can make the more brittle rocks buckle. These *buckle folds* are helpful in showing fold extensions.

Where a small-scale fold plunges steeply, it can imply more than one period of deformation. The rotational dip event occurred either before or after the smaller folds formed. To decide the sequence of events you need more evidence to reach a conclusion. Knowing the folding sequence, along with other data such as hydrothermal fluid flow from a nearby intrusive can point to possible deposits very near the hingelines.

Next examine even the tiniest faults. Are the faults filled with breccia or gouge? More importantly, is the fault younger? If the minerals are kinked or pulled apart, they had to be acted on by a later force. Wall rock breccia included in the vein indicates that the vein was later-formed. (The wall had to be there first.)

Now determine whether that *later force* was an ore-bearing intrusion or fed by hydrothermal solutions.

Broken, weathered and eroded rocks along a fault usually disguise it. Faults are rarely seen in a natural outcrop exposure. However, outcrops can suggest a fault. Check for rock parts obviously out of place, or rock patterns indicating sudden and unlikely changes. Is there a change in bedding, banding, or foliation patterns? Clues are breccia, altered rock forms, or increases in jointing.

If a natural fracture surface shows striations or scratch lines along the face or side, it is a squeezed intrusion. Think of toothpaste lines issuing from the squeezed tube-mouth.

ROCKPECKER TIP # 89 - Slickensides are polished striations, scratches or flutings found on wall rocks in a fault plane. They can indicate the direction of relative movement of the fault. By feeling the fault surface striations, the opposite wall has moved parallel to the scratch direction. However, a close inspection may be called for to determine if the scratches were only the latest episode in multiple faulting.

ROCKPECKER TIP # 90 - Multiple faulting raises the question as to whether a faultline is of pre-ore or post-ore faulting. If pre-ore faulting, the mineral stream may possibly be in the latest faultline. If post-ore faulting, the unseen continuation of a folded mineral deposit is a more direct exploration target.

ROCKPECKER TIP # 91 - Where the polished wall-rock slickensides wont match what would be the opposite rockwall, look for displacement of the outcrop VERTICALLY. Faults in dipping beds displace their outcrop patterns. The amount of movement may vary from less than an inch to many thousands of feet vertically, to more than 100 miles horizontally.

Some fractures are not faults, but they carry features of a fault. Your task requires you to follow a single rock layer up to the fracture. If the fracture crosses the single rock layer without an offset, you have a fault. If you find a rock offset in the layer you are following, you have a later fracture. Remember, faults can carry mineralization. If you visualize a rubber ball cut in half, but laid together as an uncut ball, the cut is a fault and may be mineralized. Now lay the ball halves together with an out of round lie and you have a later fracture. Later fractures occur after a mineralized intrusive event.

Once you have discovered a fault, look for rock exposures in stream banks or road cuts. Collect rock fragments alien to the wall rocks. Make a notebook entry of your findings. Regardless of how good you believe your memory to be, when you need to know, you'll have forgotten.

Faults come in many configurations. In the main, either one side slipped up or down to leave the striations. No matter how thin the fault, strata layers retain the deposition order. The strata can be matched to a hanging wall or foot wall for continuity identification. Again, "looking" for the match is your principal task. A match will give you the other side of the fault. If you're lucky, you'll find mineralization.

NARROW STRIP GEOLOGIC MAPPING

You have found what appears to be a fault, or at least a fracture. The definitive answer is still open.

SHOULD YOU FIRST VISIT ALL THE STRIKE OUTCROPS AND THEIR CONTACTS?

Yes, but with another objective also in mind and on paper. *THAT IS THE ROCK CONTACT TRACE MAP.* Start at the initial outcrop between two distinctly different rock exposures. Your mapping tools will be your compass, feet, paper and pencil. (Altimeter and clinometer to measure elevation and slope angles, if you have these instruments.)

Take a bearing or azimuth reading from your starting point along the strike to a clearly visible object. Now take a second reading at right angle to that line to establish a reference point. Mark the starting point with ZERO bearing and step-off the distance to the first clearly visible object you sighted in your first bearing. MARK THAT POINT #1. Continue your survey to points #2, #3, #4, etc. with bearings. Make any additional notes you may need for later clarification. If your fault carries mineral values, you may need professional help. Your trace map will cut the pro's time to save you money.

While the step-off method has a degree of inaccuracy, particularly on slopes, you now have a readable map of your fault-line.

At this point you can develop your mineral inspection along the bearing lines. Note any geologic features next to the traverse course. You should note folding, rock sequence, slope angles, and changes in strike and dip.

With elevation and slope angles, you can profile the area. With rock notes you can project the profile downward as the rock character changes along the step-off course.

ROCKPECKER TIP # 92 – Where the contacts you are following are concealed, walk a ZIG-ZAG course predicted by the strike and dip of the exposed outcrop.If you have difficulty locating the contact, search for float fragments of one or both rock units you are following. Other evidence can be obtained from animal burrow float diggings.

Don't overlook soil properties such as texture, color, mineral grains, sediment compaction and vegetation to get you back on track.

CHAPTER 29

SAMPLING

You have a likely prospect; perhaps even a significant discovery of ORE-GRADE MINERALIZATION. Whether the effort expended is a technical success will depend on your next two steps. They are SAMPLING and RESERVE ORE DETERMINATION. When you look at your claim as a mine delivering a broken ore stream, it is the recovered or extracted value that counts.

The first step is to collect samples which represent a snapshot of the entire ore body. Sure, you can "eyeball assay," but precision and accuracy are vital to mineral grade. The tenor of your deposit is the unbiased "real", not an item on your wish-list.

Surface samples are suspect, as not showing the ore body at depth. Surface leaching or secondary enrichment can introduce a bias or inaccuracy into the sample. Even drill core samples, where some of the core is unavailable, is suspect. The missing core segment may have none or most of the values.

The opposite bias comes into play when samples are "salted". That is, too much "high grade" ore, or too much waste material in the sample. You can anticipate the test results in advance.

If you want to falsify the results, you may as well explode gold dust into the rock crevices, or take a sealed sample sack and inject silver nitrate into the sample. You may fool the pros, but not for long! The real fool is the one you see in a mirror.

PROPER SAMPLING

1. *CHANNEL SAMPLES* – In your outcrop, use your hammer and chisel to cut a channel or slot across the area to expose fresh

material. Now take your sample. Be sure to collect the channel fragments. Take your time to get a uniform sample. Place this sample in a clean, zippered, plastic bag to avoid contamination.

2. *CHIP SAMPLES* - After you clear the weathered material from the outcrop, cut and chip a broad face at right angles to the ore zone for a distance of 15 to 20 inches. You must include wall rock samples too. This is repeated for each separate zone.

3. *GRAB SAMPLES* - You take enough fragment samples over an area large enough to mirror the grade of mineralization.

4. *DRILL CORE SAMPLES* - They are beyond the scope of this chapter. They fall into the area of "reserve determinations".

YOUR NEXT STEP IS TO TAKE THE SAMPLES TO A LAB FOR ANALYSIS.

However, before you take this step, you need to make a spending decision as to the type of analysis you are prepared to buy. If money is no object, then a "QUANTATIVE ANALYSIS" is your obvious choice. You can order a search for every element present, and the percentage or value of each. If you draw a blank, you must be prepared for a "NO VALUES" report.

You may opt for a "QUALITATIVE ANALYSIS". This means that the element tested for is either present or absent. In buying an assay, you need to specify an element certain. The assayer will determine the percentage or value of that specified element in your rock or ore. If you had called for gold and silver as the search elements, your assay report might read: 0.25 oz. Au and 9.5 oz Ag/ton. From the current market prices you can then determine the value of your ore or rock.

There is a third option that you should seriously consider. That is, run some pre-assay tests. Visual identification of native elements is possible for a few elements with the aid of a hand lens. Gold, silver, platinum, copper and iron come easily to mind. However, these elements, when chemically combined, can look like ordinary country rock. For this reason cookbook chemistry tests should be part of every serious Rockpecker's kit bag.

Don't overlook the physical identification tests of powder streak, specific gravity, luster, and hardness. They can point you to your mineral's identification. Mineral color is the least reliable and the most ambiguous of the physical tests.

Referring to cookbook chemistry, there is one simple test that can be thought of as the "it ain't there" test. If no precipitate forms, these metals are ABSENT: silver, mercury, iron, bismuth, copper, cobalt, cadmium, manganese, nickel and magnesium. If your test mineral forms a colored precipitate, the colors indicate these elements PRESENT: Blue – copper or cobalt; yellow – mercury; red-brown – iron; brown – silver. Remember, these are pre-assay tests. Other simple tests can indicate or positively confirm the presence of about 40 elements. Isn't it smarter to know in front, that the element you specify in your assay order, will be found? The assay will positively confirm your element and tell you how much of it is present. More importantly, you are getting value for your assay dollar.

ROCKPECKER TIP # 93 – Always take a DUPLICATE sample from each location so you can have another lab run a second assay. This increases the level of confidence and operates as a *CHECK ASSAY.*

ROCKPECKER TIP # 94 – Always ask the lab to retain your sample. This allows for a *repeat assay*, or for an *additional element* assay. (Remember, the gangue may be the ore.)

ROCKPECKER TIP # 95 – As a whole, professional assayers are honest and will treat your assay results as confidential. However, it is probable they will be location-curious. Common sense suggests that you take precautions so that greed will not get a chance to rear its ugly head.

It is good practice to keep your location secret. If you hear a comment such as: "That looks like ore from XYZ", don't acknowledge it! Simply smile and say nothing. If you feel pressed for an answer, locate the ore source in another state, or county, or even better, Canada. Make the sly comment that it's only a check assay.

NEVER LOCATE THE SOURCE SPECIFICALLY, you may find claims by a *paper hanger* plastered all around yours, all because of a slip of the lip.... Yours or the Assayers!

ORE RESERVES

Assuming your assays show a tenor worthwhile exploiting, your next step would encompass the determination of ore body size. At this point, think drilling, professional expertise, and investors to cover the expenses.

If you ore falls into the Federal Minerals A or B classes, you may be eligible for government exploration assistance.

A successful exploration program will block out reserves labeled *PROVEN, PROBABLE, and POSSIBLE.* The U.S. Bureau of Mines will use the terms *MEASURED, INDICATED, and INFERRED* to give the same meaning.

Just what is the value of a geophysical or geological survey as it applies to your ore deposit? Only the costly exploratory operations like core drilling, trenching, and sinking a shaft can provide the ultimate proof of ore volume. These are major information expenditures without which an investor would keep his wallet zipped. There is no practical way to avoid these costs.

However, there is a way to minimize exploratory costs. A core drill hole in barren ground costs the same as one striking ore. Since an ore body is rare, compared to barren ground around the body, the success of a well conducted geophysical survey is measured by the time, effort, and money saved in eliminating the more costly methods on barren ground. The ore body portions selected for further work are the masses identified by geophysical data.

Even the barren ground eliminated is only a *probable barren.* The ore body could extend deeper. It could be out of range of the method's detection capability.

Geophysical data interpretations can give reliable estimates of the probable depth, length, breadth, and the attitude at which the mass lies. This is valuable information needed to plan the next exploratory decision.

If you decide that you are unable to develop your mine as a personal investment, you have at least two other options. Sell out, if you can find a buyer; do nothing and sit on your claim. If you are prepared to "hold" your claim for 25 years, and if it is strategically located near a large operating mine, a strategy you might use is a ploy known as *SCHUMAKERING.*

Schumakering is a Canadian term used to represent a doubling of asking price, each time the other mildly interested party rejects your offer to sell.

Schumaker paid $8,000. for his strategic gold prospect around 1910. In 1911 he offered it to the nearby mining company for $75,000. The offer was rejected. He told them that if they became interested again, his asking price would increase to $150,000, then to $300,000, then $600,000.

As the mining company's workings neared the Schumaker, each selling offer that failed, was doubled. In 1936, the property was finally acquired for $2,000,000. Hence, the term "Schumakering" to describe this negotiating ploy.

The bottom line objective for most Rockpeckers is to discover and identify a mineral target, sell out, and let private enterprise, (or government) develop the ore body.

ROCKPECKER TIP # 96 - If your ore deposit discovery results in a "maybe" sale, that is, you grant an exploration OPTION on your claim, get a MINERAL-SMART LAWYER to represent your contractual interests. In addition to front money for the option, bargain for all the exploration data results as additional consideration, should the option be dropped or not exercised. You will have then pocketed the front money plus exploration development data for the time your claim was tied up.

An aside: Prior to 1950, 85% of all Canadian ore deposits were found by conventional prospecting methods using exposed mineralization guides. The most widely accepted field evidence of hidden mineralization is still the location of shallow deposits.

Always be guided by this old prospectors rule:

NEVER QUIT, UNTIL YOU PUT IN ONE MORE ROUND!

CHAPTER 30

MINING GRADE ORES

REVISITING OLD MINING DISTRICTS

In many of the old mining districts, mining operations were directed at the most accessible, most valuable, high grade ores. It was not uncommon for the operator to follow the vein until it ran into a marginal ore grade, then cease work there.

My admonition to you is: STAY OUT OF OLD MINES. LIFE IS TOO SHORT! For example: Uranium mines produce radon gas that attaches to dust particles. Breathing it into your lungs can result in lung cancer. Drinking water from these same mines can produce bladder cancer. How sound are the timbers today?

ROCKPECKER TIP # 97 – If you find a mine tailing dump that ceased work before 1930, take some grab samples for assay. At today's precious metal prices, a gold ore that was low grade at $20. to $30. the recovered ounce, is perhaps now high grade ore at $400. the ounce.

If the location is open to claim, you may have a heap leaching mine prospect. The ore is already above ground and crushed. Now all it needs is the modern extractive touch.

You should be aware that major mineral producing districts of the past are being revisited quite heavily today.

Remember too, the gangue or waste rock may be the ore.

ROCKPECKER TIP # 98 – You have located and assayed your mineral find. Is it of *MINING GRADE?* Remember, recovery of *CO-PRODUCTS and BI-PRODUCTS* from your ore can make a much lower grade economic.

The following listings can give you some ideas.

MINING GRADE ORES

METAL	MINERAL	METAL CONTENT	MINING GRADE - % OR GRAMS PER TON
BERYLLIUM	BERYL	14% BeO	0.7 - 3%
	BERTRANDITE	42% BeO	0.7 - 3%
	CHRYSOBERYL	20% BeO	0.7 - 3%
CHROMIUM	CHROMITE	33 - 58% Cr_2O_3	12.0 - 50.0%
COPPER	CHALCOPYRITE	35% Cu	.5 - 3%
	BORNITE	69% Cu	.5 - 3%
	ENARGITE	49% Cu	.5 - 3%
	CHALCOCITE	80% Cu	.5 - 3%
	COVELLITE	66% Cu	.8 - 3%
	TETRAHEDRITE-TENNANTITE	30 - 50% Cu	.8 - 3%
GOLD	CALAVERITE	39% Au	5 - 30 Gms/Ton
	SYLVANITE	24% Au	5 - 30 Gms/Ton
	PELTZITE	25% Au	5 - 30 Gms/Ton
LEAD - ZINC	GALENA	86% Pb	8 - 30% Pb
	SPHALERITE	60 - 67% Zn	8 - 30% Zn
LITHIUM	SPODUMENE	4 - 8% Li_2O	2 - 4% Li_2O
	LEPIDOLITE	3 - 6% Li_2O	2 - 4% Li_2O
	AMBLYGONITE	8 - 10% Li_2O	2 - 4% Li_2O
MANGANESE	PYROLUSITE	55 - 63% Mn	18 - 40% Mn
	MANGANITE	50 - 62% Mn	18 - 40% Mn
	PSILOMELANE	35 - 60% Mn	12 - 50% Mn
	RHODOCHROSITE	40 - 45% Mn	8 - 15% Mn
	RHODONITE	32 - 36% Mn	8 - 15% Mn
MOLYBDENUM	MOLYBDENITE	60% Mo	.05 - .8% Mo
NICKEL - COBALT	PENTLANDITE	10 - 40% Ni	.5 - 5% Ni/Co
	NICCOLITE	44% Ni	.5 - 5% Ni/Co

MINING GRADE ORES

METAL	MINERAL	METAL CONTENT	MINING GRADE - % OR GRAMS PER TON
NIOBIUM	COLUMBITE	45 - 78% Nb_2O_5	.5 - 3.5% Nb_2O_5
	PYROCHLORE	23 - 73% Nb_2O_5	.5 - 3.5% Nb_2O_5
PLATINUM GROUP	SPERRYLITE	56% Pt	5 - 15 Gms/Ton Pt
SILVER	ARGENTITE	87% Ag	300 - 1500 Gms/Ton Ag
	PROUSTITE	65% Ag	300 - 1500 Gms/Ton Ag
	PYRARGYRITE	60% Ag	300 - 1500 Gms/Ton Ag
	STEPHANITE	69% Ag	300 - 1500 Gms/Ton Ag
	POLYBASITE	64 - 72% Ag	300 - 1500 Gms/Ton Ag
TUNGSTEN	WOLFRAMITE	60 - 75% WO_3	.5 - 5% WO_3
	SCHEELITE	80% WO_3	.5 - 5% WO_3
TITANIUM	ILMENITE	53% TiO_2	15. - 50% TiO_2
	RUTILE	92 - 98% TiO_2	15. - 50% TiO_2
	TITANITE	41% TiO_2	15. - 50% TiO_2
URANIUM	URANINITE	47 - 88% U	.05 - .4% U308
VANADIUM	CARNOTITE	12% V_2O_5	2. - 5% V_2O_5

ROCKPECKER TIP # 99 – Even if your ore shows mineable grades in terms of the above percentages or grams recoverable, it has only an "in place" value, if due to other expense factors, it cannot be marketed at a profit. To that extent, the mining grades listed are presented only as a rule of thumb.

CHAPTER 31

THE SEARCH - A RECAP

METALLIC ORE VEINS

Imagine an igneous intrusive the size of a small city pushing a massive path into the overlying sedimentary rocks. This batholith crumples its way upward to send out off-shoot stringers to flow through openings in the crust produced by earlier tectonic deep earth events. Thanks to earthquakes that faulted and folded the upper crust, the magma has multiple passageways that fill as dikes, veins, laccoliths, and sills.

The dike crosscuts existing sedimentary strata. Dike stringers begin to feel their way through available cracks to reappear as veins and veinlets.

Magmatic liquids and solution vapors flow with the magma and race with and in front it. The colder country wallrocks cool the liquids to cause the precipitation of any mineral burden. This may be the first magmatic event; or a later episode; or it may have been continuous for months or years.

The vein filling materials of silicon, gold, silver, and other deposit metals are building up in favorable settling sites. The mineral streams may be sucked into the pore spaces of permeable host rocks, or become disseminated in the permeable sedimentary strata. In vapor form, the metals may squeeze between sedimentary shale layers to deposit in paper-thin sheets as pure metal such as gold, or as electrum, a gold-silver amalgam.

Consider that the vapor state of silicon occurs at 2355 C; aluminum at 2467 C; copper at 2595 C; gold at 2966 C; silver at 2212 C. That slight a difference in degrees centigrade, makes their deposition commonality easier to comprehend.

Now, 50 million years later, the overlying sedimentary rocks have eroded and passed their way to the ocean; thence to recycle, yet again, into magma. Portions of vein material are visible at the surface as various sized outcrops. They may be in close proximity to the intrusive in the form of vein swarms or may appear miles distant. Your geologic map indicates intrusive rock patterns and evidence of area outcrops. Generally the map question at this point, boils down to whether the intrusive was emplaced before or after the faultline was born.

You see, if the intrusive was emplaced before the faulting, then how could vein filling have occurred; there were no solution channels present until later. Magmatic pressure? Well, maybe!

ROCKPECKER TIP # 100 – Prospecting map tip – Look closely at your geologic map, particularly at the intrusive rock shown. If the map shows the faultline running through an intrusive body which shows that movement of the body took place along the faultline, it is a sign that faulting took place after the intrusive hardened. Magmatic fluids had no fault channels to flow through during emplacement.

ROCKPECKER TIP # 101 – Cut an apple in half. Reposition it so that the halves are offset. The faulting came second. Take the same apple halves, but now place them as they were before being cut. The faulting came first and may have acted as a channelway for rising metallized solutions or vapors.

You conclude that the faulting came first, and now you are prowling the intrusive outcrop. You next step would be to search for faults surrounding the intruded area, but in the sedimentary rocks. Bear in mind that rich veins may have been formed within the intrusive, but that occurrence is not as commonplace as veins found in sedimentary formations.

The last few paragraphs have introduced you to the concept of what is known as vein and replacement mineral deposits.

Within a range of several hundred to a few thousand feet, hot water solutions percolate under high pressure laterally and upward from magmatic origins. The hot intrusive body-solutions relieve pressure by searching for a fault, joint, shear zone or other cracks and fissures. These relief pathways become the sites for mineral deposition. *This is the significant process in the formation of many metallic ore deposits.*

Simple cavity filling is the dominant vein-forming process. Crystals grow inward from the cooling fissure walls. Magmatic material, deposited as a sheet structure between the fault-walls of the passage, are called *veins.*

Where the fissure crack is not an internal clean break, or where through fault movement, the headwall grinds against the footwall, pieces of angular rock material or a fine gouge are included in and around the invading solutions. These pieces may be dissolved and replaced, or may become part of the vein. Oftentimes, the fine gouge acts as a stopper to terminate the vein. These are called *pockets or brecciated zones.* The word breccia comes from the German "brechen", meaning "broken" or "to break".

The hot solutions are referred to as *hydrothermal solutions.* The term includes not only very hot water, but also steam. It is my view that these fluids are mostly *vapor states.*

To explain variations in mineral deposits, the pressure behind the metal-laden vapors that are forced into rock pore openings is described as high, medium or low temperature-pressure hydrothermals. This reflects the idea, that as the vapors and solutions travel away from the heat source, they will lose both heat to the confining walls, and pressure reductions due to the increased travel distance.

Another phase of hydrothermal solution travel embodies the concept of the interaction of the solution with the invaded rock. This process involving removal of one or more minerals from the host rock, with a replacement-exchange or substitution of a solution mineral into the vacated host rock space. This replacement process is called *metasomatism.*

Minerals such as quartz, fluorspar and barite, alien to the rocks in which they are found, are called introduced minerals. Carbonate rocks are susceptible to the replacement process. That process accounts for barite in dolomite; fluorspar in limestone.

The replacement-exchange process is oftentimes found at a distance from the intruding magma. However, it can operate with rocks in contact with the intrusive, and even within the intrusive itself. Where replacement is in contact with the magmatic intrusion, by definition, it would be considered metamorphic rock in a technical sense.

ROCKPECKER TIP # 102 - When you find an intrusive outcrop, the metallic veins can lie at a far distance, perhaps even miles away. You might try the *unwinding coil walk* with the outcrop as the centerpoint beginning. Remember that the vein will rise about 60 degrees from the intrusive. If you are lucky, you will spot a small outcrop head poking through the sediment.

Rather than assume familiarity with an intrusive igneous rock outcrop, let's explore the probabilities and expectations.

SILL, DIKE AND VEIN OUTCROP MATERIALS

This is intended as a general guide to igneous rock terms as used in the literature, but more importantly, the stuff of veins.

In your readings you will frequently run across rock terms like *acid or basic rocks, mafic or felsic rocks.* It gets even more confusing for the prospector when *ultrabasic or ultramafic* are used as descriptive terms. It is my wish to clarify, if possible, some distinctions.

ACID vs BASIC ROCKS - In the long ago, it was thought that when quartz or silicon oxide (silica) dissolved, it formed siliceous acids. Such acids in solution, when forming rocks, would be high in silica. They were the acid rocks. Those low in silica were termed basic rocks. Then it was proven that siliceous acid did not exist in nature. The terms were in use so long that they simply became ingrained.

Acid rocks are light-colored because of the high quartz or feldspar content. They are now usually referred to as *felsic.*

Basic rocks are dark-colored and are termed *mafic.* (Mafic stands for ferromagnesium. You can remember the dark rocks as mafic, if you think of the black hand or mafia.) Ultrabasic or ultramafic rocks are very dark, nearly black, and usually consist wholly of the dark-colored mineral olivine, with perhaps some pyroxene present. These are all igneous rocks which are identified by their overall general appearance. Remember that a rock which has volcanic origins, whether rising from depth, or blown into the sky, or flowing over the surface, has the same composition. The difference is simply texture and name. Quick cooling extrusive rocks have a tight or no-visible internal grain structure. Underground, slow-cooling intrusive rocks have a coarse grain structure. Grain size is the important distinction within the term *texture.*

Geologists have named these rocks and their counterparts as follows:

Igneous rocks are classed as intrusive when they have never left the crust and have hardened under the surface. The heat exposure over time produced a texture of large grains of various minerals.

GRANITE is the most common. It is a light-colored rock with a mix of dark flecks of equal grain size. While overall grain size can vary, the average is about one quarter inch across. A light-gray granite is actually *quartz monzonite.* The metamorphic equivalent is a rock with alternating light and dark wavy bands.

When granite is extrusive, that is flows out of a volcano, it is light-colored with tiny or no visible grains. It can show flow marks that appear as bands or streaks.

SYENITE rock is similar to granite, but has no quartz content. Grain size is very fine.

PEGMATITES are granites with very coarse grains.

DIORITE is a dark-colored rock with a coarse-grained texture like granite. Quartz is usually absent, its feldspar is plagioclase.

TONALITE is a diorite containing quartz.

GRANODIORITE looks like granite, but is mostly plagioclase feldspar with subordinate potassium feldspar.

In commerce, granite includes true granite, syenite, quartz monzonite, granodiorite, quartz diorite, and diorite. These are the light-colored granites.

The so-called black granites range from anorthosite gabbro with a bluish tint, caused by the primary labradorite feldspar content, which gives it the optical shiller effect, to diabase, and pyroxenite.

DIABASE occurs chiefly in sills. Being near the surface of the crust when it cooled gave it a fine grain structure from the faster cool-down. This is a hypabyssal rock of medium depth.

PYROXENITE is composed entirely of pyroxene. If the rock is chiefly olivine, it is dunite. If composed primarily of the dark minerals pyroxene and olivine, the rock is called peridotite, which in altered form is kimberlite, the host rock for diamonds. A gabbro with abundant pyroxene and olivine is called norite.

All these name distinctions are due to chemical analysis and are important to geologists when writing technical papers. As a prospector, just think of colored granite and black granite when thinking of intrusive outcrops. I included these granite distinctions to give you a broad-brush glossary of rock words you may hear in a conversation or read in an investors prospectus.

You know, talking "prospecting" can be less strenuous, and at times more fun than looking, picking, packing and roughing it.

While we're at it, let's recap volcanic extrusive rock.

RHYOLITE is a light-colored lava flow of granite. If it moved while still molten, it can have flow bands or streaks.

OBSIDIAN - Sometimes RHYOLITE cools so quickly that crystals have no time to form. In that case the grain structure will be invisible to the eye, and the rock will have a glassy texture.

PUMICE is a rhyolite violently expelled by volcanic explosion. It has a glassy texture with many trapped gas bubble pores.

BASALT is a dark, fine grained lava flow rock. It can be black, greenish, gray or brown in color. It weathers to a skin deep tan.

ANDESITE is the volcanic equivalent of diorite. It often resembles a dark-colored rhyolite. Andesite is common in porphyries.

LATITE is the fine grained equivalent of MONZONITE. Monzonite is often associated with mineral deposits.

TRACHYTE is a fine grained porphyritic textured equivalent of syenite. Makes an excellent flooring material because it does not become shiny from rubbing.

PHONOLITE is a trachyte with more than 10% feldspar. The white ash that smothered Pompeii is of phonolite composition.

PORPHYRY is a rock that contains various mineral grain sizes in a fine grain matrix. The large grains are called phenocrysts; the matrix, the groundmass. The large grains are half-cooled deep-in-the-earth crystals that were forced to the surface along with fresh lava which became the groundmass. Think of raisin bread.

Is the rock porphyry, conglomerate or breccia? The distinction rests with identification of the large grains. Rounded pebbles means conglomerate; angular pieces define breccia; mixed large crystal grains are probably phenocrysts of porphyry.

The order of *younging* can be helpful in locating the direction of the younging, so as to locate the ore source rocks. For example: You find a piece of float that looks promising. Sure, you can charge off in all directions to perhaps "luck-out", or you might employ the *law of included pebbles.* When one piece of rock is included WITHIN ANOTHER ROCK, the included rock is always older.

How can this help you? Imagine that you picked up some float that shows a blue, rounded pebble, probably quartz. The blue quartz is wrapped in a layer of pumice rock, or basalt. You know the blue pebble was deposited earlier than the pumice because it is included within it. Your search direction should follow the pumice path. If you find it, you may have found a buried lead, (pronounced "leed") also known as a tertiary channel. It is said that quartz pebbles are stained blue in fossil channels.

However, since the pebble is blue, it may be fluorite that was radiated that color by uranium proximity. Test the pumice for uranium compounds.

As to included pebbles in sedimentary rocks, the constituents of the sediment are always older than the sedimentary rock. However, pebbles in a sediment that are associated with a nearby adjacent sediment must be younger than the enclosing rock.

As a closing note, remember that only igneous intrusive material can form a vein. Extrusive igneous material is just another sedimentary strata waiting to weather.

CHAPTER 32

MINERAL ORES AND CRYSTALS

We can see the process of crystallization every day in our ice cube tray. Water, (a mineral) at freezing point, alters its form and structure to ice. All mineral deposits were developed by some earth process which works to physically or chemically separate, then concentrate particular elements.

Falling rain and snow, which we call meteoric waters, penetrate rock fractures and fissures then, begin to dissolve the contacted rock materials at the higher levels. As the mineral laden water trickles downward, it redeposits metallic minerals, in enriched amounts, at the lower ground water level. Underground waters percolating through limestone joint systems, pick up carbon dioxide to yield weak a carbonic acid that is strong enough to be an effective solvent for limestone. As this underground chemical erosion continues, ultimately a large cave system is developed. A reverse process fills the cave with calcite or aragonite crystals.

When meteoric waters descend to the levels of magmatic activity, the heated water rises to carry with it the dissolved minerals. As the solution temperature drops to the saturation point, precipitation of calcium or silicon in the form of *tufa or sinter*, coats the hot water channels with crystalline mineral crusts.

Where the hot solutions, as fluids or gases, have a magmatic origin, faults and fissure cracks provide an easy escape path to the surface to create hot springs. Minerals in the fluids deposit in the inverse order of solubility. The most easily dissolved minerals are ejected with the hot water; the lesser solubles deposit lower down in the fissure system. Fissure veins of quartz, galena, sphalerite, pyrite, chalcopyrite and others result from hot solutions. Chemical reactions form new chemical substances where various solutions mix, then deposit on the fissure walls.

In arid regions, evaporation of inland water basins deposit a layer of bottom salts. Repetition over long periods build up minable thicknesses. Boron salts, chlorides and sulfates form the minerals of colemanite, borax, ulexite and glauberite in what are termed *playa lake deposits.*

Near areas of volcanic action, vapors or gases composed of water, carbon dioxide and hydrochloric acid together with compounds of boron, fluorine and metal chlorides (known as volatiles) react to form new products. These *pneumatolitic* mineral deposits may form topaz, fluorite, tourmaline, scapolite and phlogopite.

Magmas containing heavy minerals separate into enriched masses. Gravity and solubility are the separating factors. Minerals of the platinum group, chromite deposits, dark course grained rocks of the pyroxene and amphibole class (dark granites) are formed by this magmatic segregation. Accumulated segregations can crystallize in place to form ore bodies grading into other surrounding igneous rocks, or be forced into faults to form a vein-dike system of metallic minerals.

Originating the same way as way as vein-dikes are pegmatites. The difference being their association with the light colored granites and syenites. Pegmatites are the "jewel factories" of light granitic rocks. Diamond, emerald, ruby and sapphire are gemstones associated with basic (light colored) rocks. Granitic veins and vug cavities formed by cooling-shrinkage can contain smoky quartz, microcline feldspar, albite, clear quartz, micas and topaz crystals several inches long.

Metamorphic processes produce changes in rocks, especially limestone in contact with magma. *Contact metamorphic* mineral deposits of magnetite, chalcopyrite, lead-zinc, gold and cassiterite form. The minerals of bornite, pyrite, pyrrhotite, galena, sphalerite, gold, and silver are common metal associations. Silicate associations are garnet, wollastonite, epidote, tremolite, diopside, zoisite, vesuvianite, quartz, calcite, axinite, tourmaline, fluorite and scapolite.

Regional metamorphism, chiefly due to pressures of tectonic plates crashing into and folding the earths crust, at the high grade level, take fine grained shales and turn them into schistose rocks that develop garnet, staurolite, rutile, cordierite, ilmenite and magnetite. Clay and other impurities in limestone change into phlogopite, scapolite, oligoclase (moonstone), tremolite, actinolite and diopside.

While ore minerals can form through a variety of earth processes, the composition of a mineral consists of particles so minute they are invisible to powerful microscopes. These tiny particles are atoms of one or more substances called elements. These elements unite with each other to form chemical compounds through electron sharing. This produces a molecule with properties entirely different than those of its constituent elements.

HALOIDS are simple compounds of chlorine, fluorine, bromine or iodine. The name "haloid" comes from the mineral "halite" which we know as table salt. Haloids are "salts".

SULFIDES are the most important metal carrying ores. Sulfide ore compounds, usually metals and sulfur, are one of the most important classes of all chemical compounds

SULFATES are formed because of the strong affinity of sulfur with oxygen, where with one or more metals gypsum and barite result.

OXIDES are compounds of oxygen with another single element. The oxide of silicon is called silica by chemists, but quartz by mineralogists.

CARBONATES are compounds of carbon combining with metals or non-metals. In the mineral world,the metal calcium is found as the mineral calcite, or as limestone rock. Both are calcium carbonate.

SILICATES are compounds of oxygen, silicon and one or more metallic elements.

PHOSPHATES are compounds of phosphorous, oxygen and a metallic element. In nature it is the rock "phosphorite". The farmer using phosphate on his field is fertilizing with calcium phosphate.

ARSENATES and VANADATES are compounds of arsenic or vanadium combined with oxygen and a metallic element. This class produces minerals of great colorful beauty. *Erythrite,* known to Rockpeckers as *cobalt bloom,* is a common alteration product of cobalt arsenides. It's pink color and dusty appearance is a prospectors guide to cobalt minerals.

Chemists have a shorthand formula for describing a complex chemical compound. This text avoids the shorthand by verbalizing the mineral name. For example, in the mineral chart, *polybasite [201]* is a sulfosalt where antimony, arsenic or bismuth join with a metallic element to form a double salt. The chart describes it as: *silver copper antimony sulfide.* This is a silver antimony sulfide where 9 to 12% copper can substitute for the antimony.

Mineral crystals usually form by accretion, one atom at a time, much like building a brick wall. The crystal of a specific mineral usually has a specific face-shape which is called its *habit.* All solids are, by definition, crystalline. A crystal is merely a mass of crystalline material bounded by natural growth surfaces. A crystal is a crystal because of the way its atoms are arranged. The external shape of a crystal is what is known as its *habit.*

In the real world, the surfaces of a crystal do not have to be plane, smooth and shiny. That is simply an accident of growth in a *euhedral* (bounded by plane surfaces) crystal. *Anhedral* crystals have no plane faces at all; *subhedral* crystals have crude geometric shapes. Crystal symmetry results in the repetition of crystal faces, which are flat surfaces that *sometimes* form the outer boundaries of well-formed crystals.

ROCKPECKER TIP # 103 – A mineral crystal can have several habits. Don't let surface shape alone, cloud your conclusions. Habit is only one of the ways to identify a mineral. For example: Gypsum, hydrous calcium sulfate, is found in several different forms. As selenite, satin spar, alabaster, or the rock gypsum, yet it is the same hydrous calcium sulfate with a new shape.

It is helpful to know and identify some of the CRYSTAL GROUPING descriptions of rocks and minerals. A few *descriptive* terms are:

ACICULAR: Slender needle-like crystals, randomly oriented, found within rocks, cavities, or upon open fissure walls.

AMYGDALOIDAL: Mineral filled almond shaped cavities; occurs in volcanic rock flows; cavities contain many kinds of minerals.

ARBORESCENT: Tree branch mineral structures, sometimes seen in native gold, silver and copper specimens.

ASTERIATED: Needle-like inclusions create starlike reflections. Phlogopite can display asterism, as well as polished star-sapphire gem stones.

BLADED: Long knife-like blades. One of the stibnite habits.

BLOCKY: Cube-like small block appearance; much like galena.

BOTRYOIDAL: Bubbly mineral surface that appears as a cluster of grapes.

CAPILLARY: Hairlike, thin, very slender crystals. Millerite, a nickel sulfide is commonly called "capillary pyrites." Found in limestone, dolomite or ore veins.

CELLULAR: Many open cavities caused by gas bubble escape from glassy surface magmas such as basalt.

CLASTIC: Clearly visible rock fragments or minerals in a fine grained matrix such as metamorphosed conglomerate.

COCKSCOMB: Describes the arc with triangular points of a rooster comb. Often found in marcasite, a polymorph of pyrite.

COLUMNAR: Crystals in parallel column structures such as aragonite.

COMPACT: Same as "massive".

CONCENTRIC: Arcing bands or layers of different colors, textures, or minerals. Malachite displays this banding.

CRUSTIFIED: Minerals growing in fissure walls and rock cavities which point inward toward the open area. Also know as COMB STRUCTURE.

DENDRITIC: Fern-like branching within a mineral such as moss agate or pyrolusite.

DRUSY: Wall or rmicrocrystal growths.
Druses can be found in the vug openings of all kinds of rocks.

EARTHY: Material resembling ordinary soil, often the alteration product of earlier minerals such as limonite after weathered iron minerals.

FEATHERY or PLUMOSE: Feather-like roughly parallel aggregates of mineral grains such as muscovite.

FELTY or LEATHERY: Fibrous intergrown minerals such as actinolite, criss-crossing in a mat to fill a small rock fracture.

FIBROUS: Fine, silky, flexible fibers with a thread-like separation such as chrysotile.

FILIFORM: Applies mostly to native metals in wire-like, thread-like shapes such as wire silver.

FOLIATED: As in micas and chlorites, a leafy, platy or bookpages-like structure.

GLOBULAR: Round or nearly so sphere-like pieces of any size such as gravels in sandstone.

IRREGULAR: The usual and common arrangement of crystals in mineral groups. The cause is impeded or restricted growth space.

LAMELLOR or LAMINAR: Fine striations or plates as parallel laminations at the cleavage planes.

MAMMILLARY: Large rounded bust-like masses.

MASSIVE or COMPACT: As in clays, uniform, solid masses without visible crystal structure.

MICACEOUS: Thin scales or plates of mica as applied to schists and found in biotite or muscovite.

MIRROR-LIKE: Some opaque minerals such as pyrite have smooth surfaces that will reflect.

OOLITIC: Small, rounded, BB-like pieces sized like fish eggs. From the Greek "oon" meaning an "egg".

PISTOLITES: Like oolites, but about the size of peas as in bauxite.

RADIAL: A wheel-like pattern where spoke spacing is varied as in goethite.

RENIFORM: Many smaller rounded shapes bunched together in kidney form as in hematite kidney ore.

RETICULATED: Bladed crystals in a criss-cross network as in cerussite.

SCHISTOSE: An easily splittable metamorphic rock where micas are arranged in parallel layers.

STELLATE: A radiating group of crystals with distinct wheel-like separation of crystal groups as in pectolite.

STRIATED: Straight parallel lines or ridges where two crystals intersect as in pyrite. Also those lines in plagioclase feldspar, when cleaved, to distinguish it from orthoclase feldspar without lines.

STUBBY: Short, pyramidal, prismatic crystals as in quartz.

TABULAR: Thin tabletop crystals with large faces as in barite.

VESICULAR: Surface bubble-like cavities in igneous rock flows.

VUGGY: In any kind of rock, a small, open cavity often lined with mineral crystals.

WARTY: Small, round, rough-surfaced groups of minerals.

ZONED: Colors concentrated in layers throughout the crystal as in amethyst crystals with varying shades of purple.

CHAPTER 33

OVER 30 GOLD ORE ASSOCIATIONS

Gold ore has been found associated with the following minerals which in this recap may be considered as gangue. These associations deliberately omit the locations where the find occurred. This is for the reason that it is essential to keep an open mind when Rockpecking.

GOLD HAS BEEN ASSOCIATED WITH:

1. Dikes and sills of Tertiary rhyolite in altered sediments.

2. Beds of carbonate rocks with thin bands of shale and quartz sand.

3. Beneath impervious slate.

4. With white quartz, chlorite and minor mica, pyrrhotite, pyrites, and arsenopyrites.

5. With chalcedony, clay minerals and dolomite.

6. With arsenic, realgar, orpiment, cinnabar, stibnite, pyrite, and tennantite.

7. In a gangue of dolomite, chert, and montmorillonite.

8. With organic carbon, in that it is thought to have absorbed and reduced the gold. The carbon later oxidized to leave native gold. (Investigate any sandstone black line for gold. Algae, now carbonized, may have trapped placer gold.)

9. In porous minute quartz crystals, carbon, and with

realgar and orpiment.

10. In quartz, pyrite, alunite, kaolinite, marcasite, bismuthinite, and tennantite.

11. With granite, gneiss, ash-flow tuffs, and volcanics.

12. As gold-silver sulfides or tellurides with gangue of crustified quartz, carbonates and fluorite.

13. As gold telluride ores in Tertiary volcanics with rocks of phonolite, syenite and basalt.

14. Crystalline rocks with gold, sulfides, and molybdenite in a gangue of carbonates, chlorite, and garnet.

15. As gold, slight silver, metal sulfides and ankerite.

16. As gold, some silver in massive sulfides.

17. Gold with quartz, pyrite, sericite, tourmaline, carbonates with considerable carbon.

18. Native gold with quartz, pyrite, tellurides, calcite, metal sulfides, and tourmaline.

19. Gold in quartz veins in carbonized volcanic rocks.

20. Gold and silver with massive zinc and copper ore.

21. In thin sedimentary conglomerate beds alternating with quartzites and volcanics.

22. In well-graded conglomerate rocks in old stream channels of probable Tertiary origin.

23. As gold in hornblende schists, iron-bearing quartzite, and porphyitic granite.

24. In quartz veins in granite rocks, veins with sulfides,

arsenopyrite, and tourmaline. Gangue of quartz, barite, and carbonate rocks.

25. Gold along shear zones in quartz metamorphic rocks, with mineral associations of pyrite, metal sulfides, sheelite, vanadinite, and bismuthinite.

26. In dike rocks of quartz, iron sulfides, molybdenite, wulfenite, and crocoite.

27. With greenstone as gold-pyrite-tellurides in quartz.

28. In schistose diabase rock containing quartz, gold, and metal sulfides.

29. Gold-bearing quartz in saddles localized at the crest of an anticline between slate and sandstone beds, with small amounts of pyrite, arsenopyrite, and pyrrhotite. Minor galena, sphalerite, stibnite, molybdenite, and bournonite in dolomite and ankerite gangue.

30. As an indicator, a thin band of black slate replaced by quartz and pyrite.

31. Gold in quartz in crusty sulfides in tuffs that also carry copper with gold.

32. In fissure veins associated with granite intrusions.

33. Fissure veins in sediments carrying stibnite, sheelite, and arsenopyrites.

34. Gold quartz veins in andesite carrying arsenic, antimony, mercury, copper, lead, zinc, and iron.

MOST, IF NOT ALL, OF THESE GOLD ORE BODIES WERE IN SEVERELY FAULTED , FOLDED AND FRACTURED ROCK ENVIRONMENTS.

35. *LOW-GRADE MINERALIZATION OF WALL ROCKS:* Fine-grained, inconspicuous, or invisible gold values can be generated in large

volumes by the out-flow of *hot springs.* These deposits represent the concept of travel and reprecipitation of gold along thrust faults. Special porosity-permeability of host rocks, adjacent to the outflow, spreads values in the host rock. These deposits are termed *epithermal fluid deposits* to distinguish them, as a special classification, from replacement and cavity fillings.

36. *CARLIN TYPE:* A sub-type of epithermal deposit has now come to be known as "Carlin Type", after the Carlin Gold Mining Co. ore body. The metallic gold particles are so small, they are invisible to the eye. The values are detectable only by assay. These gold deposits are disseminated bodies containing 7 to 10 PPM gold ore.

The Carlin Type are characterized as disseminated in thrust faults, of domed and fractured *carbonate rocks.* Fine-grained gold, associated with arsenic-sulfur minerals with antimony or mercury relatively uniform throughout, but with very little quartz. The gangue is a sulfate mineral.

These deposits are *spatial,* in that they are spread throughout the limestone, along an older dike system. Silver is absent. It is believed that leaching carried off the silver during formation of the deposit. Because of relative price increases in gold, these bulk, low-grade, gold deposits have become economic.

Carlin Type deposits illustrate a basic economic principle. If you find a low-grade ore in bulk quantity, that can be extracted cheaply at a profit, you have a valuable mining property.

Remember an uneconomic, low grade gold mineralization can be extracted profitably, if bi-products and co-products in the gangue have sufficient extraction value.

CHAPTER 34

SILVER & ASSOCIATED MINERALS

There are 55 well known silver minerals. The most important are the tellurides and tetrahedrides. The common gangue materials are quartz, calcite, dolomite, and rhodochrosite. Silver is associated with lead; gold is rarely free of it; many copper and zinc ores carry some of it. Cobalt is also an associated metal.

The origin of most silver-lead deposits have been hydrothermal solutions filling cavities and fissure veins. Replacement silver mineral deposits are the most valuable, when replacement is massive.

FOUND SILVER ORES AND THEIR ASSOCIATED MINERALS AND GANGUE:

1. In quartz veins in quartzite.

2. Associated with siderite, quartz, galena and tetrahedrite.

3. In hanging wall chambers of crushed quartz, calcite, sulfides, gold, electrum, argentite, and polybasite.

4. As replacements in limestone, silver-lead, silver-copper, zinc, or gold ores.

5. In greenstone rock formations overlain by cobalt veins.

6. In sedimentary uranium ore veins associated with native silver, cobalt, nickel, arsenides, copper sulfides, bismuth, quartz, and rhodochrosite.

7. In stocks and fissure veins in slate, graywacke, and altered vocanics with metal sulfides, quartz and calcite.

8. Veins in red shale and conglomerates as crusty fissures with quartz and pyrite.

9. In folded limestone beds separated by shale, near a monzonite intrusion. Associated with barite, dolomite, pyrite, galena, tennantite, and sphalerite with a zone grading from copper to silver lead.

10. Minor amounts of metals in unweathered rock may be concentrated into worthwhile deposits during weathering. Water charged with carbon dioxide and oxygen will oxidize, hydrate, and carbonize rock-forming minerals. The sulfides are converted to soluble sulfates or more stable oxides.

Lead, zinc, and copper are preserved as carbonates. Native gold, silver, and copper may remain stable in the weathered zone or leach down and concentrate. Silver is especially soluble in sulfate solutions, going from solution to reprecipitation.

SILVER BEHAVES MUCH LIKE COPPER IN THE ZONE OF OXIDATION, BUT WITH FEWER OXIDATION PRODUCTS. (THINK GOSSAN.)

Acanthite, the silver sulfide, releases the silver downward. In arid climates, cerargyrite may be left in the oxidized zone. Native silver is common. (This is probably due to lack of water.)

Silver can occur as a nugget of copper-silver alloy which is called a *halfbreed.* While silver is more abundant than gold, it is usually obtained as a refining by-product of lead, zinc, and gold ores. AGAIN, SILVER IS MOST COMMONLY ASSOCIATED WITH LEAD.

THE CHIEF MINERALS OF SILVER ARE:

Argentite, - 87% Ag.; Cerargyrite, - 75% Ag.;

Polybasite, - 80% Ag.; Proustite, - 65% Ag.;

Pyrargyrite - 59% Ag.

CHAPTER 35

PLATINUM METALS GROUP

The platinum group consists of platinum, osmium, palladium, ruthenium, iridium, and rhodium. These metals, when pure elements, range in specific gravities from 12 to 22. Osmium is the heaviest element by actual density measurement. However, theoretical calculations of density from the iridium space lattice seems to indicate that iridium may be the heaviest element.

These metals are not only associated together, but are generally alloyed together. If you have found one, you have found what are called *platinum group metals,* or PMG's for short.

Platinum, as a native alloy, has a specific gravity range of 14 to 20 as found in your goldpan. Native platinum is never pure. In addition to the alloyed PMG's, it can form a natural alloy with iron and copper. Dependent on the amount of alloyed iron, PMG's can be slightly magnetic. In placers, a brown iron oxide coating makes it difficult to pick platinum out of black sands.

TO GIVE YOU AN IDEA OF THE RELATIVE PHYSICAL PROPERTIES OF PMG's: THEY ARE AMONG THE RAREST OF METALLIC ELEMENTS.

ELEMENT	SPECIFIC GRAVITY	HARDNESS	MELTING POINT	BOILING POINT	COLOR
	Density	MOHS	Degrees Centigrade	Degrees Centigrade	
PLATINUM	21.45	4.0 - 4.5	1772	3827	Silver-white
PALLADIUM	12.02		1544	2970	Steel-white
RHODIUM	12.41+		1966	3727	Silver-white
RUTHENIUM	12.41		2310	3900	White
IRIDIUM	22.42		2410	4130	Silver-white, yellow hue
OSMIUM	22.57		3045	5027	Blue-white, high luster

SPOT PRICE: RHODIUM 12/89 = $ 1275.00 OZ.; 4/92 = $ 2680.00 OZ.; 9/93 = $ 860.00 OZ.

PGM's were recognized in California as a bi-product of placer operations shortly after the Gold Rush days. PGM's have been found in the American, Merced, Mokelumne, Tuolumne, Yuba, Cosumnes, Klamath, Salmon, Scott, Smith, and Trinity Rivers. The Trinity River near Junction City has been the most productive. As an oddity, rhodium and palladium are nearly absent in California placer gravels. Very few nuggets have been found. Most PMG's are finely divided as rough grains. They do not flatten into gold-like flakes.

MOST PLATINUM RECOVERED TODAY IS A REFINING BI-PRODUCT OF NICKEL ORES.

Platinum is associated with mafic (dark) and ultramafic (very dark) rocks. It is intimately associated with chromite and nickel sulfides.

In South Africa, platinum has been found in carrot-shaped pipes in olivine-dunite rock along with phlogopite and hornblende. A strata-like dike layer of mafic norite rock was below and seemed to encase the pipe.

The nickel sulfide ores of Australia occur in greenstone belts. The platinum is in low-grade disseminated sulfides such as pentlandite, pyrite and millerite.

PMG's are most likely to be found in chrome-like pods rather than veins. Associated metals are chrome, nickel, copper, and cobalt. Associated rocks are dunite, peridotite, and serpentine.

A typical nickel sulfide assay sample IN PARTS PER BILLION:
PLATINUM = 200; GOLD = 110; RUTHENIUM = 220;
OSMIUM = 110; IRIDIUM = 6.

A prospecting boon has been the ELECTRON PROBE. This instrument can detect platinum down to 0.6 parts per million. As a result, more rock and mineral associations are coming to light.

In 1989 the National Defense Stockpile goals were amended for PGM metals. The stockpile targets for platinum, palladium and iridium were reduced. Stockpile goals for rhodium and ruthenium were initiated. This does not mean that these elements are no longer in need by the Department of Defense. The reality is that the old goals were never met. The targets were simply reset to meet a lower kilogram aim with rhodium and ruthenium added to the strategic metals program.

PGM deposits have been found in a variety of geologic environments. Many deposits are simply not economic in that the tenor of the PGM's are not cost effective to exploit. There are three general categories of PGM occurrence.

1. Stratiform deposits where PGM's are found in layered zones in Precambrian rock, such as the Montana Stillwater Complex.

2. Nickel and copper bearing dikes and sills near rift structures as in the old U.S.S.R.

3. A geological freak such as the Canadian, Sudbury, norite rock intrusion. (This one is thought to have been caused by a meteorite impact.)

Other PGM associations that supply the world's needs are:

1. Placer deposits.

2. Nickel-sulfides associated with extrusive rocks.

3. Metal sulfide base metal deposits hosted in shale.

4. Sulfide deposits in zoned dark black rock intrusions.

5. Chromite deposits in both stratiform and pod forms.

6. Porphyry copper deposits & hydrothermal deposits.

ROCKPECKER TIP # 104 - PGM's found in placer deposits in the Junction City area of the Trinity River have been past-productive. The PGM source has never been found. LOOK AGAIN !

CHAPTER 36

SERPENTINE ROCKS

GEOBOTANICALS & DIAMONDS

Ultramafic rocks, largely peridotite and serpentinite weather to exceptional chemical properties. The soil produced thereby is responsible for the descriptive term *serpentine barrens.*

Serpentine can be mistaken for shale. Both rocks readily cleave and fracture into a flake-like division. The soil is dry and gray with high magnesium content. Some plant life has an addiction for this soil to the point where it can be called *serpentine specific,* or endemic.

Some plants that are specific to serpentine are the *LEATHER OAK, BUCKTHORN – MUSK-BUSH, and MOSS PHLOX.* These are faithful indicators of serpentine.

Leather Oak – (Quercus Durada) is a scrub-oak in chaparral with a low spread two to five feet high with small leaves about an inch long.

Buckthorn – musk-bush – (Ceanothus Jepsonii) stands four to five feet high with a gray bark. Its flowers are white or blue and give off a musky odor.

Moss Phlox – (Phlox Subulata) is a low herb with a slightly woody base with two opposing leaves. It is know as the *Pink Hills,* for it blooms with a pink flower that covers the entire serpentine area.

The soil over serpentine outcrops is thin and dries quickly. It lacks phosphorous, calcium and nitrogen. The high magnesium concentration is poisonous to most plants and dwarfs others. This habitat selects specialized plants that act as *indicators of serpentine.*

Ultramafic rocks, like serpentinite, dunite and peridotite, have an iron-magnesium silicate as the chief constituent. Serpentines occur in sill-like sheets, squeezed up through overlying rocks invading sedimentary rocks. They are structurally associated with faulting patterns.

Metamorphosed serpentine rock contains chrysotile asbestos. Contact zones contain rodingite which is rich in calcium and shows a plant life change which delineates the serpentine zone that is visually obvious from a distance. Both soils and rocks contain nickel and chromium in trace to moderate concentrations, and are commonly associated with quicksilver ore deposits.

Serpentine barrens show a sparseness of timber and brush growth. They may be nearly bare with a gray ash-like covering you see after a forest fire landscape. Serpentine gets its name from the snake-like mottled pattern on a serpent's skin. Sort of a Boa Constrictor pattern. Nickel, mercury, chromium, asbestos talc, soapstone, magnesite and jadite have been taken from serpentine deposits. Nickel extraction has not yet proven profitable. However, Strepthanthus Polygaloides hyper-accumulates nickel. Tissue accumulations usually 1000 PPM can run to 30,000 PPM.

ROCKPECKER TIP # 105 - In searching for nickel ore deposits, collect the mustard plant, S. Polygaloides. Subsequent nickel analysis may prove to be a short-cut to valuable nickel claims.

ROCKPECKER TIP # 106 - Where you find underlying rocks of greenstone, serpentine, peridotite or one of their rock-relatives, look for a peculiar *red clay* which is likely to form as soil. This clay is called *laterite*, from the Latin for brick-red. Laterite is rich in iron which colors it red. Sometimes it is also rich enough in nickel to qualify as an ore.

GEOBOTANICALS

There have been assertions of plant associations with metallic ore deposits; the gold weed with gold; the different colors a plant leaf absorbs from copper deposits; certain moss types on arsenic-bearing minerals. Even microscopic gold in the dung heaps of cattle feeding on range pasturage.

There may be some truth or coincidence in these tales. I present the idea of geobotanicals only for a broader prospecting viewpoint.

The rattleweed or locoweed (Astragalus) has been positively associated with serpentine soils, and indirectly with vanadium and uranium deposits.

ROCKPECKER TIP # 107 - Only a few species of locoweed grow where the soil is rich in selenium. Selenium is strongly associated with uranium - vanadium ore. Therefore, watch for rattleweed or locoweed *in strong concentrations*. You may have a uranium-vanadium deposit. Serpentine rocks and selenium go hand in hand.

Uranium deposits are of many types ranging from hydrothermal to sedimentary. Use of a Geiger counter or scintillation counter allows the radioactive emissions from daughter elements of uranium to be detected from a distance. Uranium as an element is not radioactive, however it is radiotoxic.

We are concerned here with stratified sandstones with nearly horizontal clusters of ore deposits. Deep-rooted tectonic events disturbed the surface in places with moderate folds and high angle faults. The deposits lie within the uplifted regions. Sandstone ore bodies tend to follow buried stream courses. Uranium and vanadium. when present, commonly fill the pore spaces of sandstone and small fractures in shale and limestone.

The principal ore mineral is carnotite, the oxidation product of uraninite. Vanadium, as a vanadate, is a principal associated metal. Native selenium and native arsenic is strongly associated with uranium - vanadium ores.

ROCKPECKER TIP # 108 - A very thin line of carbon in the sandstone seems to be characteristic of uranium or gold content.

DIAMONDS

The occurrence of diamonds in kimberlite naturally suggests they may be found in allied basic and ultrabasic rocks. The Arkansas diamonds are sourced in a group of small peridotite intrusions. CHROMIUM DIOPSIDE IS A DIAMOND INDICATOR MINERAL.

The best examples of ultrabasics are ore bodies of serpentized peridotite and pyroxenite that produce asbestos and talc in crystalline rocks. They have proven to be of no importance as diamond sources, although gravels derived from them have yielded scattered stones in Tennessee, Alabama, Georgia, North Carolina, Virginia and the asbestos belt of Quebec, Canada. Small stones of similar occurrence have been found in British Columbia and the gold gravels of California's Sierra Nevada's.

Based on the findings of the firehose miners at Mokelumne Hill, late Tertiary events occurred at least eight times. It was here that the water canons uncovered a series of eight long vanished river courses. Many hundreds of diamonds wound up in the sluice boxes. Since the miners were not looking for diamonds, how many diamonds escaped downstream or were discarded as rock crystal or crystal quartz? Don't overlook them!

ROCKPECKER TIP # 109 – In the past, both platinum metals as well as diamonds were recovered with the placer gold. While diamonds have been found in the goldpan, it is more probable that the stone will be crystal quartz. Save all such finds!

An easy way to distinguish a diamond from crystal quartz is to rely on the index of refraction. (Both scratch glass.)

***WHAT** TO DO:* Get some kerosene and pour it in a clear glass vessel. Drop your probable diamond into the kerosene. If the stone edges appear to disappear, it is crystal quartz. The reason rests on the fact that crystal quartz and kerosene each have nearly the same index of refraction and will blend into each other.

CHAPTER 37

PLANTLIFE MINERAL GUIDES

Suppose that you are building a cabin on your claim. There are no running streams as a water source; you know you need to dig a well for your supply, or else haul every drop from miles away. Desperate to find an underground source, you try water witching or dowsing and get no reliable response. What next?

ROCKPECKER TIP # 110 - In desert regions seek a plant that you know has a deep root system that penetrates the zone of saturation of the water table. You would look for mesquite, greasewood and perhaps alfalfa.

In non-desert regions look for plants that like water-logged soil. You would look for willows, ferns, and alders.

As long as you are looking for water, keep an eye open for large, shallow, earthen, depressions that could indicate a gossan precipitation of ore metals from shallow ground water. Recall, you will be looking for limonite or iron traces within the depression.

Geobotanical prospecting employs visual inspection of plants as a guide to buried ore. This method has the advantage of providing results that are immediately available. This is for the reason that direct observation of plants provide the clues; even the absence or distribution of plants can be meaningful.

The method is based on the idea that root systems are mighty sampling systems that siphon solutions containing dissolved minerals (including metals), which may deposit in the *UPPER* parts of the plant, or modify the vegetation's normal growing habit, or produce a frail, unhealthy plant that may not survive. Live Juniper tree roots have been found in mines at the 300 foot level.

The geobotanical prospector looks for plants that show an

excess of a toxic element that interferes with the plant's health. "Toxic element" in this sense means more of that element than the plant can normally tolerate. Much like a child eating too many chocolates. Think of it as plant-indigestion. With no antidote handy, it shows the harmful results in its fruit, leaves, twigs, bark, and roots.

It must always be borne in mind that grasses, herbs, shrubs, and trees whether with leaves or needles, require mineral elements for their normal health. The metallic elements of chromium, cobalt, nickel, copper, zinc, molybdenum, lead and zinc are some of the minor elements that plants take-up in *unmineralized* ground. Each plant has its own requirements; the minor element take-up varies with each species as their environment changes.

ROCKPECKER TIP # 111 - You may see a clue that leads you to believe the plant is directly rooted in ore. Remember that leaching of the ore body will spread the included elements adjacent to the ore body. You need to think of it as like eluvial gold laying near but close to the ore source.

ROCKPECKER TIP # 112 - Select for inspection, a common, but uniformly distributed plant species. Deep-rooted plants are preferred to shallow-rooted species. Be sure to inspect plants at the outer edges of what you perceive to be the orebody.

ROCKPECKER TIP # 113 - There are only a few UNIVERSAL indicators such as the ASTRAGALUS (rattleweed) which favors selenium in serpentine rocks. This can point to URANIUM, NICKEL, VANADIUM, and CHROMIUM ores. It does not grow elsewhere. The yellow violet, known as the CALAMINE VIOLET, which grows only on ZINC-RICH soils and nowhere else, are the two best known indicators. Most others can be thought of as LOCAL indicators which prefer mineralized areas, but can grow in non-mineralized ground in other regions.

There is an attempt to place ASTRAGALUS (rattleweed) on the endangered plant list. This plant is well known in Northern California and the Colorado Plateau. It is a member of the locoweed family. As forage, it is toxic to sheep and cattle as well as to other grazing WILD animals.

COPPER FLOWERS OF THE LABIATE FAMILY

Yes, it will require some special training to distinguish members of this herb family, but it can pay-off big if you take the time, and have an interest in COPPER prospecting. Your research need be only as far as the library.

A small annual mint known as HAUMANIASTRUM ROBERTI has been reported to be resistant to copper toxicity and thrives over COPPER ores.

Another copper flower that shows strong growth in a copper ore environment is GYPSOPHILA PATRINI. It is a copper indicator.

One of the best geobotanical indicators of copper is a copper flower known as BECIUM HOMBLEI DE WILD. It is a low herb with characteristic flower and leaf form. It must be carefully examined, for it is very similar to B. OBOVATUM which grows more rapidly in normal soils, but is unable to grow in high COPPER or NICKEL soils. An interesting point about BECIUM HOMBLEI is that its growth distribution has been found to outline the underlying COPPER deposits and leach-off areas.

The CALIFORNIA POPPY has been associated with COPPER deposits in Arizona. Seems that the more vigorous and dense the flower stand, the richer the copper deposit. The problem with this guide is that the California Poppy can grow almost anywhere. Never-the-less, if you see a densely populated field of this flower, take some soil samples for later cookbook chemistry tests.

A useful, though NON-SPECIFIC guide to NICKEL, COPPER, COBALT, CHROMIUM, and ZINC ORES has been the yellowing of plant leaves, known as chlorosis. It is understood that this is caused by interference with plant chlorophyll production leading to leaf discoloration. It seems that those mentioned metals mess up the iron intake that is essential to plant growth. While chlorosis can be caused by other factors, leaves with abnormal coloring are important clues for the Rockpecker to follow. Special attention should be given to small flowering plants and grasses in moist areas. CHLOROSIS IS COMMON IN HIGHLY MINERALIZED SOILS.

METALLIC ELEMENTS

TOXIC PLANTLIFE EFFECTS

METALLIC ELEMENT	GROWTH CHANGE CLUE
CHROMIUM	GREEN VEINS IN SHADES OF YELLOW LEAVES.
COBALT	WHITE PATCHES ON NORMAL LEAVES.
COPPER	PURPLE STEMS, LOWER LEAVES WITH DEAD PATCHES STARTING AT THE LEAF TIP, GREEN VEINS IN SHADES OF YELLOW LEAVES.
MANGANESE	CURLING LEAVES WITH DEAD EDGES, GREEN VEINS IN SHADES OF YELLOW LEAVES.
MOLYBDENUM	YELLOW-ORANGE COLOR OF FOLIAGE, STUNTING.
NICKEL	LEAVES SHOW WHITE, DEAD PATCHES.
URANIUM	FRUITS WITH DEFORMED SHAPES.
ZINC	LEAF TIPS APPEAR DEAD, STUNTED PLANT, GREEN VEINS ON SHADES OF YELLOW LEAVES

Geobotanical methods can have a diminished prospecting confidence level because drainage, sunlight, plant age, and other variables complicate the conclusions. The method's main advantage rests in the speed of a survey and the ease of inspection.

ASIDE - It is well known that metals in an ore environment will leach out through chemical actions. These metals are carried along with the transporting waters to form, over time, deposits known as "disseminated ores". These are the same waters used by plantlife to sustain their growth.

Gold is one of the metals, that in trace amounts, appears in solution in some water tables. Based on that fact, some innovative merchandising person incorporated it in a real estate promotion to move some Western ranch land.

The only practical use for the ranch land was in cattle ranching. It was remote, desolate, and in point of fact, nearly worthless.

How could this imaginative marketer distinguish this offering from all the other similarly worthless ranches available? How could he impart some desirability into the undesirable? How could he make this property memorable enough to attract a buyer, or at least a looker?

He advertised the offering as a gold property! His ad-copy ran with many descriptive words of delight to someone interested in gold.

In the ad he stated that gold had been discovered, in trace amounts, in the cattle manure. That some enterprising owner could collect and process the manure for its gold content.

We know from our geobotanical knowledge that this could be possible, in that the grasses could have accumulated enough gold for it to show up in steer dung. After all, that's the basis for geobotanical prospecting.

The author supposes, that after a potential buyer or looker was in hand, the rest of the sales talk involved an unknown, hidden, undiscovered gold deposit that would not only pay for the ranch, but make the buyer wealthy. HOW'S THAT FOR GEOBOTANICAL MARKETING ?

CHAPTER 38

GEOCHEMICAL PROSPECTING

Geochemical techniques bring to prospecting the detection and mapping of geological structures hidden beneath the overburden of soil or barren rock. It has the ability to detect concealed ore bodies over a large area in a cost-effective manner.

Essentially the geochem prospector is measuring trace elements in a qualitative and quantitative way over a wide expanse. The tested material may be water, soil, vegetation, or sediments. Where the element tested-for shows up in an abnormal amount, it is called a "geochemical anomaly". That finding may be a clue to hidden ore.

There are two types of geochemical surveys:

FIRST TYPE - You own a ranch composed of 5,000 acres of mostly worthless ground. You want to know the mineral possibilities of your ranch, particularly any areas that should be looked at more closely. This is called a *reconnaissance survey.* Samples would be taken for chemical testing over large spacings, such as one sample per 100 acres. Once the element tested-for is determined for the overall ranch, an excess over the average; may indicate a mineral prospect.

Stream waters, stream sediments and wells are commonly tested locales. Soil sampling and stream sediment sampling give the best confidence level for most metals. Water samples are good for the more soluble elements.

SECOND TYPE - Your ranch shows higher levels of the trace element at two general, but non-specific sites. You now need to make a decision as to whether you want more detailed information as to mineralization specifics. At this time a *detailed survey* would be taken for the purpose of outlining the mineralization, then to localize the source.

Samples are taken at five to one hundred foot distance intervals. The samples are usually soil samples that can run the cost of the survey up significantly. Cost determines sample spacing schedules. Soil samples, particularly deep samples, give the most dependable results, in that leakage of metals from an orebody show up more strongly in the adjacent soil residues.

Geochemical surveys are used in all major exploration efforts. They are used to confirm findings made by other methods.

Copper, lead, zinc, silver, gold, nickel and molybdenum are the major trace elements sampled. Earth materials chemically change as they move, mix and redeposit with other materials. This can occur both at or below the surface. The mobility of elements form *dispersion patterns* away from the orebody. This allows for a larger discovery target when employing geochemical means.

An *indicator element* is an element surveyed for in order to detect an orebody. For example, the element copper is an indicator element for copper ores. The same can be said for uranium. But what if the mineral component is hard to analyze, or difficult to interpret? What then?

Whatever the geological process, elements of similar mobility tend to be associated. Therefore, a constant finding of a high content of one element in association with another, means the probability of finding the associated element is also high. Such an element is called a *pathfinder element*.

For example, in a surface environment, molybdenum is more mobile than copper. It can be used as a pathfinder search element for locating porphyry copper deposits. This is for the reason that porphyry copper deposits usually contain some molybdenum. Copper is a pathfinder element for nickel-copper ores. A pathfinder element is used because it is easier to detect than the element being sought.

YOU LOOK FOR ONE ELEMENT TO INFER THE FINDING OF THE ASSOCIATE.

SOME COMMON ELEMENTS IN GEOCHEMICAL ASSOCIATIONS ARE:

PATHFINDER ELEMENT	TARGET ELEMENT SOUGHT
Potassium	Rubidium
Calcium	Strontium
Aluminum	Gallium
Silicon	Germanium
Zirconium	Hafnium
Niobium (columbium)	Tantalum and rare earth metals.

IN BLACK SHALES -

Aluminum	Silver, arsenic, gold, bismuth, cadmium, molybdenum, nickel, lead, vanadium, zinc.

Before you can effectively recognize dispersion patterns that relate to ore, you must first determine the *background* of indicator elements that would normally occur in non-mineralized ground. (See ELEMENT chart "E", column "K".) When parts per million of a given element are known, it is more realistic to understand "background" as a range on either side of the PPM value.

An anomaly is simply a "deviation from the normal". An orebody, being a rare occurrence, thus deviates from the normal, is an anomaly; as is the geochemical dispersion pattern related to erosion or genesis of an orebody.

When samples are taken and analyzed, if the concentration of the indicator element is greater than the top threshold of the background range, it can be considered a *significant anomaly*, and as such used as a guide to ore.

ROCKPECKER TIP # 114 - It is obvious, that if known ore deposits are in the geochemical search area, background readings will be higher than normally would be expected. Therefore, it is cost-effective to use local background thresholds at the higher level to detect and map an orebody. It is also meaningful to graph the normal background threshold.

Once the geochemical anomalies are described, they can be pictorially graphed as a ratio of anomaly strength to background strength. This allows for visual comprehension of the orebody.

The effectiveness of a geochemical exploration survey greatly depends on a sensitive procedure that will detect elements present in small concentrations. The method used must be reliable, cost-effective, and above all simple, so that multiple samples can be easily prepared for trace element analysis.

There is a choice of many chemical methods to extract the trace element. The best method is the one that gives the clearest contrast between background and anomaly. Some of the methods in use are:

1. EMISSION SPECTOGRAPH - The sample is vaporized, with the trace element directly read by optical methods.

2. FUSION - the sample is mixed with an organic salt as flux, such as potassium bisulfate which acts like an acid to decompose the sample.

3. HOT CONCENTRATED ACIDS - Acids are used to decompose the sample. It has been found that trace elements in pyroxenes and amphiboles are only partially released. Chromium, manganese, molybdenum elements may be lost to volatization from the action of concentrated acids. The method requires great care.

4. Weak aqueous acid solutions and oxidation-reduction agents are also often used to decompose samples for trace element release.

After the trace element has been released, it must be separated from other liberated elements. Two methods often used are:

A. Separation in liquids through solvent extraction of the trace element. This is the basis for many color tests.

B. Often employed is solid phase separation by precipitation.

Once the trace element is released and captured, the question of "how much of the element is present" must be answered.

Colorimetry relies on the color intensity to estimate trace element quantity. Colorimetric tests are available for nearly all elements of interest to geochemists. This method provides a low cost simplicity and portability of equipment as a major advantage. However, there is a draw-back. It can only check one element per test.

Some other methods of determining the quantity of a trace element are:

1. - EMISSION SPECTROMETRY
2. - ATOMIC ABSORPTION
3. - X-RAY FLUORESCENCE
4. - RADIOMETRIC DETECTION
5. - ELECTROANALYTICAL MEASUREMENTS
6. - CHROMATOGRAPHIC METHODS
7. - MASS SPECTROMETRY

CONCLUSION - In most geochemical surveys, sample precision is more important than sample accuracy. Sample accuracy can be checked by a second lab. Precise measurement of trace elements in a single sample, over a range of samples, gives a greater confidence level in describing and graphing the weak anomaly. It is that slight variation in trace element readings that hint at the presence and location of an ore body.

TECHNICAL? YES ! QUICK AND GOOD? YES !

At the information gathering stage of exploration, a detailed geochemical survey is faster and much less costly than bringing in major trenching and drilling equipment to arrive at the same information confidence level.

(See LaRune's COOKBOOK CHEMISTRY Notes, PART IV)

CHAPTER 39

NATURE'S HEAP LEACHING

SEDIMENTARY CHEMICAL WEATHERING - SUPERGENE ENRICHMENT

SECONDARY ENRICHMENT - GOSSANS - COLORADO'S - EISENER HUT

Can you believe that metallic elements dissolve in water; that at surface conditions, metals oxidize and are washed by slightly acid solutions to redeposit deeper in the earth? Well, iron rusts, doesn't it? Some chemistry examples will explain.

These chapter sub-titles all reach the same end result. Earlier, we discussed formation of alluvial deposits through sedimentation, transportation, deposition and lithification into sedimentary rock. Primarily, then, emphasis was placed on physical weathering and physical break-up of prior rock forms. Those break-up pieces are called *clastic.* We now deal with chemical weathering as an erosion process.

Here, we will deal with minerals that dissolve in water to form solvents, that in turn dissolve metals, so as to be transported in a pregnant solution that redeposits to form new minerals.

When you think about it, ground water (as water) can only move in three directions, up, down and sideways. Our interest will concern downward water movement with some sideway options. The water, of course, is precipitation in the form of rain, snow or melting ice. On its fall to the crustal surface, the rain mixes with carbon dioxide from the air to form a mild carbonic acid mixture solution. Geologists call rain *meteoric* water to distinguish it from fossil water trapped in sediments, rising from below, which they term connate water. In time, these acidic water solutions can dissolve most rock minerals, even quartz, one of the rock forming minerals.

As existing rock weathers, erosion reduces landforms to

expose the deeper-lying rock structures. Where ore deposits are bared, they weather along with the enclosing rocks. The surface exposure oxidizes many ore minerals to yield solvents that dissolve other minerals. The mineralized bodies themselves become oxidized and leached of their values. The leaching stops at the water table or at a depth where oxidation cannot take place. The depth of oxidation is called the *oxidation zone.*

As the cold, diluted leaching solutions migrate downward, they lose all or part of their mineral cargo within the zone of oxidation, to form oxidized ore deposits.

The oxidation and leaching of the upper part of the weathered ore deposit, leaves the upper portion barren of mineralization. (except gold) The effect of supergene enrichment has made the uneconomic ore above, an enhanced, enriched ore grade in the water table. The process of supergene sulfide enrichment is the genesis of some of the great copper deposits.

In recap, the worthless, exposed mineralized deposit has solution weathered through chemical action, to transport the metals in a precipitate, to form valuable economic ore deposits in the water table.

ROCKPECKER TIP # 115 – Where you see a rusty color over a large, circular, surface area that gives the appearance of a halo laying over the ground, you have found a *gossan.* The values lie below the halo and can extend more than 300 feet. The brown, rusty surface color is iron, usually goethite altered to limonite. That's why the German's call it an "Eisener Hut" or *iron hat.*

Dissolved oxygen in water alone is a powerful oxidizing agent. Carbon dioxide, chlorides, iodides, and bromine in water can yield strong solvents. Sulfuric acid reacting with sodium chloride (salt) yields hydrochloric acid that, when added to iron, creates a strong oxidizer known as Ferric chloride. While these solutions are admittedly dilute, solution chemistry with oxidation can reduce many mineral deposits.

Pyrite (iron sulfide) is the most common sulfide in ore deposits. When it oxidizes, it generates sulfuric acid to leave limonite or hematite; hence the brown, rusty color, along with tiny amounts of native, yellow sulfur in the outcrop.

Over time, the percolating of these charged waters through the marginal sulfide ore deposit will leach out copper, silver, lead and zinc. Where pyrites is absent, the oxidation products of other sulfides will remain in place. This is also true when carbonate wall rocks neutralize the acidic water to cause dissolved metals to precipitate.

Silver behaves much the same as copper in the oxidation zone, but has fewer oxidation products that are stable there. Silver sulfide, acanthite, is oxidized and made soluble in ferric sulfate solutions. The silver content is then carried downward. In an arid climate, the halogens attack the weakly soluble silver salts to form silver chloride, cerargyrite; silver bromide, bromyrite; silver iodide, iodyrite; which deposit in the oxidized zone for lack of enough water to dissolve them. Native silver is commonly found here.

Zinc sulfates are very soluble. Ore deposits at the surface oxidize to leach out into the ground water system. In more arid areas, a shortage of leaching solution can leave zinc in the oxidation zone to form smithsonite, hydrozincite and hemimorphite.

Gold, chlorine and manganese present an interesting geochemical concept. Field evidence indicates that gold and chlorine in solution is capable of short transport, when manganese oxide combines with it to form the stable gold chloride. Manganese is present in many gold ores.

Gold enrichments in surface areas have been found. Coarse gold found in cracks and wall rocks in higher vein locations, point to short distance solution and redeposition. This gold was not in the physically weathered, eluvial deposits.

Oxidation of sulfides can produce acidic waters strong

enough to dissolve tram car rails in a mine environment.

Gossans are weathered surface sulfide deposits where leaching and bleaching leaves a tell-tale halo. A gossan can be thought of as a large diameter pipe extending downward. Supergene or secondary enrichment can move the valuable sulfides from upper to lower pipe depth levels. Silver, copper, and zinc sulfides are the desired targets at depth. But how can you tell what's at depth? If metals are present, is the mineralization economic?

Are there economic values in the pipe? Is the iron-stained halo a *false gossan?* Has a tectonic event interrupted the oxidation process by diverting the ground water into a newly formed fault?

Remember that in a gossan, descending waters charged with dissolved minerals, tend to seek a density level for concentration in the water table, where the sulfides are stable. No water, no oxidation, no solution, no reprecipitation in recap.

ROCKPECKER TIP # 116 - Chemical weathering, supergene enrichment, secondary enrichment, gossan, colorado, Eisener Hut, presuppose a large igneous intrusion such as a stock or sill, deeply buried in the crust, erosively, now poking out of the crust. The exposed outcrop carries ore values that are chemically attacked by solvent water solutions, which reprecipitate in the water table as valuable ore concentrations, *or mineral alteration products.*

LEACHED OUTCROPS OF MINERALIZED BODIES HAVE BEEN A LONG SOUGHT TARGET. MANY HAVE BEEN DISCOVERED AND MINED.

ROCKPECKER TIP # 117 - A gossan zone of mineralization need not be presently exposed. Evidence can show up in a drill core sample, as long as the oxidized ore minerals reside in the zone of oxidation. However, alfide mineral, such as pyrites, and the rusty limonite color at the surface are essential components.

Gossan associated minerals are the copper-bearing sulfides of chalcopyrite, chalcocite, covellite; sphalerite is the zinc-bearing sulfide; galena, the lead sulfide. Silver chlorides-bromides-iodides are the silver-bearing halogens.

ROCKPECKER TIP # 118 – Jasperoid is a prime indicator and prospecting clue in gossan-search. The oxidation process converts the feldspars to clay, while leached-out silica alters to iron-colored jasper.

Where mineralized solutions pass through limestone, the acidity is neutralized. Sphalerite migrates as zinc carbonate; the lead in galena stays put; silver migrates; gold is separated from the sulfide, to deposit nearby as free-milling gold.

ROCKPECKER TIP # 119 – The rusty limonite blanket lying on the surface is a "here I am" marker.

ROCKPECKER TIP # 120 – The supergene zone (super means above) has been known to produce *black copper called sooty chalcocite.*

ROCKPECKER TIP # 121 – We tend to think of a gossan as a *thing.* Rather, our concept should embody the idea of a carpet-like, rusty, limonite outcropping mass, that usually overlies concentrated iron-bearing sulfide deposits.

ROCKPECKER TIP # 122 – If there is no iron, there is no gossan. Iron, as limonite, is universally present. To determine the underlying minerals, it is visually more important to determine the color of the limonite and the color of the boxwork skeleton shapes. *Limonite boxworks* is a descriptive phrase for the skeletal remains of the honeycomb patterns, as the surface remains of the leached-out sulfide mineral. Where limonite has moved to the fringe of the sulfide mineralization, a bleached halo may form as another gossan indicator.

ROCKPECKER TIP # 123 – Sulfide ores may be expected to underlie the boxworks. But what if erosion has destroyed the boxworks? Then, the source-sulfides are apt to lie upward and lateral of the halo.

FIELD EXPERIENCE, BASED ON THE WEB-LIKE ARRANGEMENTS OF THE BOXWORKS, TOGETHER WITH THE LIMONITE COLOR CAN INDICATE THESE UNDERLYING GOSSAN MINERALS:

SURFACE BOXWORKS COLOR	*UNDERLYING SULFIDE*
1. Halo - no box - BRICK RED	PYRITES
2. Halo - no box - BRICK RED	PYRRHOTITE
3. Halo - no box - YELLOWISH-BROWN	SPHALERITE
4. YELLOWISH-ORANGE	BORNITE
5. BROWNISH-MAROON	CHALCOCITE
6. YELLOWISH	CHALCOPYRITE
7. ORANGISH-YELLOW or ORANGISH-BROWN	GALENA
8. YELLOWISH-BROWN	SIDERITE
9. MAROON	MOLYBDENITE
10. CHOCOLATE	TETRAHEDRITE

The minerals redeposited in the zone of oxidation are the carbonates, silicates and oxides of: COPPER, ZINC, LEAD, SILVER, VANADIUM, and URANIUM. All have been found in economic quantities, but in the *lower* section of the zone of oxidation.

Generally, the water table follows the contour of the surface. Supergene sulfide enrichment starts at the top of the water table where reduction and reprecipitation can begin. Obviously, the lower portion of the water table would be most heavily mineralized. This would not be true where a fault provides an easy passage downward, or where a *perched water table* overlies the normal water table.

The zone below the water table is known as the *primary or hypogene zone.* This is the point at which secondary enrichment stops. Therefore, the ore below the water table is lean, uneconomic and unaltered. Mining engineers call this *protore.* That is, mineralized rock too lean to yield a profit. Had there been no water table, and no secondary enrichment, it would have all been the primary ore zone, and as such, worthless.

CHAPTER 40

GEOPHYSICAL PROSPECTING

The one blanket, one jackass prospector, when searching for outcrops, was using a tried and true method of ore discovery. The theory he was following was quite simple. If the mineral discovery shows in a surface outcrop, it likely extends at depth for a possible mine.

Compared to barren ground, an orebody is rare. Over time, many of the surface outcropping ores have already been found and mined. Therefore, any method that can 'see' below the surface, or provide under-surface information that can be interpreted to discern concealed earth conditions, has a prospecting value.

The application of physics and its physical principles, as applied to the earth, is known as geophysics. Geophysics provides some methods that can give us a glimpse of sub-surface formations. Within a particular method's limitations, large areas can be eliminated as being barren of specific ore types. It gives an "it ain't there" answer.

Orebodies exhibit specific physical properties that differ from its encapsulating country rock. The important differences are magnetism, density and electrical conductivity. Geophysics uses these ore properties to eliminate barren ground; it can also direct more detailed searches to areas that are more favorable ore sites. *The primary mission of a geophysical survey is to separate the barren ground from the more promising ground.* More to the point, the success of a geophysical survey is measured by the money, time and effort saved by subtracting barren ground from an exploration schedule that could only have been eliminated by the more expensive methods of drilling, trenching and shaft sinking.

ROCKPECKER TIP # 124 – It is important to realize that even the finding of *barren* ground is only a *probable*, and not a *positive* conclusion. The capability of the equipment and/or the expertise in data interpretation may fall short of an ore target to miss it.

Geophysical methods have been devised for ore exploration that use magnetism, electric conductivity and density of earth materials as search tools.

I. MAGNETIC METHODS

With the earth's natural magnetic fields, a compass can be used to guide your way. When you pass over a magnetic, concealed, orebody, that compass will show a variation in inclination of the magnetic field. This was the basis for the old "Spanish Needle". If you can measure those variations in the magnetic field, you can use the variations to locate magnetic ores.

Magnetite, ilmenite and pyrrhotite-bearing sulfide deposits, as well as manganese and chromite ores produce distortions in the earth's magnetic field. These distortions can be employed to locate those ores.

ROCKPECKER TIP # 125 - Manganese and chromite ore's magnetic response can vary from strong to none depending on their iron content. This magnetic irregularity means that magnetic prospecting for these ores is, at best, hit or miss.

Magnetic field is the mechanical force that surrounds a magnet, as well as an electrical current. The lines of magnetic force in a magnetic field are known as *flux density.* The lines of magnetic force in a bar magnet start from the north pole, flow outside the magnet as a force field to the south pole, then continue within the magnet to form closed-loop force fields. The current in the loop multiplied by the loop area is called the *magnetic moment*. The moment can be measured and expressed in gauss units of one ordinary ampere.

For convenience, geophysics divided the gauss unit to create a sub-unit of 1/100,000th of a gauss called a *gamma.* The strength of the earth's total magnetic field is 60,000 gamma or .06 gauss.

(100,000 gamma = 1 gauss = 1 maxwell per cm^2)

Orebodies increase the earth's natural magnetic strength by thousands of gamma. Therefore, when your metering equipment reads 60,000 gamma, then jumps to 75,000 then 100,000 gamma, you are probably over a magnetic ore body... or under high tension wires. This reading is called *intensity of magnetism.* As you move away from the intensity, your meter readings decrease. The increase in intensity of magnetism is said to be *induced* by the field.

Iron, cobalt, nickel, magnetite, pyrrhotite, chromite, manganese, and a few other ores are magnetic in a permanent way.

If the intensity of magnetism is acted on by induced intensity, the flux density increases. If an ore deposit develops north and south fields on its surface, the field either remains constant or north in the field direction, and south on the opposite surface. South is said to be positive with north the negative. This +/- is thought of as *susceptibility of the body.*

Substances with high susceptibility are rocks with ferromagnetic mineral content of iron oxide or iron sulfides. Basalt, skarn, diabase rocks are more magnetic than slate, sandstone or limestone. Pyrite and hematite intensities of magnetism are so low that the earth's field cannot distort them enough to register with the current magnetic detection instrument technology.

MAGNETIC DETECTION OF AN OREBODY DEPENDS ON THE DIFFERENCE IN THE SUSCEPTIBILITY OF THE OREBODY AND THE SUSCEPTIBILITY OF ITS HOST ROCKS.

On the whole, the permanent magnetism of *igneous* rock is dependably stable. In the field, a fluxgate magnetometer is taken to a reference point, set to zero, then moved to another reference station where the needle deflection angle is measured in gamma and recorded. Depending on the terrain, one man can take 100 to 200 reference station readings over a 50 foot spacing per day. This is most cost-effective.

An orebody resembles a large bar magnet. From the surface, the ore anomaly's magnetism is directed downward for a positive susceptibility. The north pole develops on the lower surface with the south pole on the top surface of the orebody. With proper expertise, pole theory interpretation of magnetic intensity readings, can produce *estimates* of depth, width, strike, length, and attitude of the deposit. The data is also pictorially graphed for visual comprehension of the magnetic anomaly.

II SELF-POTENTIAL METHOD

Natural voltages due to differences in chemical activity in the ground are used to measure electrical conductivity. Think of the earth as a giant battery. If two non-polarized metal stakes are driven anywhere into the ground, then connected by cable with a voltmeter in the line, voltages can be read on the meter. A normal voltage reading in non-mineralized ground would show 50 or less millivolts.

However, above sulfide bodies containing pyrite, chalcopyrite or pyrrhotite, you can get 400 to 500 millivolt readings. Graphite, pyrolucite, psilomelane, magnetite and coal also produce higher readings. THIS IS CALLED THE SELF-POTENTIAL OR S-P METHOD. When the surface layer is a weak conductor, this method cannot be effectively used.

III ELECTRICAL METHODS

A. Resistance

If we send an electric current into the ground by means of two electrodes driven into the earth, a cable / voltmeter connecting the electrodes can show the distribution of voltage and the results can be mapped. Where the measured ground is of uniform composition, the voltages, known as *equipotentials*, have the same constant value with no voltage difference. The current path traveling through the ground becomes distorted in the neighborhood of an orebody that is a better electrical conductor. The better conductor offers an easier current path (less resistance) through the orebody. VOLTAGES TAKEN OVER THE OREBODY WILL BE UNIFORM, BUT READINGS INCREASE AS YOU REACH THE PERIPHERAL REGIONS OF THE ORE.

OREBODIES THAT ARE POORER CONDUCTORS THAN THE SURROUNDING COUNTRY ROCK WILL GIVE HIGHER READINGS BETWEEN ANY TWO POINTS OVER THE ORE, WHILE THE PERIPHERAL ORE-REGION READINGS WILL DECREASE.

The resistance of an ore deposit depends on the content of conducting minerals. The better the ore conducts electricity, the lower the resistance readings will be.

OTHER GEOPHYSICAL PROSPECTING METHODS CURRENTLY IN USE ARE:

1. ELECTROMAGNETIC

2. INDUCED POLARIZATION

3. GRAVITY

4. SEISMIC

7. RADIOACTIVE

8. AIRBORNE MAGNETIC

9. COMBINATIONS OF TWO OR MORE METHODS

As technology for positive prospecting advances, we can expect improved results through the use of geophysical methods.

CHAPTER 41

ELECTRONIC PROSPECTING

GEIGER COUNTERS
SCINTILLATION COUNTERS
METAL DETECTORS

If you consider uranium as *radioactively* emitting "here I am" signals on your electronic equipment, this is simply NOT so.

URANIUM IS NOT RADIOACTIVE! It is, however, RADIOTOXIC. Where uranium has had time to produce daughter-elements, it is the daughter-element radioactivity that points the way to associated uranium.

Elements are composed of a nucleus with orbiting electrons. The number of electrons affects the chemical form the element takes.

If you can visualize a nucleus that broadcasts its presence with a property we call radioactivity, then we may be able to tune-in on its homing signal and thereby locate it. These spontaneous emissions, in the form of radiation, disintegrate the nucleus, which in turn becomes the nucleus of a new element. Call it a daughter-element. It may be radioactive.

The emissions are electrically charged particles. The *alpha* particle is positively charged; the *beta*, negatively charged. The *gamma* emitter is an X-ray of short wavelength.

There are over 50 elements with natural radioactivity. Neptunium and thorium are but two of them.

The alpha, beta and gamma emissions are all able to ionize air or gas, so that the air or gas is able to conduct electricity. The alpha and beta rays are easily stopped by the slightest obstruction. The gamma emissions, in theory, can be detected across any thickness. However, gamma intensity reduces in a gradual way, to fall below the sensitivity and background noises of detection devices.

When gamma rays strike a crystal such as sheelite, the crystal produces small flashes or *scintillations* of light that can be measured. The radioactive search, then, would be for sites with high gamma radiations. Since not all radioactive elements emit gamma rays, they cannot be detected, unless a daughter element emits the gammas. Uranium is such an element. Therefore, the search for uranium and thorium is actually a search for the daughter element radium, polonium or radon.

Granite, shale, limestone, sandstone, and basalt all contain small gamma ray emitters. These make up the *background noises* that mask uranium and thorium deposits.

GEIGER COUNTERS AND SCINTILLATION COUNTERS

In a Geiger Counter, the ionized gas in its tube completes a circuit when hit by a gamma ray. This causes a pulse which can be heard with earphones and read on a meter as *counts per minute.* A Geiger Counter is relatively inefficient in that only 1 in 100 gamma rays ionize the gas in the tube. This is a 1% sensitivity.

When a certain crystal absorbs a gamma ray it flashes with EACH hit. Each scintillation can be counted, heard, and read on a meter. This makes the Scintillation Counter about 100 times more sensitive than a Geiger Counter. Is it any wonder that the Geiger Counter is a poor second choice for the uranium prospector? There is a trade-off though. The Geiger Counter has a greater sturdiness with better electronic stability.

In using either of these electronic devices, you must first take a background noise reading. This reading must be adjusted from place to place because the background counts will vary from different locations. TO ADJUST FOR THE BACKGROUND NOISE, THE BACKGROUND COUNTS MUST BE SUBTRACTED FROM YOUR LOCATION READING

COUNTS.

ROCKPECKER TIP # 126 - A COUNT IS NOT MUCH GOOD UNLESS IT IS THREE TO FIVE TIMES THE BACKGROUND NOISE LEVEL.

ROCKPECKER TIP # 127 - Certain tin, copper, lead or silver veins can contain uranium. If you find a vein indication with your counter, be sure to take at least one sample. Be sure to note it in your field book.

ROCKPECKER TIP # 128 - When your counter reads twice background noise level during a slow walk survey, walk a grid pattern of about 20 feet square. Take notes of the adjusted readings. Now place your counter on the ground and take that reading as the area count. Continue your survey-walk for more indications. Plot the counts on your map. Draw line-curves through *equal intensity readings* to give your map a visual radiation picture. Your map should look like contour elevation map lines.

After you have plotted readings on your radiation map, you would normally take the area with the highest readings count, as the CENTER OF RADIATION. Pay particular attention to your plotted curve counts. A pegmatite outcrop can give higher counts than one at greater depth. Conversely, strata rubble can cause a significant count to fail to register.

If your radiation map looks spotty, it may be due to *SKATTER.* Probable cause: Radioactive decay with consequent loss of screaming radon daughters.

ASIDE: The use of a metal detector, Geiger Counter or Scintillation Counter falls within the area of Geophysics. You are now a Geophysical Prospector.

Surface deposits of uranium deposited in the past 10,000 years may show little or NO radioactivity. Your radiation detector will miss those deposits.

Uranium forms highly soluble compounds with carbonates and phosphates. In solution with surface or subsurface waters, the uranium can travel for miles from the granite or volcanic tuff sources. Discovery of non-radioactive surface deposits have been found in recent years. The uranium content must be confirmed by chemical analysis.

GENESIS: Uranium in a soluble compound will precipitate in a reducing environment in the presence of organic materials such as peat, mud, muck, or black clay.

ROCKPECKER TIP # 129 - Organic materials in a wet environment STINK and show black color. A peat bog, marsh, or other wet, black area bears investigation. Stinking, black, organic mud or clay, especially that which is located near granites or volcanic tuff, could be a young uranium deposit. The more stink, the better your odds.

It is said that young surface deposits of ELEMENTAL uranium, being non-radioactive, have an extraction value over and above conventional uranium mining. That is, the tailings will not be radioactive, thereby eliminating those environmental concerns.**

** The principal ore of uranium is uraninite. The theory of the pure element uranium, combining with phosphates or carbonates to then precipitate, assumes that the radioactive thorium isotopes are absent, or nearly so. This is highly unlikely, but theoretically possible. ** 100 % extraction recovery is equally unlikely with a nil probability. From an environmental concern standpoint, RADIOTOXICITY is a much greater risk.

METAL DETECTORS

In its most simple form, a metal detector emits a downward signal that rebounds from a metallic object to translate into an audible beep. It is the beep sound that pinpoints the target. These units can detect metals on the surface as well to a certain depth. Some detectors have a ground canceling tuner so that, in mineralized areas, the metal target will be more audible with the background noises neutralized. The detector head coil varies in diameter to the purpose for which it was manufactured. Most units are water safe. They can be employed without harm to the instrument as under water scanners.

Placer gold prospectors favor the metal detector as useful in finding the magnetic black sands that most often deposit in the same trap locations as alluvial gold. It is the author's experience that the favorable goldbearing streams carry the black sands to deposit them at most all streambank locations. Because of this experience, that particular use for a metal detector results in a lot of ambiguous readings. To be fair though, with the instrument correctly set to eliminate normal background sand reflections, an experienced operator can efficiently distinguish the echos by sound intensity.

Don't overlook searching for paystreaks forecast by black, magnetic sands that lie higher up on the banks. Your target is a bench formed by a younger streamcourse or even bank locations made in high flood. Detector sound intensity can clue you to hidden sandtraps and older gold drops.

A detector use favored by many is called *nugget shooting.* Here, likely looking ground is systematically scanned. Most detectors will locate the larger gold pieces, but smaller gold will not give a *hit*, and thus miss or overlook those possibilities.

The author believes that one of the most favorable places for nugget shooting is on the TOP of one of those old dredge gravel piles that you see along many gold country streams. Why the top, and not some other place on the heap? And why a dredge?

These are two sophisticated questions. The answers ground the reasons for designating an old dredge site as favorable.

Those massive gold panning machines worked day and night for 365 days a year. One would think that from an efficiency standpoint, little, if any, gold escaped the collection system. You would be right too! However, you should realize that there were system design flaws that did permit the escape of gold nuggets.

One design problem was the trommel separation type. "Trommel" is the German word for drum. These drum separators passed through the larger gravels on the way to the tailings dump. The drum had holes drilled around the circumference, so that as it rotated, the gold and smaller gravel could pass through for further processing by the system. The gold sized smaller than the drum holes was caught, nuggets over the hole size escaped. Trommels usually had holes smaller than one inch. This means that the larger nuggets encountered by the dredge fell into the tail. Now you know the reason for exploring old dredge tailing piles with your detector.

Some dredges, that were not trommel equipped, were rigged with caged sluice boxes to trap all sizes of gold. They were caged to keep the gold in the sluices and out of the dredge worker's pockets. Sluices were cleaned at regular intervals. It was not unusual to have captured many thousands of ounces before the next scheduled cleanout. The problem that the sluice dredges could not control was "clayballing". The dredge would pass into and collect stream clays. That clay passed through the system and through the sluice boxes. When it hit the boxes, it collected the trapped gold, balled around the gold, to be carried to the tail as a mudball. A gold mudball could not be distinguished from any other mudball in the tail. During the night as operations continued, the sluice box could be emptied of its treasure.... and nobody knew when. That gold rests within the tail today. It also could be near the top as well as within the pile. Over the years, the erosion process has dissolved the mud on top of the pile. Now it's waiting for you and your detector.

We know that gold, being heavier than most anything else dropped by stream sediments, will eventually work down to rest on bedrock. The dredges were designed with continuous bucketlines that could dig down to 50 feet or deeper. That's were the richer gold was to be found, and lost to trommel design and mudball clay contamination. Obviously, the deepest alluvial material wound on top of the tailings pile. That's why the top of the heap is the avenue of best nugget shooting.

CHAPTER 42

LODE & PLACER GOLD TRIVIA

When we look at rockbound ore deposits we divide them into *syngenetic*, the ore formed at the SAME time as the host rock, and *epigenetic*, the ore formed LATER than the host rock, as in hydrothermal fluid additions. Sedimentary strata-bound rocks that contain gold, are an in-between. That is, the gold was formed and captured as an unconsolidated sediment to lithify and become part of a sedimentary rock. They are sometimes called *fossil* placers. When gold or other minerals lie in unconsolidated sediments, we call them *placer deposits.*

Placer deposits are the result of weathering, erosion, transport usually by water, and deposition at favorable sites, along a present or past watercourse. Besides gold, other resistant minerals in placers are: Platinum, cassiterite, magnetite, hematite, monazite, ilmenite, diamond, garnet, feldspar, and quartz in the form of gravels and sand grains.

The four types of placers are *eluvial, alluvial, marine or beach, and fossil.*

Eluvial placer deposits are found in the immediate area of the source-deposit. They are physically separated from the main ore deposit, but usually lie close to the source. When the minerals are moved downslope through water action, the eluvial rubble develops into the most important placer type. It is known as an *alluvial* placer. During its water travel path, abrasion, leaching, and other forces, cause rounding of the transported material. A chemical dressing takes place to leach silver from the gold rubble. The same is true with platinum, where iridium and osmium are reduced as an inclusion in the platinum.

ROCKPECKER TIP # 130 - When your goldpan displays a silver-gold mix (the silver will probably be black), sample the stream in its upward course. If the silver content of the pan continues to increase, you are nearing the eluvial source. If the silver content decreases, you have passed it by.

Also remember that there may be more than one silver source showing in your pan. At this point, you are interested in all origins, but the nearest source would be that point upstream where the silver content is the greatest. (It is a two-source feed.) Obviously, the same search-logic applies to gold.

Silver can be another search-clue. If it is bright and shiny, it was probably a recent deposition. It may not have had time to build up the black oxide tarnish. If what you suspect is silver with a dull gray coating, think platinum!

If you find a heavy mineral that acts like gold in your pan, but the color is brown or black, realize that while gold itself does not tarnish, it can pick up a coat of iron oxide to appear brown. A coat of manganese oxide can give gold a black coloring. The presence of black or brown gold tells you that it was there long enough for the passing iron or manganese minerals to rub off. Test you mineral by scratching it with your pocket knife or nipping it with your tweezers. The gold color will be readily visible beneath the coating.

An interesting hypothesis has been made as to the formation of gold nuggets. One states that fine and micron gold combines chemically to unite and form the lumps we call nuggets. Impact welding of fine particles is another idea presented to form a nugget. My own view rests simply on a nature-made-it-that-way in the ore. Bouncing along the streamcourse reduced its shape and size a little.

Just what is a *nugget*? There is no hard rule. You can define it anyway you like. We think of a nugget as a lump of gold, but of what size? If you can pick it up with a tweezer, or pick it up with your fingers, is it a nugget? Probably, yes.

In passing, let me comment on marine or *beach* placers. Incoming surf tosses the heavier materials toward the shore, the outgoing water action washes the lighter material toward the water. Gold, ilmenite, and diamonds have been found in beach placers. They are also called *marine* placers.

Finally, *fossil* placers: They are the most ancient. The placer gold was deposited so long ago, that the material that surrounded the gold turned to rock. It can even show subsequent metamorphism to turn gold-bearing sandstone into gold-bearing quartzite. The fossil placer distinction is made to distinguish it from the hydrothermal method of ore deposition.

Gold as a metal has little practical application other than ornamental, electronics, dentistry, and a few industrial uses. Gold's value, historically, has been as a forerunner of civilization. The search for gold quickly populated an area. As the gold ran out, depopulation followed with like speed.

However, gold is an international storehouse of value, a concentrated form of wealth. It is looked upon as a stable inflation asset, in that it appreciates in value as other forms of money depreciate to become worthless. Basically, gold retains its value because everyone wants to possess it. Since it was first known by man, it is estimated that total world gold production is less than 100,000 tons.

For the past 60 years, it appears to be U.S. policy to discredit gold in the eyes of the ordinary citizen. The "bad mouthing" continues today! Yet, the U.S. is the largest gold holder in the world. The world's bankers control one-third of all gold ever mined, probably because gold is their best reserve asset. The fact is, gold-holding has a calming effect in times of financial adversity.

In both ancient and modern times, gold has acted as a financial stabilizer. It is a protection against inept, destabilizing government. After U.S. controls were lifted from private ownership of gold, active search for the metal avidly accelerated, propelled by market price increases. Therefore, it behooves the prospector to know about lode and placer gold, particularly its mineral associations. I define *lode* gold as that ore still in the vein or disseminated in other materials as submicroscopic or micron gold. Gold deposits range in geological age from Precambrian to late Tertiary. It is usually found throughout the world where igneous activity occurred. Most of the Western U.S. gold is related to the Tertiary Period. Tertiary volcanics, as a belt of shallow lode deposits, run from Central Mexico to Southern California, Arizona and Nevada.

ROCKPECKER TIP # 131 - Watch for a thin volcanic ash or tuff strata in sedimentary rocks. This is the layer that identifies the Tertiary Period.

In an economic recovery sense, the only worthwhile gold ore minerals are the gold tellurides, calaverite, sylvanite and petzite. Of course, native gold and electrum is a favorite target. Electrum is a mix of gold-silver ranging from yellow to white in color. As the electrum silver content increases, the color tends to sharper whiteness.

In the absence of new technology, where arsenic is the associated mineral with gold, it can quickly make extraction unprofitable.

ROCKPECKER TIP # 132 - If your gold ore is highly mineralized with arsenic, stake your claim anyway. A new cheap extraction process may be on the verge of discovery.

Gold that is termed *free milling* is usually extracted by cyaniding, amalgamation, and now the newer, cheaper *heap leaching* method. *Refractory* gold ore may require fine grinding, roasting, and if associated with copper, smelting. Mostly, these gold ores are concentrated and smelted.

Heap leaching, also known as *hydrometallurgy*, is the extraction of metals by leaching with an aqueous solvent that dissolves the valuable metal without attacking the gangue. It permits subsequent recovery of the dissolved metals from the solution. It is really a large-scale application of a chemical laboratory process known as extraction; the separation of a soluble substance from an insoluble one by means of a solvent.

CHAPTER 43

MINERAL RECOGNITION CLUES

*This chapter is designed to acquaint you with mineral classes and how the related **MINERAL I.D. CHARTS** can feed you recognition clues.* Simple identification tests such as hardness, streak, color, and specific gravity measures are dealt with in later chapters. *For field work, pay particular attention to the Found:, Associated with:, Looks like:, Alters TO/FROM: categories. They speak to ore genesis as explained in ensuing chapters.* ABOVE ALL, DON'T LET THIS CHAPTER PANIC YOU. If you recognize a pattern in mineral relationships, you've grasped the content. The rest is simple determination.

MINERAL GROUP CHARACTERISTICS

The earth's crust is the location of many minerals of value to man. It is the Rockpecker's job to locate and identify those that are demanded by an industrial society. That same crust shares its space with minerals of yet unrecognized value or of minimal industrial demand. Whatever you find, your day will come, if for only to supply the needs of mineral collectors who desire to build up fine display collections. This group of potential purchasers should never be neglected. They pay generously, and by the gram, for our product. They deserve the best we can supply.

Within the mineral kingdom, let's distinguish between rocks which are a mixture of mineral substances and the pure mineral where each part is the same as any other portion. That is, it must be solid and homogeneous with dimensional stability under normal temperature and pressure conditions.

By *homogeneous* we mean the substance has the same chemical composition throughout all its parts. This means the chemical identity is the same wherever found, be it Georgia, California, Italy or Africa. The internal structure of a mineral links atoms together in an ordered space lattice, as in a cube, where they are in specific triple dimensions. This we call *crystalline*. It is the nature-formed geometry of the crystal lattice structure.

A mineral is a substance, that when heated to a specific temperature, goes directly from the solid state to its liquid state. There is no plastic, softening stage between the solid and liquid phase. Every mineral body has a specific temperature at which this occurs.

There are a few accepted minerals that technically do not fit within these definition parameters. As defined, mercury is not a solid; opal is not crystalline. Yet they are unquestionably minerals. Above all, a mineral should be a natural substance from the earth's crust. A purist would exclude man-made minerals, even though they are identical in all respects. He would class a synthetic gem as a "laboratory chemical", even though the gem was more in perfection-beauty than the natural gem material. This same thinking prevails with man-made gold nuggets. "Laboratory-Made" is enough to void the name mineral. You decide !

The classification of minerals progresses from the simplest native elements as Class I to the progressively complex silicates as Class VIII. This is essentially a chemical classification as used in Dana's 1837 system, elaborated on by Strunz in 1938, modified in 1970. (Dana's scheme has 12 classes; Strunz has 9 classes, but for our purposes we use 8 classes. (Organic substances, such as amber, belong to the 9th class)

To give you a feel for the mineral classes, CLASS I will detail information as gleaned from the MINERAL CHART. Where a set of numbers is shown such as: (000/000), THE FIRST NUMBER IS THE MINERAL NUMBER ALPHABETICALLY; THE SECOND NUMBER IS THE MINERAL CLUE-CODE NUMBER. Where only one number (000) is shown, it is the mineral alphabetical number.

CLASS I - NATIVE ELEMENTS

Native elements are found in nature in the free, chemically uncombined state. With some exceptions, it is rare to find an element as a natural mineral. Elements are grouped as a metal, (gold, copper, silver, platinum); semi-metal, (arsenic, bismuth, antimony); or non-metallic, (carbon, diamond, sulfur). Metals have a metallic look and are malleable; semi-metals have a metallic look but are brittle; non-metals are nearly everything else.

COPPER - Cu (083/108): copper-red color often hidden by a green patina of malachite (162/085). Associated with silver (230), malachite (162), cuprite (092). Alters to cuprite (092), malachite (162).

SILVER - Ag (230/123): silver-white color, tends to blacken in air. Looks like platinum (200). Associated with calcite (056), quartz (213), barite (036), fluorite (112), copper (083), uraninite (260). Natural alloys of copper and silver are known as *halfbreeds*.

GOLD - Au (124/244): yellow, tending toward white when silver in content. Associated with pyrite (207),arsenopyrite (026),chalcopyrite (068), quartz (213), galena (116), sphalerite (234), pyrrhotite (212), molybdenite (173), silver (230), copper (083), chalcedony (066). Looks like chalcopyrite (068), pyrite (207).

MERCURY - Hg (167/263) silvery liquid drops. Found in low-temperature veins in volcanic or hot spring regions.

ARSENIC - As (025/073) dark gray or brownish-black. Found in crystalline rock hydrothermal veins and metamorphosed dolomite.

ANTIMONY - Sb (019/058) metallic gray luster. Found in crystalline rock ore veins as hydrothermal lodes.

GRAPHITE - C (125) black, greasy to the touch. Found in metamorphic rocks, pegmatites and hydrothermal veins.

DIAMOND - C (097) transparent, colorless, yellow or green. Found in kimberlite breccias and placers.

SULFUR - S (243) bright yellow, brown-yellow, reddish-yellow. Found in sedimentary evaporative deposits and fumaroles.

CLASS II - SULFIDES

This class includes compounds of metals with sulfur (sulfides), selenium (selenides), tellurium (tellurides), arsenic (arsenides), antimony (antimonides) and bismuth (bismuthides). Most are easily recognized by their metallic luster. Brittleness is characteristic. Many alter to secondary ore minerals. Those that are composed of a metal element with sulfur are called *sulfides;* those with the metal element and sulfur plus the semi-metal arsenic, antimony or bismuth are known as *sulfosalts.*

THIS IS A VERY IMPORTANT INDUSTRIAL GROUP

CHALCOCITE - (067/062) COPPER SULFIDE - 79% Cu.
Found: As secondary enrichment of sulfide ore veins. An important copper mineral. Known as copper glance. lead-gray color.
Associated with: (034), (083), (092), (068), (248).
Alters TO: (086), (034).

COVELLITE - (086/068) COPPER SULFIDE - 66% Cu.
Metallic blue crystals in thin layers; a rare mineral.
Found: In hydrothermal copper veins.
Associated with: (067), (048), (103).
Alters FROM: (068), (103).

BORNITE - (048/075) COPPER IRON SULFIDE - 63% Cu.
Bronze color, turns to purple and violet on a fresh break. Known as peacock ore. An important copper ore.
Found: in high-temperature hydrothermal veins and pegmatites.
Associated with: (067), (068), (103), (207), (026).
Alters TO: (067), (069).

ARGENTITE - ACANTHITE - (024/121) SILVER SULFIDE - 87% Ag.

Lead-gray color, metallic luster; associated with silver metals or disseminated in lead. Known as silver glance. A main ore of silver. At normal temperatures argentite is really acanthite.
Found: in low temperature hydrothermal veins.
Looks like: (067), (230).
Associated with: (230), (203), (206), (116).
Alters TO: (201), (238).

SPHALERITE - (234/253) ZINC IRON SULFIDE - 67% Zn.
Color yellow to brown, red or black. Known as zinc blende. Main ore of zinc. Fluoresces orange under ultra-violet light.
Found: in hydrothermal veins with galena.
Looks like: (116), (228).
Associated with: (116), (207), (068), (056), (164), (126), (136), (232), (024).
Alters TO: (232), (136).

ROCKPECKER TIP # 133 - Sphalerite is an important guidepost mineral to silver.

GREENOCKITE - (126/100) CADMIUM SULFIDE - 77% Cd.
Greenish-yellow coating on sphalerite. A replacement mineral of cadmium-rich sphalerite. IT IS THE ONLY ORE OF CADMIUM.
Found: As crusts on sphalerite, smithsonite, prehnite, natrolite, and in calcite veins.
Associated with: (202), (177), (056), (232), (234).

CHALCOPYRITE - (068/094) COPPER IRON SULFIDE - 34% Cu.
Brass-yellow color. Can be confused with gold or pyrite, but produces a green streak. A widely found copper ore. Fools gold.
Found: in hydrothermal veins and metamorphic volcanics.
Looks like: (124)
Associated with: (207), (116), (234), (067), (056).
Alters TO: (067), (086).

OTHER SULFIDES OF INTEREST

TETRAHEDRITE - TENNANTITE - (248-065) COPPER IRON SULFIDE - In crystals or granular form, these substances with a gray or black

metallic luster look alike. Arsenic and antimony substitute for each other to create a new mineral. Both contain some iron. When the substitution is silver it is called freibergite (30% Ag.); with mercury the name is schwartzite; with nickel it is called frigidite. Perhaps your find will contain a new substitution chemical form. You can then call it YOUR-NAME-ITE.
Found: in hydrothermal veins on copper-silver ores.
Alters TO: (034), (162).
Associated with: (068), (024), (116), (234), (203), (203), (206), (056), (213).

PYRRHOTITE - (212/002) - IRON SULFIDE - 36% S.
Bronze color in thin flat plates. Also called magnetic pyrites.
Found: in mafic and ultramafic extrusive rocks; in hydrothermal veins; in pegmatites; in contact metamorphic deposits; in sedimentary deposits associated with siderite.

ROCKPECKER TIP # 134 - Pyrrhotite is a guidepost mineral for gold, platinum, cobalt and nickel.

MILLERITE - (171/093) - NICKEL SULFIDE - 64% Ni.
Thin needle-like brass-yellow crystals with metallic luster. Rare - the richest nickel ore. Also known as capillary pyrites.
Found: in cavities of iron minerals and calcite veins.

GALENA - (116/059) - LEAD SULFIDE - 86% Pb.
Lead-gray, bright metallic on freshly broken surface. The main ore of lead. Cubic cleavage in 3 directions.
Found: It is a medium-temperature hydrothermal mineral deposit.

ROCKPECKER TIP # 135 - Galena is a guidepost mineral for silver and zinc.

CINNABAR - (075/120) - MERCURY SULFIDE - 86% Hg.
Color vermilion red tending to brown. Associated with realgar. An important ore of mercury. Easy to confuse with realgar.
Found: in hydrothermal veins; replacements in sedimentary rock.

NICCOLITE - (171/093) - NICKEL ARSENIDE - 43% Ni.
Bronze with a touch of pink color. Also known as nickeline.

Associated with silver. Also known as Kupfernickel. (G)
Found: in hydrothermal veins of red copper deposits.

STIBNITE - (240/066) - ANTIMONY SULFIDE - 71% Sb.
Steel-gray distinctive crystals with length-wise scratches. Main antimony ore. Often altered to red-yellow powder antimony ochres.
Found: in hydrothermal veins on lead silver ores.

PYRITE - (207/092) - IRON SULFIDE - 53% S.
Pale brass-yellow color, metallic luster, black streak. One of the "fools golds", but can contain gold, and then is known as "auriferous pyrite". Granular variety weathers to limonite.
Found: in rocks of igneous, metamorphic and sedimentary origin.

ARSENOPYRITE - (026/001) - IRON ARSENIC SULFIDE - 46% As.
Silver-white prismatic crystals, semi-metallic luster. A widely occurring arsenic mineral. Can contain some gold. Sparks when struck with hammer. Also known as mispickel.
Found: in high-temperature hydrothermal veins and pegmatites.
Associated with: (271), (223), (124), (173), (060), (230), (116), (207), (068).
Alters TO: (225), (108).

ROCKPECKER TIP # 136 - Arsenopyrite is a very important guide-post mineral to gold, cobalt, silver and nickel.

MOLYBDENITE - (173/026) - MOLYBDENUM SULFIDE - 60% Mo.
Lead-gray color in flexible, but brittle, leaf-like plates. Main ore of molybdenum. Looks like graphite, which is blacker.
Found: disseminated in igneous rocks; pegmatites; contact meta-morphic limestone; in cavities of granitic rocks.
Associated with: (223), (271), (026), (068), (207).

SKUTTERUDITE - (231/009) - COBALT ARSENIDE - 28% Co.
Crystals are small, cubic, silver-white with iridescent film, metallic luster, black streak. Ore of cobalt, nickel, arsenic. Also known as smaltite. Guidepost mineral to native silver.
Found: in hydrothermal veins with silver, cobalt and nickel.
Looks like: (026), (120).
Associated with: (056), (213), (036), (078), (180), (230), (043).

ROCKPECKER TIP # 137 - Skutterudite is a guidepost mineral to silver.

PROUSTITE - (203/119) - SILVER ARSENIC SULFIDE - 65% Ag.
Scarlet color crystal prisms, sparkling luster, brittle, transparent to translucent, scarlet streak. Known as ruby silver.
Found: in low-temperature hydrothermal silver sulfide veins.
Looks like: (206), (092).
Associated with: (056), (213), (230), (024), (206), (201).
Alters TO: (024).

PYRARGYRITE - (206/107) - SILVER ANTIMONY SULFIDE - 59% Ag.
Prismatic crystals with different terminations in black with dark-red tints, purple-red streak, translucent with sub-metallic luster. Also known as dark ruby silver. Has specimen value.
Found: in silver mineral hydrothermal veins.

BISMUTHINITE - (044/067) - BISMUTH SULFIDE - 81% Bi.
Tin-gray crystals with striations, sometimes woven structure. Opaque with metallic luster; MOHS hardness 2.0; streak gray; specific gravity range of 6.80 to 7.20. The best ore of bismuth.
Found: in medium-temperature hydrothermal veins; pegmatites.
Associated with: (043), (060), (026), (213), (271).
Looks like: (240).

REALGAR - (214/102) - ARSENIC SULFIDE - 70% As.
Red or orange compact masses or small crystals. Orange-yellow streak; MOHS 1.5-2.0; specific gravity 3.5-3.6; garlic smell when heated. Destabilizes in the light. Can be mistaken for cinnabar.
Found: Hot springs; low temperature hydrothermal veins.
Associated with: (186), (240), (101),
Alters TO: (186) when exposed to light.

ROCKPECKER TIP # 138 - Realgar is a good gold and silver clue.

ORPIMENT - (186/248) - ARSENIC SULFIDE - 61% As.
Fibrous or leafy masses colored lemon-yellow, gold, brownish-yellow, orange-yellow; yellow streak; MOHS 1.5-2.0; specific gravity 3.48. Destabilizes in light.

Found: in low-temperature hydrothermal veins with native arsenic.
Associated with: (214), (075), (056), (036), (025), (240).
Alters FROM: (214)

OTHER SULFIDES OF INTEREST:
SYLVANITE - (244/069) - SILVER GOLD TELLURIDE - 24% Au.
STEPHANITE - (238/011) - SILVER ANTIMONY SULFIDE - 68% Ag.

CLASS III - HALIDES

These are compounds of metals combined with halogens. (fluorine, chlorine, bromine, iodine) The luster is generally vitreous. The water-soluble members can be recognized by taste. Confusion with water-soluble sulfates and nitrates is possible. The best known member is halite. (common salt) The water-insoluble members crystallize in the isometric or cube-form.

CERARGYRITE - (062/233) - SILVER CHLORIDE - 60-70% Ag.
Small gray or gray-green crystals in crusts or masses which tarnish purple-brown when exposed to light. Waxy to sub-metallic luster; MOHS 1.0-1.5; specific gravity 5.5-5.6; Also known as horn silver. Called chlorargyrite in Europe.
Found: in oxidation zone of silver deposits as an alteration product of argentite.
Associated with: (230), (024), (206), (116), (036).
Alters FROM: (024).

FLUORITE - (112/169) - CALCIUM FLUORIDE - 48% F.
Has a wide range of colors from green, yellow, sky blue, pink, colorless; fluoresces blue or violet in ultraviolet light. MOHS 4.0; specific gravity 3.1-3.3; transparent to translucent; luster vitreous. Used in production of hydrofluoric acid.
Found: in hydrothermal veins and pegmatites.
Associated with: (056), (116), (234), (271), (173).
Looks like: (056).

OTHER HALIDES OF INTEREST:
SYLVITE -(245/166)- POTASSIUM CHLORIDE - 52% K.

CARNALLITE-(058/125)-HYDRATED POTASSIUM MAGNESIUM CHLORIDE -14%K.
CRYOLITE -(90/202)- SODIUM ALUMINUM FLUORIDE - 54% F.

CLASS IV - OXIDES AND HYDROXIDES

These are compounds of metallic elements with oxygen. Many have a secondary origin formed as a result of changes in other minerals such as sulfides upon surface weathering. Gold does not combine chemically with oxygen. Therefore, "gold oxide" is an impossibility. Some of the most abundant ores are simple oxides.

CUPRITE - (092/110) - COPPER OXIDE - 88% Cu.
Dark ruby-red crystals; bright red streak. Known as ruby copper.
Found: in oxidation zone of copper sulfide deposits.

SPINEL - (235/201) - MAGNESIUM ALUMINUM OXIDE - with a hardness of 7.5-8.0 in colors of white, red, blue, pink, black are called "gem spinels". The Royal Crown of England has a large, red transparent spinel which for many years was thought to be a ruby.
Found: in metamorphic marly dolomites and placers.

MAGNETITE - (061/010) - IRON OXIDE - 72% Fe. Strongly magnetic.
Is an iron-rich spinel. Most important ore of iron.
Found: Common in sedimentary, igneous, metamorphic rocks.

CHRYSOBERYL - (072/273) - BERYLLIUM ALUMINUM OXIDE - 19% BeO.
An important gem material with a MOHS hardness of 8.5. Known as cats-eye. A clear, bright yellow-green is highly valued.
Found: in pegmatites, mica schist, intruded granite.

HEMATITE - (135/112) - IRON OXIDE - 70% Fe.
Blood-red streak; color iron-black; often colors rocks and minerals shades of red and red-brown. A major ore of iron. Also known as bloodstone and kidney ore.
Found: in lava rock, pegmatites, hydrothermal veins, sediments.

ILMENITE - (142/015) - IRON TITANIUM OXIDE - 31% Ti.
Looks like hematite but with black streak. A major titanium ore.
Found: in Igneous and metamorphic rocks, pegmatites and placers.

SILICA OXIDE GROUP
QUARTZ, TRIDYMITE, CRISTOBALITE, OPAL

QUARTZ - (213/143) - SILICON OXIDE - coarse crystals; an essential constituent of volcanic, metamorphic and sedimentary rock; glassy or milky in color. Brown is Morion quartz; yellow is Citrine quartz; violet is Amethyst quartz. MOHS 7.0. When quartz has rutile inclusions, it is known as rutilated quartz; with mica, it is Adventurine quartz; with crocidolite (089), blue alternating yellow, it is a Falcon's eye or Tiger eye gem.
Found: Igneous magma injected in veins and fractures.

CHALCEDONY - (66-144) - SILICON OXIDE - micro-crystalline quartz.
AGATE - Concentric, banded, fibrous; gem cutting use.
JASPER - Massive, fine-grained, strong shades of red.
PRASE - Grayish-green agate in basaltic rock cavities.
ONYX - Black and white parallel layers; carved cameos.
SARDONYX - Red and white parallel layers; carved cameos.
CARNELIAN - Orange-red, shades of brown; used for seals.
CHRYSOPRASE - Translucent, yellow-green, apple-green.
HELIOTROPE - Opaque green with red spots; bloodstone.
CHERT - Dark gray to black flint.
These are all used, as an easy to polish, ornamental stone.
Found: as water solution precipitate.

TRIDYMITE - (256/153) - SILICON OXIDE - Above 870 degrees centigrade, the stable form of quartz structure rearrangement.
Found: A sublimation product in felsic volcanic rock cavities.

CRISTOBALITE - (087/165) - SILICON OXIDE - Above 1470 degrees centigrade, the stable form of quartz structure rearrangement.
Found: In igneous, felsic rock cavities.

OPAL - (185/207) - HYDROUS SILICON OXIDE - This is the hydrous silica colloidal state in compact and earthy masses. A quartz with a variable water content. Precious opal shows fine colors.
Found: in sedimentary rocks precipitated from silica-rich solutions. Occurs as small veins and crusts. Never found in crystals.

CASSITERITE - (060/217) - TIN OXIDE - 78% Sn.
Also known as tin stone. Occurs in granular masses or crystals colored yellow, red-brown, brown-black. Greasy luster.
Found: in pegmatites and sedimentary placers.

Manganese oxides and hydroxides are both numerous and difficult to distinguish when in earthy masses. *WAD* is a field term for a rock-like mixture of different hydrous manganese oxides.

PYROLUSITE - (208/024) - MANGANESE OXIDE - 63% Mn.
A highly valued manganese mineral used in the steel industry. Iron-black or steel-gray color in compact masses.
Found: as a surface alteration of manganese deposits.

RUTILE - (219/029) - TITANIUM OXIDE - 60% Ti.
An important commercial ore of titanium. ANATASE (012/249) and BROOKITE (054/250) have the same chemical formula with a different crystal systems. Colors are brown, black, yellowish, bluish, or violet crystals. MOHS 6.0-6.5. Light brown streak.
Found: in quartz intrusive veins and metamorphic rocks.

OTHER OXIDES OF INTEREST:
ZINCITE (274), CORUNDUM (085), THORIANITE (250), URANINITE (260), FERGUSONITE (111), HAUSMANNITE (132).

CLASS V - CARBONATES, BORATES, NITRATES

The salts of carbonic acid are called carbonates. Carbonates usually form excellent crystals. They consist of a metal combined with the carbonate radical CO_3. They can be highly colored, transparent or translucent crystals. Carbonates are tested with a dilute solution of hydrochloric acid which causes a carbonate to fizz like seltzer. Sometimes the acid must be first warmed; sometimes the mineral must first be powdered. In the main, carbonates are secondary minerals.

CARBONATES OF INTEREST:

MAGNESITE - (160/167) - MAGNESIUM CARBONATE - 28% Mg.

RHODOCHROSITE - (215/104) - MANGANESE CARBONATE - 47% Mn.

CALCITE - (056/154) - CALCIUM CARBONATE - 40% Ca.

DOLOMITE - (101/161) - CALCIUM MAGNESIUM CARBONATE - 13% Mg.

CERUSSITE - (063/147) - LEAD CARBONATE - 77% Pb.

SMITHSONITE - (232/228) - ZINC CARBONATE - 52% Zn.

MALACHITE - (162/085) - HYDROUS COPPER CARBONATE - 57% Cu.

HYDROZINCITE - (141/189) - HYDROUS ZINC CARBONATE - 75% ZnO.

CLASS VII - PHOSPHATES, ARSENATES, VANADATES

This class contains over 300 minerals, many with brilliant colors. Phosphates carry the PO_4 radical; arsenates - the AsO_4 radical; vanadates - the VO_4 radical. Many in this class are secondary minerals in that they form through alteration processes from other minerals. Their MOHS is to 6.0 or less; specific gravity is high due to the heavy contained elements.

PHOSPHATES OF INTEREST:

MONAZITE - (174/230) - Rare earth metals phosphate. 59% Ce., 12% ThO_2. Yellow to brown-red, translucent, stubby crystals. Radioactive; main ore for cerium and thorium.

TURQUOISE - (258-090) - HYDRATED COPPER ALUMINUM PHOSPHATE. Usually a light blue or green microcrystalline mass. A valuable gem stone. Can be confused with "bone turquoise".

Other phosphates of interest: AMBLYGONITE - (010/206) - 10% Li_2O; BRAZILIANITE - (051), PHOSPHOPHYLLITE - (198), VARISCITE - (263), WAVELLITE - (268), TORBERNITE - (253), AUTUNITE - (032), LAZULITE - (153), APATITE - (020), PYROMORPHITE - (209).

ARSENATES OF INTEREST:

MIMETITE - (172), ANNABERGITE - (017), ADAMITE - (003), ERYTHRITE - (108).

ROCKPECKER TIP # 139 - ERYTHRITE, with distinctive pink, dusty color, also known as cobalt bloom, is an important guide to cobalt mineral deposits. Sometimes bluish-pink to raspberry-red.

VANADATES OF INTEREST:

DESCLOIZITE - (096/099) - HYDROUS LEAD ZINC VANADATE - 22% V_2O_5.
VANADINITE - (262/255) - LEAD VANADATE CHLORIDE - 19% V_2O_5.
CARNOTITE-(059/243)-HYDRATED POTASSIUM URANIUM VANADATE- 63% UO_3, It is one of the main ores of uranium and vanadium - 20% V_2O_5.

CLASS VIII - SILICATES

This group includes about one third of all known minerals. Several are used as gemstones, other have industrial use.

The first seven mineral classes were classified by their *chemical* composition. Silicates are classified on their *structural* bases. Specifically, on the way the basic tetrahedral building block SiO_4 is linked with neighboring ions.

In silicates, the rigid structure is caused by the great strength of the silicon-oxide bond. The easy substitutability of ions produces complex solid solutions in their crystal habits. Rare in other classes, *isomorphism* is a characteristic of silicates. Isomorphism is a property that creates a mineral series. That is, the chemical composition varies slightly from the first to the last mineral in the series so that each is entirely different. In the case of olivine (182), for example, between the beginning of the series (rich in magnesium), to the opposite extreme (rich in iron), there is a wide range of minerals with an intermediate chemistry.

SILICATES, SUB-CLASS 1 - NESOSILICATES
(Isolated tetrahedral groups)

EXAMPLES OF THIS SUB-CLASS: (095), (251), (237), (252), (149), (013), (229), (275), (183), (269), (194).

The garnet group illustrates both an isomorphic series and the sub-class called NESOSILICATES. Expect a compact crystal form habit with high hardness, high index of refraction, and high specific gravity. Members of the garnet group are found in igneous and metamorphic rocks.

a) PYROPE - (210) - S.G. - 3.5; bright red, usually transparent and quite rare; much valued as a gemstone.
Associated with: (097), (183), (227), (235).

b) ALMANDINE -(007) - S.G. - 4.3; orange-red, wine-red, brown-red usually opaque, (sometimes transparent) sparkling luster.
Associated with: (013), (149), (237), (229).

c) SPESSARTINE - S.G.-4.2; is an orange or yellow-orange garnet.

d) GROSSULAR - (127) - S.G. 3.5; color is colorless, pale green or milky when pure. The Hessonite variety is orange to cinnamon-brown. Clear, finely colored crystals are cut and sold as gemstones of excellent quality. Has high collector demand.
Associated with: (272), (256), (099), (222).

e) ANDRADITE - (014) -S.G. 3.8. Color ranges from warm yellow in TOPAZOLITE; to emerald green in DEMANTOID; to brown, red or black in MELANITE.

Demantoid and Topazolite are highly valued gemstones associated with serpentine rock bodies.

Melanite is found in lavas and nepheline-syenite rocks.

f) UVAROVITE - (261) - a rare, dark-emerald-green, transparent garnet mineral used as a gemstone.
Found: in serpentine rocks rich in chromite.

SILICATES, SUB-CLASS 2 - SOROSILICATES
(Tetrahedra grouped in twos)

Sorosilicates may also contain isolated silicon tetrahedra as in the epidote group. Epidote (106/072) is usually green or brown due to iron content. Clinozoisite (077/071) contains aluminum rather than iron; zoisite (277/053) is the same mineral in a different crystal lattice. When manganese substitutes, pink piedmontite (199/111) is formed. Allanite, (006/061) forms when rare earth minerals, or thorium and uranium are in substitution. It can be slightly radioactive. Blue gem-quality zoisite crystals are sold as tanzanite. As you can see, extensive chemical substitution is possible in the epidote group.

HEMIMORPHITE - (136/177) - An ore of zinc.

EXAMPLES OF THIS SUB-CLASS: (143), (106), (265), (199).

SILICATES, SUB-CLASS 3 - CYCLOSILICATES
(Tetrahedra joined into rings)

EXAMPLES OF THIS SUB-CLASS: (033), (073), (100), (170).

BENITOITE - (038/276) - Sapphire-blue gemstone.
CORDIERITE - (084/184) - Transparent known as "water sapphire".
TOURMALINE - (254/238) - Transparent are semi-precious stones.
BERYL - (040/188) - Sought after in green, transparent, blue, yellow, or dark rose. An ore of beryllium.

Most natural beryl has an actual beryllium oxide content of 12% versus a 14% theoretical maximum. Crystals often show tapering toward one or both ends. Beryl can easily be confused with quartz in specific gravity and fracture. Look for slightly greasy luster and more highly colored than milk quartz.

Beryl occurs in pegmatites, sometimes in veins of commercial quantity. Hand cobbing makes handling very labor-intensive. 300 tons must be mined to get one ton of marketable beryllium oxide.

As a consequence, beryl is usually recovered as a co-product along with mica, feldspar and other pegmatite material.

Beryllium metal is used as an alloying agent in copper, nickel and zinc. In the atomic energy field it is used as a reflector of neutrons. Mists, fumes and dust of beryllium are highly toxic to the lungs.

Pegmatite beryl was the only known source for beryllium metal until 1960. Around that date BERTRANDITE (039/274) was found at Spor Mountain, Utah. The mineral was finely disseminated in volcanic tuff. Technically, the deposits are called "epithermal beryllium deposits in water laid tuff'. Other finds in Nevada, Alaska, and elsewhere, provided high tonnage in low grades, but difficult and expensive to process.

Beryllium material is included as one of the metals on the strategic stockpile list. Ore output levels are withheld to avoid disclosing individual company confidential data.

SILICATES, SUB-CLASS 4 - INOSILICATES
(Tetrahedra joined in elongated chains)

The pyroxenes represent the single open chain. The double or closed chain is characteristic of amphiboles. The pyroxenes are the water-free equivalent of amphiboles. Prismatic cleavage planes lie at 87 and 93 degrees to each other.

JADEITE -(144/055)- White or yellowish-white with green markings, occasionally translucent, has a vitreous luster. Usually in waxy, felt-like, compact masses. *JADE*, IN THE GEM INDUSTRY, REFERS TO BOTH JADEITE or NEPHRITE. The nephrite variety of actinolite (002) is also called "jade", but is an amphibole. PRECIOUS JADE IS ALWAYS JADEITE, A PYROXENE.

DIOPSIDE - (099/195) - Transparent crystals faceted into gems.
SPODUMENE - (236/227) - Important lithium source.
RHODONITE - (216/208) - Highly prized by collectors.

SOME PYROXENES OF INTEREST: (001), (030), (184), (104), (053).

The amphiboles are more common than pyroxenes, which they greatly resemble with their felt-like, needle-like habit. Prismatic cleavage planes lie at 56 and 124 degrees to each other.

ACTINOLITE - (002/196) - Color dark to light green, translucent, vitreous luster. The tough, felted, greenish variety is known as nephrite and often confused with jadeite (144).

SOME AMPHIBOLES OF INTEREST: (091), (255), (217), (139), (088).

OTHER SUB-CLASS EXAMPLES: (189), (216), (035), (039), (179).

SILICATES, SUB-CLASS 5 - PHYLLOSILICATES
(Tetrahedra joined together by 3 vertices)

This structure produces hexagonal and sometimes flat sheets. Orientation of the bonds are in the same direction which results in a leaved habit. The Greek word "phyllon" means "leaf".

MICA GROUP:
BIOTITE - (042/133) - Black, brown or dark green platy aggregates; transparent or translucent brownish sheets with vitreous luster. It is widespread as a sedimentary mineral, seen as small golden-colored particles that are the bane of the goldpanner, who refers to them as: "damn yellow mica".

OTHER EXAMPLES OF THIS SUB-CLASS MICA GROUP: (176), (121), (196), (156), (165), (176).

OTHER EXAMPLES OF SUB-CLASS 5: (NON-MICA) - (264), (175), (076), (147), (227), (074), (073), (226).

SILICATES, SUB-CLASS 6 - TECTOSILICATES
(Tetrahedra are joined together by all 4 vertices in cages)

In this sub-class, aluminum substitutes for silicon at the center of the tetrahedra to produce two extreme group limits, as in olivine, pyroxene, amphibole, feldspar. The substitution of cations produce a SI-O bond of rigid structures with great strength. The tecto's are grouped for microscopic inspection and petrographic reasons, not systemic ones. Most minerals in this sub-class tend to be pale-colored or white, relatively hard , and low in specific gravity.

The feldspars are aluminum silicates of soda, potash, lime and barium. When the feldspars carry chlorine and sulfur anions, they are known as feldspathoids. The scapolite and zeolite groups are members of this sub-class. Many of these minerals are of interest only to scientists and collectors.

PETALITE - (193/044) - Ore of lithium.
NEPHELINE - (178/205) - Important industrial mineral.
ORTHOCLASE - (187/239) - Important feldspar industrial mineral.
ALBITE - (005/214) - Important feldspar industrial mineral.
SODALITE - (2323/050) - Bright-blue used in jewelry trade.
LAZURITE - (154/022) - Lapis lazuli, a highly prized gem.

OTHER TECTOSILICATES OF INTEREST TO COLLECTORS: ANALCITE (011), LEUCITE (157), SANADINE (221), MICROCLINE (168), LABRADORITE (150), DANBURITE (094), NATROLITE (177), HARMOTOME (130).

****

In the opening of this chapter, it was suggested that you not panic over an inability to grasp clearly the mineral contents that followed. You should understand that mineralogists spend their entire careers researching this area of study.

****

Review chapter 13 for minerals that the U.S. must import to meet its domestic needs.

CHAPTER 44

HARD ROCK MINING SLANG

BACK — The ceiling over your head in a tunnel.
BAR — Long steel rod to pry overhead loose rock.
BELLY ROBBER — The bunk house cook.
BLOWPIPE — Pipe to blow out stick drill hole debris.
BREAST — The blank end wall at tunnel termination.
BULL COOK — The bunk house custodian.
BURN CUT — Blast drill holes, 1 center & 8 around.
CAGE — An elevator to transport men from vertical shaft.
CAP — Large overhead timber supported by side timbers.
CROSSCUT — Left or right shaft from main tunnel.
DEEP ENOUGH — The miner quits to move on.
DOUBLE JACK — 8 pound hammer for driving steel.
DRIFT — When being driven, a crosscut tunnel.
FIRE IN THE HOLE — Ready to blast warning.
FOOTWALL — The wall opposite the hanging wall.
GAD — Like a moil with a flattened end.
GOB — Distributed waste rock as in a stope.
GRAVEYARD — Midnight to morning work shift.
GRIZZLY — Heavy rail grates to catch oversize rocks.
GYPO — A piecework production contract miner.
HANGING WALL — The exposed wall after blasting.
HI—GRADER — The man who steals rich mine ore.
HI—GRADE — Rich ore of great value.
HITCH — A notch in rock to contain a timber stull.
HORSE — Country rock around the ore material.
JIGGER — Work area foreman.
LAGGING — Timber shims above main timber.
LEVEL — Any horizontal tunnel.
LOOK OUT BELOW — a down-climbing warning.
MAIN — The horizontal tunnel into the hill.
MANWAY — A built-in ladder for climbing.

MUCKER — The man who loads waste or ore.
MOIL — 10 inch sharpened, pointed steel.
MOP RAG — Canvas over steel to prevent spatter.
MUDSPOON — Removes muddy cuttings from drill hole.
NOTHING DOWN — An up-climbing warning.
OLD MAN — The big boss of entire mine.
ORE CHUTE — A gated, wooden chute over ore car.
PIE CAN — Miner's lunch bucket.
POWDER MONKEY — Delivers explosives to crews.
PRIMERS — Blasting cap with slow burning fuse.
RIB — The side walls in a tunnel or shaft.
ROCK IN THE BOX — Mineral dust lung silicosis.
ROUND — Series of holes for blasting.
RUSTLING — Waiting around mine for work.
SHIFTER — The top underground shift boss.
SHOOT THE BREEZE — Leisure conversation.
SINGLE JACK — Double faced 3 to 4 pound hammer.
SLICK SHEETS — Shovel easy sheet iron under ore.
SLICKER OUTFIT — Water shedding coat and pants.
SMITH — The mine tool blacksmith.
SQUARE POINT — Square nosed shovel used on slicks.
SQUARE SET — A style of mine support timbers.
SPLIT — Cut end of fuse to expose fresh powder.
SPITTING — Igniting one end of a fuse.
STATION — A room blasted out of solid rock.
STEEL — 3/4th inch tool steel, chisel point rods.
STICK — Stick of dynamite.
STOPE — (Cut & Fill) Ore removed, waste dump area.
SWING — Afternoon to midnight work shift.
TAMPER — Wooden stick to press dynamite into hole.
TITTY BOTTLE — Compressed air line oiler.
TOMMY KNOCKERS — Mine noises ascribed to ghosts.
TRAMMER — Loaded car to chute dumper.
WINDLASS — A cranked hoisting device over a hole.
WINZ — A down-hole in a tunnel for ore search.

NOW'S THE TIME TO "DRIVE THE WEDGE".

CHAPTER 45

REVISITING OLD MINING DISTRICTS

242 PRECIOUS METAL LOCATIONS.

130 GOLD PLACER LOCATIONS

33 STATES WITH PLACER GOLD

A journey of a thousand miles starts with the first step. For your prospecting adventure, that important first step involves a simple decision. That is, where to begin?

A tried and tested idea is to research past memories. Ghost towns which are no more; towns with waning population or with a stagnant present, yet history-rich in prospecting lore. Bear in mind that the townsite was the focal point of any surrounding mining activity. This is your starting point. Your thrust will be to answer the question: Where is the treasure overlooked by the old-timers? With my new-found expertise, can I do better?

The following list of towns, ghost or otherwise, qualify with a rich history of the colorful mining past. This should make your research a bit easier. When you select your target, respect and treat the past with dignity. It deserves our honor. Think prospecting, not souvenir collecting. Let sleeping legends survive! Above all, trust your prospecting instincts.

ARIZONA:

BUMBLE BEE- Nearby Bradshaw Mountain placer gold exhausted by the early 1900's. During the depression of the 30's, many made their survival-living reworking this area.

CALABASAS - Near Tucson, also known as Pumpkinville. Lode copper and gold.

CHARLESTON - located across the San Pedro River from Mill-ville, where the Tombstone ore was milled.

CHLORIDE - Turquoise mines and lode silver.

CLEATOR - In the foothills of the Bradshaw Mountains, lode gold. Placer gold in Turkey Creek.

CONTENTION CITY - East side of San Pedro River, lode gold.

CROWN KING - Mount Wasson slope of the Bradshaw Mountains, lode gold at the 7000 foot elevation.

DOS CABEZAS - In the foothills, lode copper in veins.

EHRENBERG - Placer gold in Weaver and Antelope Creeks. Nearby lode gold in galena.

GLEESON - Blue-green turquoise in the Dragoon Mountains.

GOLDROAD - Lode gold at Sitgreaves Pass near 3500 foot elevation.

HARSHAW - South of Durazno in center of Patagonia Mountains, lode silver.

JEROME - Mingus Mountain - copper with gold and silver values.

McCABE - Placer gold and lode gold in Bradshaw Mountains.

MINERAL PARK - Turquoise and gold in the foothills; lead-silver in the Cerbat Range.

MOWERY - Lead-silver in the Patagonia Mountains.

OATMAN - Lode gold.

QUARTZSITE - Rich deposit of tin in Plomosa District. (probable myth)

SALOME - Glory hole lode gold in veins.

STANTON - Placer Gold.

TOMBSTONE - Placer gold. Lode silver and gold mines.

TUBAC - Silver lodes in the Diablito and Santa Rita Mountains.

TWIN BUTTES - Lode copper 26 miles S-W of Tucson in Santa Rita and Sierrita Mountains.

VULTURE CITY - Lode gold in pure quartz.

WEAVER - Lode and placer gold.

WHITE HILLS - Lode and placer gold.

COLORADO:

ALMA - Supply town for surrounding placer gold areas.

ANIMAS FORKS - Lode gold at 11,300 foot elevation.

APEX - Placer gold in Pine Creek. Lode gold in veins.

BLACK HAWK - Placer gold on Gregory Creek. Lode gold in veins.

BONANZA - Lode gold at 8,000 foot elevation.

BRECKENRIDGE - Placer gold in Swan and Blue Rivers. Gold, silver, lead, zinc mines nearby.

BUENA VISTA - Supply center for Leadville lead-silver mines.

CENTRAL CITY - Placer gold in Little Dry Creek, lode gold same area. Area called the "richest square mile on earth".

CREEDE - Placer gold in Willow Creek. Silver-lead mines at 8800 foot elevation.

CRESTED BUTTE - Gold placers on Slate and Coal Creeks. Silver-lead mines at 9000 foot elevation. Coal mines built town.

CRESTONE - Free-milling gold in gash veins in Sangre de Cristo Mountains.

CRIPPLE CREEK - Lode gold on slopes of Mount Pisgah.

EUREKA - Gold, lead, silver, zinc, copper in lode mines.

FAIRPLAY - Placer gold in South Platte River.

GEORGETOWN - Placer gold in streams; lode silver in veins.

GLADSTONE - Lode gold on Clement Creek.

KOKOMO - Gold, silver, lead mines at 10,600 foot elevation.

LAKE CITY - Lode silver in San Juan Mountains.

LEADVILLE - Lead-silver in carbonate sands.

OHIO CITY - Gold placers in Ohio and Gold Creeks. Lode gold nearby.

OPHIR - Lode gold on Howard Fork of San Miguel River.

PITKIN - Galena (lead-silver); gold lodes.

PONCHA SPRINGS - Hot mineral springs.

SHAVANO - Above the timberline - galena (lead-silver) mines.

SILVER PLUME - Silver-gold ore on Columbia Mountain.

SILVERTON - Silver-gold in San Juan Mountain Range.

TELLURIDE - Gold-Silver tellurides in San Juan Mountains.

VICTOR - Lode gold on town lots. Road between Victor and Cripple Creek paved with low grade (for that era) gold ore.

WARD - Placer and lode gold in Left-Hand Canyon.

WHITE PINE - Hot springs area. Silver glance, silver wire, silver pyrites on Contact Mountain.

UTAH:

ALTA - Silver-bearing galena and quartz in Wasatch Mountains.

BINGHAM CANYON - Placer gold in canyons, gold-silver in galena in Wasatch Mountains.

CORINNE - West bank Bear River - site of inefficient smelter. Worthwhile values still found in smelter slag that paved the streets.

EUREKA - South in Tintic and Oquirrh Mountains, gold, silver and copper mines

MAMMOTH - lode silver mines.

MERCUR - Cinnabar found which carried high gold values. Placer gold in Lewiston Canyon. Lode gold and silver in Oquirrh Mountains and Horse Thief Canyon.

OPHIR - Horn silver on Silverado Hill.

PARK CITY - Gold-silver in quartz.

SEGO - Coal mining area. Said to have Anthracite coal.

SILVER CITY - Tintic mining district - ancient silver mines reopened - later flooded and abandoned.

SILVER REEF - Silver in both red and white sandstone.

MONTANA:

BANNACK - Placer gold in Grasshopper Creek.

BEARMOUTH - Lode gold in local mountains.

BEARTOWN - Lode gold in local mountains.

CABLE - Lode gold in decomposed quartz.

CASTLE CITY - Silver-lead in galena in Castle Mountains. (domed lava protrusions)

ELKHORN - Placer gold in creeks. Lode gold-silver on Elkhorn Mountain.

GARNET - Lode gold in local area.

GILTEDGE - Lode gold - refractory type. (requires roasting)

GOLD CREEK - Placer gold in streams.

GRANITE - Sapphires in Rock Creek gravels.

KENDALL - Finely disseminated gold ores - gold not visible.

KEYSTONE - Lode silver.

LANDUSKY - Placer gold in Alder Creek. Lode gold is free-milling.

LAURIN - Placer gold in Alder Creek.

MAIDEN - Placer and lode gold in Judith Mountains.

MAMMOTH - Gold-silver lode mines in area.

MARYSVILLE - Area gold mines.

PARDEE - Silver-lead galena at Flat Creek.

PIONEER - Placer gold in area streams.

PONY - Placer gold in Pony Creek. Lode gold in granite.

RIMINI - Gold-silver-lead mines.

RUBY GULCH - Gold tellurides in clayey gangue.

SOUTHERN CROSS - Lode gold on Iron Mountain.

VIRGINIA CITY - Placer gold on Alder Creek.

WICKES - Smelter town processing gold-silver ores mined nearby.

ZORTMAN - Placer gold in Dry Beaver Creek. Lode gold in quartz in Little Rockies.

IDAHO:

BAY HORSE - Placer gold in Bay Horse Creek - Finely divided silver-lead in iron ore.

BONANZA - Placer gold in Yankee Fork of Salmon River at 6000 foot elevation. Lode gold on Custer Mountain.

BURKE - Placer gold in Canyon Creek.

De LAMAR - Placer gold in Owyhee Mountain streams. Lode silver.

GILMORE - Lead-silver lodes.

IDAHO CITY - Placer gold in Moore's Creek.

LEESBURG - Lode gold in quartz. Placer gold in Leesburg basin.

MOUNT IDAHO - Lode gold on Mount Idaho.

MURRAY - Placer gold in Prichard Creek.

PIONEERVILLE - Placer gold in Boise basin.

POTOSI GULCH - Placer gold in local streams.

SHOUP - Placer gold in Pine Creek. Lode gold in same area.

SILVER CITY - Lode silver assaying $5,000 per ton in 1870.

WARREN - Placer gold in Slaughter Creek. Lode gold-silver mines in area.

WYOMING:

ATLANTIC CITY - Placer gold in local streams.

ENCAMPMENT - Gold, copper, cobalt lode mines.

GLENCO - Coal mining area.

OPAL - Railroad center for gold mines at CORA. This area rich in minerals and precious stones.

RAMBLER - A copper boom-town.

SOUTH PASS CITY - Placer gold in Sweetwater Creek.

NEVADA:

AURORA - Lode gold on Antelope Creek.

AUSTIN - Gold and silver in quartz.

BELMONT - Lead-silver lode mines; town had a turquoise boom.

BERLIN UNION - Lode gold, but area more important for agatized marine fossils, such as Ichthyosaurs (fish lizards). Marine fossils found on top of Mount Berlin. NOW A STATE PARK.

BROKEN HILLS - Lode silver in area.

CANDELARIA - Silver, gold, lead, copper lodes. Water 9 miles distant.

EUREKA - Silver, gold, lead mines.

GALENA - Silver-lead in galena ore. Lode gold.

GENOA - Placer gold in Carson River.

GOLD HILL - Placer and lode gold in Gold Canyon.

GOLDFIELD - High-grade lode gold mines.

GOLD POINT - Horn silver lode mines at 5800 foot elevation.

GOODSPRINGS - Highly mineralized area: gold, silver, platinum, vanadium, azurite, chrysocolla, malachite, cinnabar.

GRANTVILLE - Gold and silver in veins of porphyry, limestone and quartz. Lead, antimony and copper mines.

HAMILTON - Nearly solid silver deposit found on Treasure Hill at 8,000 foot elevation.

IONE - Large gold-bearing quartz finds with later cinnabar and silver discoveries.

JOHNNIE - Lode gold in decomposed quartz - ore assayed $30,000. per ton in 1909.

MANHATTAN - At 7,000 foot elevation - jewelry-grade gold ore. Area reworked during the 30's depression by survivalists.

MIDAS - Lode gold mines.

NATIONAL - Jewelry-grade gold-in-quartz - ore assayed $75.00 per pound in 1909.

NELSON - Lode gold in El Dorado Canyon - Silver-lead and copper mines.

PINE GROVE - Lode gold mines.

PIOCHE - Lead-gold-silver ore in ledges.

RAWHIDE - Gold-silver ore assaying $26,000. per ton in 1907.

RHYOLITE - Lode gold.

ROCHESTER - Lode silver with gold as by-product.

SUTRO - Tunnel site for draining flooded Comstock lode mines.

TONOPAH - Lode gold-silver on Mount Oddie. Area contains much semi-precious gem material and petrified wood.

TUSCARORA - Lode silver on Mount Blitzen. Good water area.

UNION - Short-lived lode gold area. Very large clay deposit used for brick manufacture.

UNIONVILLE - Lode silver mines - Gold in foothills.

NEW MEXICO:

CERRILLOS - Pre-historic Indian gold mines - Blue turquoise area.

CIMARRON - Placer gold in local streams.

DAWSON - Coal mining center.

ELIZABETHTOWN - Placer gold in Humbug Gulch. Lode gold in quartz. Lode copper.

FOLSOM - Ice Age fossils in area.

GOLDEN - Placer gold in Lazarus Gulch and Tuerto Creek.

HILLSBORO - Lode gold on east side of Black Range.

KELLY - Lead, silver, copper ore in carbonates with galena. Lode silver. Smithsonite (zinc carbonate) in green rocks.

KINGSTON - Lode silver mines.

KOEHLER - coal mining area.

LAKE VALLEY - Site of chamber with horn silver walls in Black Range. Chamber produced over 2,000,000 ounces by 1883.

MADRID - Coal mining area.

MOGOLLON - Lode gold-silver in quartz.

PINOS ALTOS - Placer gold in Bear Creek. Lode gold in Santa Rita Mountains. Area reworked during 30's depression by survivalists. Silver, lead, copper mines in area.

SAN PEDRO - Placer gold - some copper lodes.

SHAKESPEARE - Lode silver assayed 12,000 ounces per ton.

TYRONE - Turquoise, copper, lead, zinc in Burro Mountains.

WHITE OAKS - Placer gold - Lode gold assaying $17000. per ton in 1880.

TEXAS:

SHAFTER - Lode silver in limestone on Chinati Mountain.

STUDY BUTTE - Cinnabar mines.

TERLINGUA - Cinnabar mines.

OREGON:

ASHWOOD - Lode silver and gold in quartz on Trout Creek.

AUBURN - Placer gold on Blue Canyon Creek.

BONANZA - Placer gold in area streams. Lode gold.

BOURNE - Placer gold in Cracker Creek. Lode gold in Blue Mountains.

CANYON CITY - Placer gold in Canyon Creek.

CORNUCOPIA - Lode gold mines in Wallowa Mountains.

GREENHORN – Lode gold mines in Blue Mountains.

JACKSONVILLE – Placer gold in Jackson Creek.

KERBY – Richly mineralized area: Gold, cobalt, ilmenite, mercury, iron.

MALHEUR CITY – Placer gold in Willow Creek.

MARYSVILLE – Placer gold in Dog Creek. Lode gold nearby.

SANGER – Placer gold and lode gold on west side Eagle Creek in Wallowa Mountains.

SPARTA – Southwest edge of Wallowa Mountains. Placer gold in Shanghai Gulch and Maiden Gulch.

SUMPTER – Placer gold in Powder River.

WASHINGTON:

BLEWETT PASS – Placer gold in streams. Lode gold and silver alloyed with copper in area.

BOSSBURG – Lead-silver in galena; gold in quartz.

CONCONULLY – Silver lode mines in area.

COPPER CITY – Lode copper-gold in area.

INDEX – High-grade bornite copper ore area – mineral traces of gold, silver and antimony.

LIBERTY – Placer gold in Swauk Creek.

LOOMIS – Gold in quartz, silver-lead in galena, copper and tungsten in area.

MONTE CRISTO - Placer and lode gold area.

NIGHTHAWK - Placer and lode gold on Similkameen Creek.

NORTHPORT - Lode silver on Silver Crown Mountain.

ORIENT - Lode gold in quartz.

REPUBLIC - Placer gold on Granite Creek.

ROSLYN - Coal mining area.

RUBY - Gold, silver, copper mining area.

SULTAN - Placer gold in Sultan Creek.

TRINITY - Placer gold; copper-silver ore.

CALIFORNIA:

ALTAVILLE - Ancient tertiary stream gold.

AMADOR CITY - Placer gold in Amador Creek.

ANGELS CAMP - Placer and lode gold.

AUBURN - Placer gold in North Fork American River at Dry Diggins.

BALLARAT - Placer gold in canyons, on Western flank of Panamint Mountains.

BEAR VALLEY - Placer and lode gold.

BIG OAK FLAT - Placer gold in Deer Creek and Moccasin Creek area.

BODIE - Lode gold under Bodie Butte.

CALICO - Lode gold.

CAMP SECO - Placer gold in Mokelumne River. Gold and copper mines.

CARSON HILL - 195 pound gold nugget found here.

CERRO GORDO - Lead-silver lode mines.

CHINESE CAMP - Placer gold in Stanislaus River.

COLOMA - Placer gold on South Fork of American River.

COLUMBIA - Placer gold on Kennebec Hill.

COPPEROPOLIS - Copper ore assayed $120. per ton in 1858.

COULTERVILLE - Placer gold in Maxwell Creek.

DARWIN - Lode silver in Argus Range.

DOUGLAS CITY - Placer gold in Trinity River.

DOWNIEVILLE - Placer gold on North Fork of Yuba River. (Unverified 427 pound gold nugget found here) 227 pound nugget also reported.

EL DORADO - Lode gold in quartz. Placer gold at Shingle Springs.

FIDDLETOWN - Placer gold in creek entering Consumnes River.

FRENCH GULCH - Lode gold in quartz near Whiskeytown.

GARLOCK - Lode gold in Iron Canyon.

GOLD CAMP - Lode gold on Tropico Hill.

GRASS VALLEY - Lode gold in quartz. Placer gold at Humbug

Flats.

HORNITOS - Placer gold in streams and gulches.

JACKSON - Gold in quartz; pyrites, galena, sphalerite, ankerite, chalcopyrite.

JAMESTOWN - Placer gold in streams. Lode gold in quartz.

KNIGHTS FERRY - Placer gold in Stanislaus River.

MARIPOSA - Placer gold in Mariposa Creek. Lode gold in quartz.

MASONIC - Lode gold in area.

MOKELUMNE HILL - Placer gold in Mokelumne River.

MURPHY'S - Placer gold in area creeks.

NEVADA CITY - Placer gold in Deer Creek. Lode gold in quartz.

NORTH SAN JUAN - Placer gold on San Juan Hill.

RANDSBURG - Lode gold in quartz in Yuma Canyon. Horn silver, tungsten on Red Mountain.

ROUGH AND READY - Placer gold in area.

SHASTA - Placer gold in foothills.

SIERRA CITY - Placer gold in North Fork of Yuba River. Lode gold in quartz on Shasta Buttes.

SONORA - Placer gold in Woods Creek. Lode gold in quartz.

TIMBUCTOO - Placer gold in Yuba River.

TUMCO - Placer gold in Jackson Creek. Lode gold in mica

schist in Cargo Muchacho Mountains.

VALLECITO - Placer gold in Coyote Creek. Lode gold in area.

VOLCANO - Placer gold at Soldiers Gulch.

WEAVERVILLE - Placer gold on slopes of Oregon Gulch Mountain.

WHISKEYTOWN - Placer gold in Whiskey Creek and tributaries of Trinity River.

SIDENOTE: These early precious metal locations were all subject to the economic realities of their time. If the cost of working an area exceeded the recovery values, it was time to move on to more fertile goldfields. As you can see, there were many options for the footloose.

Where the mining venture involved lode gold, economics was an especially important and decisive factor in folding the enterprise. Up to and including 1933, the average price of gold was $20.00 or less. This means many locations that could be profitable today, were shut down in earlier times as not being cost effective.

Early inefficient lode gold recovery methods left significant gold values in the tailings. Today, these tailings are being worked profitably with heap leach chemical extraction systems. A good rule of thumb, that can be applied to old tailings that involve this recovery method, asks the age of the tailings. If the answer is 75 years or older, then the matter can be pursued further. If not, forget it! Of course, as with any rule, it is only a guideline. Common sense should prevail.

UNITED STATES GOLD LOCATIONS

33 STATES WITH PLACER GOLD

ALABAMA - At least seven counties with lode and placer gold.

ROCKPECKER TIP # 140 - Coosa County between Birmingham and Montgomery- Try Hatchett Creek, Weogufka Creek and streams around Rockford.

ALASKA - Gold available in very extensive areas that are best defined as regions.

ROCKPECKER TIP # 141 - Try Eureka area streams, Eureka Creek, Baker, American, Sullivan Creeks west of Fairbanks.

ARIZONA - At least 14 counties with lode and placer gold.

ROCKPECKER TIP # 142 - Try Safford, Graham County, streams Northeast of Tucson. Gila River. Spring and Bonita Creeks.

CALIFORNIA - At least 37 counties with lode and placer gold.

ROCKPECKER TIP # 143 - Try Mariposa County streams, East of the Merced River, especially Mariposa Creek and tributaries.

COLORADO - At least 40 counties with lode and placer gold.

ROCKPECKER TIP # 144 - Try Golden, Jefferson County, Clear Creek and tributaries.

GEORGIA - At least 11 counties with lode and placer gold.

ROCKPECKER TIP # 145 - Try Dahlonega, North of Gainesville, Chestatee River and tributaries.

IDAHO - At least 21 counties with lode and placer gold.

ROCKPECKER TIP # 146 - Try Blaine County, Sun Valley, Ketchum, Bellevue area. Wood River and tributaries.

INDIANA - At least 3 counties with placer gold.

ROCKPECKER TIP # 147 - Try Johnson County, Martinsville streams.

IOWA - These erratic placer deposits are glacier related.

ROCKPECKER TIP # 148 - Try Northeastern county streams.

MAINE - At least 8 counties with placer gold.

ROCKPECKER TIP # 149 - Try Oxford County, Bryon Township, Swift River and tributaries.

MARYLAND - At least 4 counties with lode and placer gold.

ROCKPECKER TIP # 150 - Try Howard County, Simpsonville area streams.

MICHIGAN - At least 19 counties with lode and placer gold.

ROCKPECKER TIP # 151 - Try St. Joseph County, Burr Oak area. Northwest of Sturgis streams.

MINNESOTA - At least 2 counties with lode and placer gold.

ROCKPECKER TIP # 152- Try Scott County, Jordan East of Bloomington streams.

MISSOURI - At least 3 counties with placer gold.

ROCKPECKER TIP # 153 - Try Macon County, Elmer, South of Kirksville. Sand Creek and tributaries.

MONTANA - At least 24 counties with lode and placer gold.

ROCKPECKER TIP # 154 - Try Missoula County, Greenough, East of Missoula. Elk and Union Creeks.

NEVADA - At least 17 counties with lode and placer gold.

ROCKPECKER TIP # 155 - Try Nye County, Beatty area, North-west of Las Vegas. Amargosa River.

NEW HAMPSHIRE - At least 2 counties with lode and placer gold.

ROCKPECKER TIP # 156 - Try Coos County, Milan area streams, North of Belin.

NEW MEXICO - At least 15 counties with lode and placer gold.

ROCKPECKER TIP # 157 - Try Grant County, Northeast of Silver City, Bear Creek and tributaries.

NORTH CAROLINA - At least 40 counties with lode and placer gold.

ROCKPECKER TIP # 158 - Try Union County, Waxhaw area, South of Charlotte all area streams.

OREGON - At least 19 counties with lode and placer gold.

ROCKPECKER TIP # 159 - Try Jackson County, Ashland area, Bear, Ashland, Wagner Creeks.

PENNSYLVANIA - 1 county. Gold is byproduct of iron ore.

ROCKPECKER TIP # 160 - Try Lebanon County, Cornwall area South of Lebanon streams.

SOUTH CAROLINA - At least 18 counties with lode and placer gold.

ROCKPECKER TIP # 161 - Try Spartansburg County, all Spartansburg streams.

SOUTH DAKOTA - At least 4 counties with lode and placer gold.

ROCKPECKER TIP # 162 - Try Custer County, Fairburn area South of Rapid City streams.

TENNESSEE - At least 3 counties with lode and placer gold.

ROCKPECKER TIP # 163 - Try Blount County, South of Marysville all streams.

TEXAS - At least 13 counties with lode and placer gold.

ROCKPECKER TIP # 164 - Try Culberson County, Van Horn area all streams.

UTAH - At least 15 counties with lode and placer gold.

ROCKPECKER TIP # 165 - Try Wasatch County, Heber City area, Snake Creek.

VERMONT - At least 2 counties with lode and placer gold.

ROCKPECKER TIP # 166 - Try Bennington County, Readsboro area, West of Brattleboro streams.

VIRGINIA - At least 15 counties with lode and placer gold.

ROCKPECKER TIP # 167 - Try Orange County, Fredericksburg area all streams.

WASHINGTON - At least 27 counties with lode and placer gold.

ROCKPECKER TIP # 168 - Try Douglas County, Stevenson's Ferry areas streams.

WEST VIRGINIA - At least 1 county with lode and placer gold.

ROCKPECKER TIP # 169 - Try Tucker County, Parsons area streams.

WISCONSIN - At least 1 county with placer gold.

ROCKPECKER TIP # 170 - Try Pierce County, Rock Elm Township area streams.

WYOMING - At least 17 counties with lode and placer gold.

ROCKPECKER TIP # 171 - Try Johnson County, Buffalo area Kelly Creek and tributaries.

ROCKPECKER TIP # 172 - SOME OF THE LISTED PLACER DEPOSITS ARE ON FEE LANDS THAT WILL REQUIRE THE OWNER'S PERMISSION TO GOLDPAN. DON'T TRESPASS! WHEN PERMISSION IS GRANTED, RESPECT THE OWNER'S PRIVACY AND PROPERTY. DON'T MESS IT UP FOR THE NEXT ROCKPECKER.

CHAPTER 46

14 PRUDENT INVESTOR CAVEATS

MINING SCAMS

By
Michael N. Greeley
Mining Engineer

ARIZONA DEPARTMENT OF MINES AND MINERAL RESOURCES
Circular No. 11, January, 1986

COURTESY: H. MASON COGGIN, DIRECTOR, ARIZONA DEPARTMENT OF MINES

(MICHAEL N. GREELEY is currently with the U.S. Bureau of Mines)

Additional thanks: Anonymous - ATASCADERO, CA.

A time-honored method to bilk the public of millions of dollars is the ubiquitous mining swindle. Since an unusually rich ore deposit, or bonanza, has historically produced enormous profits for the developer, many of us believe that we too, like the '49er, can strike it rich. The glamour attached to "discovery" create, in the imagination of some people, a relatively easy way to attain fantastic wealth.

Although money can be made in mining and this Department certainly encourages mining, we also have a responsibility to urge the public to exercise prudence in its investment. Too many persons have lost their hard-earned savings on an ill-advised mineral scheme. Archives are full of outrageous examples of mining scams and swindles in which the only beneficiary was a glib entrepreneur with unbounded optimism. In most cases, he disappeared before his investors realized what happened.

When making an investment in any mineral enterprise, there are a number of factors or key features to consider. A checklist of significant considerations follows:

1. GOOD TITLE TO THE MINERALS

Prerequisite to the investment in the development of a mineral deposit is legal access to the resource. The potential investor should know who ultimately owns or administers the subject mineral property and commodity. The property may be controlled by the State, Federal Government, Indian Tribe, or a private individual or organization. Moreover, jurisdiction over the land surface may be separate from the jurisdiction of the underlying mineral resource. Where ownership or control of the mineral rights is severed from the surface rights, obvious legal problems can arise.

If a mining claim or prospecting permit for minerals is not legitimate, the money invested is wasted from the beginning. In addition, the investor should understand basic differences between leasable minerals and lode or placer deposits. These classifications determine the type of mining agreement and/or claim established on the resource. Very specific requirements must be met and procedures followed to gain the right to develop a mineral deposit. Furthermore, encumbrances against the deposit, though legal, may be detrimental to its development.

Typical examples of cloudy or illegal title to a mineral deposit include oversized claims, inappropriate claim designation and improper filing, failure to perform annual assessment or to file affidavits of labor, or location of claims on a privately held mineral estate. The investor should establish exactly what rights he has to the property in question, and what conditions are imposed before he spends one penny on exploration or development.

2. SAMPLING AND ASSAYING

Perhaps the next major consideration in evaluating a mineral investment is the sample and assay data. One sample does not make a mine. A person who brings a rock that contains two ounces of gold per ton to an investor, may be carrying the entire mine in his hand. One such high grade ore specimen is not representative of the deposit. Many samples, commonly numbering in the thousands, are required to give a reasonably accurate measure of the tenor, or quality, and tonnage of the ore.

Depending on the configuration and geologic setting of a mineral deposit, there are recommended scientific procedures to follow and methods to use to properly sample a mineralization. The investor should be satisfied that the samples referred to by the mine promoter were collected specifically from the property of interest, and also that they were collected in a proper way. The sampling method should be adequately described and each sample site precisely located, preferably on a map.

It is important not to forget the practice of deliberately salting, or adulterating samples. Ingenious ways have been devised to fraudulently enhance the grade of samples either before or after they are collected, regardless of the method of collection. The temptation to salt is particularly appealing when dealing with a mineral of high unit-value such as diamond or gold.

Once collected the samples must be properly prepared and assayed. In general, the final sample preparation and assay should be done by qualified laboratories. Assayers registered in Arizona are generally familiar with different types of ore, and are knowledgeable about the proper method to test for particular metals or other components.

All ores are amenable to rigid testing, and comments to the effect that the ore is unassayable are simply not true. Statements belittling the methods of registered assayers, complaining for example, that they never report all the gold, are immediately suspect. Modern copying devices also make it a rather simple procedure to later falsify the assayer's report. If there is any question, of course, the sample pulps (unused prepared portion) may be sent to another lab for comparison.

Spectrographic analyses do not provide an accurate test of mineral samples. This type of analysis, though relatively inexpensive and useful in providing a list of components in a sample, does not yield a reliable, quantitative measure of tenor. Often a billion or trillion dollar "ore-body" is created by simply multiplying the generalized amount of each of the metals listed in a spectrographic analysis by their current market price. An ore body, however, is not that simple. At this time, there is no commercially acceptable process known whereby each element can be recovered from a deposit.

3. TYPE OF COMMODITY

A degree of skepticism should also be reserved for ores said to contain uncommon metals or minerals. Because of their rarity, these substances may command a very high price, and are therefore extremely attractive to the investor. The platinum-group metals including platinum, rhodium, ruthenium, iridium, and osmium are the darlings of the swindler. Considering their high unit-value, even minute amounts of these metals appear to be a reasonably good bet to the innocent investor.

The problem here is usually the grade or tonnage, or a combination of both. The amount of platinum, for example, is generally too low to realistically consider extraction, or the tonnage is almost limited to the hand specimen. As a primary ore, platinum has never been mined in Arizona; its only production has come from trace amounts recovered in the final stage of refining copper ores. The geologic environment of Arizona, diverse as it is, does not encourage the search for platinum-group metals, graphite, cobalt, nickel, bauxite, diamonds and a number of other commodities.

4. THE MINING METHOD

As plans are drawn to mine an ore deposit, proposals are made which frequently are misinformed and ill-advised. There are innumerable examples of deep shafts and long adits driven to "nowhere". Many of these openings have been cut at great expense and with little or no evidence to suggest they would meet success.

An example of a mining scheme which can be described at best as "ignorant" was recently sold to a number of investors in the Chemehuevis Placer District, near Lake Havasu City, Arizona. The plan called for an investor to purchase a plot of ground 60 X 120 feet in sizes, from which 8,000 cubic yards of unconsolidated gold-bearing gravel would be dug and treated. In order to recover 8,000 cubic yards of gravel from this plot, the excavation would require vertical walls 30 feet deep!

Since loose gravel cannot be mined at a slope exceeding its natural angle of repose, approximately 45°, the maximum amount of material that an investor could ideally and safely expect to obtain from an isolated parcel is about 62 1/2% of the total, or 5,000 cubic yards. No attempt was made to explain to the purchaser that in this case, after an investment of $50,000., he would actually get less than two-thirds of what he paid for. This scenario illustrates one catch to a sales promotion involving fractional interests in a mineral property. The entire package, comprised of all individually-owned parcels, must be mined together to insure each investor's return.

An interesting twist to this story is the statement made later by the developer that only 40% of the aggregate was gold-bearing. Consequently, the investor was now entitled to 20,000 cubic yards of gravel (to yield 8,000 cubic yards of gold-bearing material) from his plot. Since 5,000 yards was the maximum he could physically dig, this is truly adding insult to injury.

5. MILLSITE AND METHOD

There is a tendency among many of us to want to build. We want others to see our accomplishments. To some degree this attitude explains why a mine tunnel is begun with little justification.

The same propensity for building might explain why a mineral processing mill is erected, or a leaching facility is frequently constructed without any obvious sign of ore. Another reason these engineering marvels are installed is due to their impressiveness. The humming, turning, grinding and screeching of equipment, and the smoke and odors of a mine plant are exciting to the potential investor. He sees industry in action - his money at work - and profits just around the corner.

Unfortunately, however, he is commonly one of the multitude who has emptied his pockets for a pipe dream. With a paltry amount of ore stockpiled, a dump laden with debris, or an old mine map showing the "lost" ore body, the developer spends the last dime of every investor getting ready to treat the mineral-rich rock. The $ 500,000 mill, designed to treat 500 tons per day, mills nothing, and the dreams of many become a nightmare.

Even when good ore exists, the treatment facility is often poorly designed. Frequently its component parts are improperly matched or not sized adequately. Materials handling procedures are commonly cumbersome and energy intensive. An adequate supply of water may be lacking. Hazardous operating conditions may be present. These circumstances are a few costly examples that can shut a plant down abruptly.

6. RECOVERY PROCESS

The recovery process is in many cases a mystery to the investor. Technological methods vary according to the metal or mineral recovered. In addition there are many variants based on the size of the mineral component, its gangue association, its state of chemical alteration, the hardness and specific gravity of the ore, and a myriad of other factors.

The milling and metallurgical treatment of ores is comprised of both physical and chemical means of beneficiation. These processes, though technically sound and well understood by the professional, are frequently vague and confusing to the lay person. An investor not familiar with basic physical-chemical laws is easily misled.

Proprietary methods utilizing secret chemicals and "black box" techniques, therefore, are often praised as technological breakthroughs. According to the developer, these so-called miraculous inventions will convert formerly worthless rock to metal-rich ore, or improve many-fold the recovery of a metal or other commodity that heretofore had been difficult to extract. One should exercise caution when evaluating such claims.

Developers often speak of 100 percent recovery. Complete extraction, however, of most constituents is essentially unknown over the long term. At the turn of the century a mining firm in Ajo built a giant retort into which ores were to be shoveled and melted. Spigots were tapped into the vessel at various locations labeled copper, lead, zinc, gold, silver, etc. All the investor had to do, after he had helped finance the operation, was to turn the spigot for the metal desired. Understandably, the entire operation fizzled.

In many cases difficult technical problems are oversimplified. The ill-informed investor is merely asked to retain faith in the management and perhaps to ante a bit more so that this "minor problem" can be speedily resolved.

7. PERMITTING AND REPORTING

A host of other factors should be evaluated by the prospective investor before spending money on a mine or beneficiation plant. Proper permitting must be obtained at various stages of development from local, state and federal agencies, internal reports generated for management and for the investors should be factual, accurate, and timely.

(Note: In 1993 it was reported that an open pit gold mining company, operating in California, was required to obtain 324 permits before operations could start, and then file continuing monitor reports with 25 different agencies.)

8. SECURITY AND SAFETY

On-site security should be adequate to protect expensive equipment and supplies as well as mine product. Of course, appropriate security measures must be taken whenever the product, especially a high-value material such as bullion, is transported from the treatment facility. Acceptable safety procedures must be implemented and must be adhered to rigidly from the start to finish of any operation. Even after termination of operations, it is imperative that hazardous materials be properly disposed of and unsafe conditions, such as open shafts, be resolved.

9. MARKETING PROCEDURE

Proposed or actual marketing of the mine or mill product should be reviewed thoroughly by the investor. There may be assessed charges for further treatment of the product. There may be by-product credits returned to the miner. The investor should also be aware of the involvement of any intermediate sales agents and their remuneration.

10. PROFIT DISTRIBUTION

Another obvious consideration is the distribution of profits. What liens, including royalties, loan payments, and rental fees, must be deducted from the gross to determine net profit? Are estimates of operating expenses and pro-forma statements realistic? The investor should be satisfied with the form of payment whether it is in cash, stock, or in-kind.

11. TAX BENEFITS

Like other high-risk investments, mine and mineral developments are often subjected to careful scrutiny by the Internal Revenue Service. Beware of accelerated tax write-offs. Such an advantage was one of the attractions in the Chemehuevis Placer scam referred to earlier. Supposedly, the investor would receive a $50,000. write-off on his tax statement the first year of his investment, by merely paying an advance of $10,000. and signing a promissory note for an additional $40,000. (The prospectus projected within four years a net income, based on gold production, of $139,000.)

In this particular case, the courts apparently upheld an IRS ruling disallowing the tax deduction. Reportedly, the original developers of this program, some $3 million richer, are now unavailable.

12. SEQUENCE OF DEVELOPMENT

In every mineral development there is a logical sequence of events with which the enthusiastic, yet uninitiated, investor may be unfamiliar. Each project can be broken into phases, the completion of which can be evaluated before expending large sums of additional funds. There is no legitimate reason for throwing good money after bad. Classic examples exist which have expensive land being purchased on the basis of someone else's assays, or a costly mill being constructed without proven ore.

13. PROFESSIONAL REPORTS

Engineering reports are useful tools that will assist the mine developer and investor. Decisions to pursue a project into the next stage, and in a particular manner, will be made logically easier after consultation with the appropriate professional engineer, whether a geologist, mining engineer, or metallurgist. Other professional assistance such as financial and legal is generally warranted.

In general, professional evaluation and advice should be sought outside of the developing organization. Principals with the firm and other vested partners, though well intentioned, may write overly optimistic reports. Statements, for example, referring to the attractiveness of a deposit because of its close proximity to a famous producer, or the historically proven improvement of ore grade with depth in the mining district may have the ring of authority, but are often pure speculation. Such reports frequently speak glowingly of questionable assets that may be virtually worthless, e.g., raw mill sites, dilapidated buildings or sheds, and rusted, dismantled equipment. Past production records may be doctored, and projected production/cost data may be presented in an unrealistic manner. Profits are often inflated or guaranteed in such company-prepared prospectuses. It is recommended, therefore, that most professional advice be obtained from consultants who have no financial connection with the company principals, the property, the mineral technology to be employed, or any part of the proposed operation.

14. HISTORICAL RESEARCH

The Arizona Department of Mines and Mineral Resources, with offices in Phoenix and Tucson, is perhaps one of the first places an investor should go seeking information. Knowledgeable, qualified engineers can provide him existing historical data on numerous properties, and discuss solutions to problems he may have.

Lists of registered assayers and engineers are available at the department. In addition, an excellent reference library is maintained, as well as a museum in which the interested individual can obtain hands-on knowledge of rocks and minerals.

While it is obvious that venture capital is needed to start a mine or mineral project, and the Department promotes the development of Arizona's mineral resources, we believe it is essential that the investor be as well informed as possible. Under the best of circumstances mining is a risky business, and we should never tolerate fraudulent practices within the industry. An informed investor, therefore, is better prepared to take the risk without being fleeced at the same time.

CHAPTER 47

SCAMS

When you think about it, every time you wash a pan of placer material, in the back of your mind is the idea of quick wealth. In a way, it's a form of greed. The only difference is that you are working in honest fashion to satisfy that greed. The scam artist also works to satisfy your greed; that's the hook that gets you to part with your hard-earned bucks. The something-for-nothing appeal. Your "something" for his "nothing". The normal victim of a successful scam is never aware that he has been taken. If later on the suspicion of being taken pops into mind, a sense of shame to confess the stupidity, deters disclosure.

No matter how naive or sophisticated you are, the scam artist has you at a disadvantage because of your greed. I define greed here as a sense of easy earnings from an investment.

HOW TO PROTECT YOURSELF.

Remember any real scam is going to be polished with all the objections and red flags anticipated and satisfactorily resolved.

IT IS ALWAYS FRUITFUL TO ANALYZE THE MONEY.

Not the money to be gained through your investment, but how the money is being spent to produce a gain. The gain is usually smoke; the funds are in the scammer's pocket. BUT HOW DOES IT GET THERE? THAT'S THE POINT OF YOUR ANALYSIS.

THE DRILLING SCAM

The first warning flag is raised if the proposal involves a

LIMITED PARTNERSHIP. This vehicle is used because it avoids scrutiny by state and federal regulators. The mere fact of being a limited partnership does not make it bad. The fact of avoiding scrutiny by people whose job is to protect the investor gives reason to wonder why. Always remember your investment must be preceded by a prospectus. If offered to you, and you decline or fail to read it, too bad for you. Now let's follow the money in a PARTNERSHIP case that was offered me in 1990.

The opening gun was an unsolicited phone call. I must have been on some sucker list. The investment was a limited partnership in the Anadarka gas field. Anadarka and natural gas are almost synonymous. Natural gas was price regulated, the pitch was deregulation would increase the partnership's profits, and being a limited partnership, all I could lose was my original investment, since the general partner carried all the liability.

I asked for a prospectus which he was to speed overnight to me by express carrier. THE SECOND THING TO WATCH: Are they using express carriers so as to avoid using the mails. In my case first class mail was just as fast as the extra expense of expressing. It's one way to avoid mail fraud charges.

I followed the money: Remember that in order to avoid fraud charges, full disclosure must be made in the offering prospectus. The problem arises when the disclosure is nearly invisible. (You had to dig these disclosures out of the prospectus, but they were there to discover through diligent search.)

DISCLOSURES MADE: It was disclosed that: The corporation acting as general partner (with all the liability) owned its own building and other assets. On the face it looked sound until you looked at the liabilities of the corporation. The business equipment was leased, the mortgage on the building and other assets gave a corporate net worth of less than $63,000 for a five year old corporation. THAT SHOULD BE ENOUGH TO DUMP ANY FURTHER CONSIDERATION OF ANY DEALS OFFERED. (A weak General Partner.)

The next disclosure was that the corporation held 10,000 acres in a lease of unexplored gas fields in the Anadarka region that they paid $1.00 per acre to lease. The 10,000 acres were to be lease-acquired by the limited partnership at $40.00 the acre. You can see a $390,000 profit for the corporation at this stage.

Disclosed also were "not-at-arms-length" other corporations controlled by the General Partner.

1. The selling corporation took a 30% commission on invested funds.

2. A controlled production company took 25% for the general partner. (Just in case gas was hit.)

THE INVESTMENT FUNDS

1. For a $100,000. investment the General Partner had a 25% share. (This left $290,000. profit on the lease sale.)

2. Each investor/partner had to invest a minimum $10,000. per share, to get a participation in the remaining 75% limited partnership share total. (2 shares were recommended.)

3. The partnership was to be capitalized at $4,000,000.00.

THE MATHEMATICS

After commissions of 30% the $4,000,000. partnership had $2,800,000 left to spend, less General Partnership salaries and overhead. Then we need to deduct the $400,000. cost of the gas lease. This leaves drill funds of $2,400,000.

It was disclosed that wells in the Anadarka were in the 12,000 foot depth range. This meant that drilling costs had to be less than $200. per depth foot, if gas were encountered at that footage. It was further disclosed that some wells were at 15,000 foot depth. At that depth there was less than $160. to spend per depth foot.

Should gas be encountered, it was disclosed that no pipeline existed in the well location to get the gas to market. When and if marketed, the not-at-arms-length production company (controlled by the corporation) would get 25%; with a maybe 75% distributed to investors.

MY FINAL CONCLUSIONS

Visualizing the partnership running out of funds before reaching gas depth, additional calls for funds were designed to run out some or all of the limited partners. I concluded that the General Partner intended the project to run out of funds. It was specified that the General Partner could buy out the limited partners if they refused to continue participation. (But at what price?)

A CLOSE READING OF THE PROSPECTUS REALLY AMOUNTED TO GIVING THE GENERAL PARTNER AN UNSECURED LOAN OF A MINIMUM $10,000. WITH A MAYBE PROMISE TO SHARE ROYALTIES.

I believed that the purpose of the project was really to dump 10,000 acres of OVERVALUED gas lease on a limited partnership that was intended to fail from the beginning.

THE LOST MINE SCAM

It was an innocent classified ad placed by a Canadian advertiser in a treasure magazine.

In essence, the ad was seeking someone interested in locating a LOST INDIAN MINE in the San Bernardino mountains.

"I HAVE THE DETAILS. AGE 75" ...... was the hook.

Variation #1 - Pay me traveling and lodging expenses. I will meet you anywhere you say. But I can't leave until after Thanksgiving Day.

Object of the scam - To get a free winter expense-paid trip to Southern California.

Variation #2 - I've suffered a heart attack and can't make the trip. Will be glad to return your expense money, or in the alternative, if you send me another $100.00, I'll send you mine location details on condition that my estate shares 15% in any finds you make.

Object of the scam - To get easy money!

THE ROADSIDE GOLDPAN SCAM

Many arterial and local travel roads in gold country run alongside gold panning streams. It is not unusual for a passing motorist with a fancy for gold panning to stop and inquire about finding "colors".

In this case, a poorly dressed man and woman purportedly husband and wife are panning at the stream. The mark stops to inquire. In cheerful fashion, the woman displays a small vial of gold dust. She also comments that this was a desperation move for them because they were about to be evicted from their $200.00 monthly rental rooms. "I think we have about an ounce, want to see?" She walks over to her old, beat-up pickup to get a cup of coffee from the thermos. She comments that tomorrow may be too late to get the gold sold.

Meanwhile the mark is fingering the gold vial. You can see gold fever overtaking him. The husband walks over to the pickup and says, "Mister, that vial is worth about 350. to 400. dollars. If you give my wife $250. now, you can make a fast profit and help both of us." The mark declines and hands the vial to the wife. She puts it in her apron pocket.

At this point the greedy mark offers $200. the amount of rent due. The offer is accepted. (The moment the mark is out of sight, they hurriedly leave the scene.)

Object of the scam: - To switch the vial with gold for one with a little gold plus mostly brass shavings. (If they are caught, they have really sold some gold.)

THE SALTED GOLDPAN SCAM

This scam: COURTESY: DR. RALPH E. PRAY - MONROVIA, CA.

This scam usually begins with a GOLD MINE FOR SALE advertisement in a reputable mining or treasure publication. The price may or may not be disclosed in the ad, but a call will make it seem like a bargain. The ad sometimes carries a reason for selling, such as an injury or terminal illness, to explain the need for a hasty close to the transaction. The paperwork for the claim is always in order, BUT CHECK FOR RECENT DATES.

A likely buyer is taken to the claim, handed a gold pan, and told to dig anywhere on the mineral claim. He brings the pan to the seller, who washes it to show the colors.

The buyer is told to get another pan of material anywhere on the claim and wash it. The buyer finds no gold in the pan, but does see the black sands. The seller asks to see the pan to make sure the buyer didn't overlook the gold. As the seller takes the pan, he holds it normally with his thumb crooked over the pan lip. As he begins to pan, he releases the gold dust secreted under his thumb nail into the pan.

Variation #2 - A co-scammer walks by and is asked to check the buyers pan for gold. In checking the pan, the co-scammer's thumbnail carries the gold dust to salt the pan.

THE FUNNIEST SCAM THAT WASN'T

Special Thanks to: Anonymous - SAN ANDREAS, CA.

MAN DELIVERS NUGGET IN 8 DAYS

This event occurred in historic Calaveras County, home of Mark Twain's "Jumping Frog" fame. At the time it made the national newswires. It was reported in detail in the Calaveras Prospect, the Stockton Record, the San Francisco Chronicle, and other newspapers in July, 1987. (Surnames have been changed)

California State Route 4 crosses the Golden Highway State Route 49 at Angels Camp which sits midway between Sonora and San Andreas on SR 49. About 10 miles East of Angels Camp on SR 4 is a captivating Mother Lode community with the unique name of "Murphys", a well preserved relic of the 19th century.

Murphys is usually quiet and uncrowded. It rests at the 2100 foot elevation; is called "home" by 1200 residents. Wells Fargo records show that in the decade following 1849, over 15 million in gold* was shipped out of the area. The main street is lined with stone buildings dating from the Gold Rush days.

*(In 1849 one ounce of gold was selling for $8.00 in coin, $16.00 in trade. This means 1,875,000 ounces of gold were shipped)

Opened in 1855 as the Sperry and Perry Hotel, the most imposing old building in town was renamed Murphys Hotel which retains its 19th century atmosphere. Within the hotel is a bar patronized by the locals, that is, except during the third week in July, when the town overflows as it hosts the Annual Homecoming Celebration.

Thomas James, of Angels Camp, was on a high as he seated himself at the Murphys Hotel well-used wooden bar. He produced a large gold nugget he had recently found, passing it around to be admired by the patrons. It's easy to ogle a freak of nature that measures over an inch long and three-quarter inch wide with a value in the $1500. range. As the nugget reached Timothy Mark, he plopped it into his beer and chug-a-lugged the beer along with the nugget. You can imagine the shock experienced by the nugget owner as it disappeared into a digestive tract. This next point may be disputed, but James is said to have pulled his hunting knife with the promise to surgically remove the nugget if it was not immediately produced. Upchuking failed. More reasonable minds called the sheriff who arrested Mark on a theft charge, then trotted him off to jail.

Mr. Mark spent the next 36 hours in County jail parked next to a portable toilet. The District Attorney declined to prosecute because he couldn't prove Mark's "intent to permanently deprive", an essential element of a theft charge. Mark maintained all along that he was just kidding when he dropped the nugget into his beer and then swallowed it.

A doctor estimated the matter would pass in five or six days. The D.A. released Mr. Mark with the warning that if the nugget was not returned in eight days, he would prosecute. With Mr. Mark, it took eight days. After the nugget was delivered to the D.A., Mr. James described it as having a ghost-buster ghost on one side with a buffalo on the obverse. James was so pleased to regain his gold nugget that he declined to prosecute. The fact that Mr. Mark returned the nugget on his own, reinforces his lack of theft intent. Mr. James determined never again to show-off to strangers a bragging size nugget. Mr. Mark learned that kidding has a price too.

This bit of fun did produce a beneficial side effect which fixes the event into gold country folklore history. Here's how Glenn Wasson of San Andreas, California relates the tale !

TOO MUCH TO SWALLOW

COURTESY: By GLENN WASSON - SAN ANDREAS, CA.
Copyright 1987
From the forthcoming book of gold country poetry,
"LEGENDS OF THE MOTHER LODE"

TOO MUCH TO SWALLOW

Many tales are told in the old Mother Lode
From the pens of Bret Harte and Mark Twain.
To this treasure of lore upon us bestowed
Is another I'll attempt to explain.

One boisterous night in the Murphys Hotel
As the patrons lined up at the bar
An event took place too dubious to tell;
'Twas unseemly, quite gross and bizarre.

A miner walked in with an obvious grin
Ordering drinks for the roistering crowd.
The bar settled down to let him begin
Which he did with a manner quite proud.

From out of his pocket, a nugget he'd panned
Was placed where everyone saw
To be fondled and hefted in each heavy hand,
An object of envy and awe.

As the drinkers inquired where the miner had dug it,
A patron slipped it into his beer,
Thinking, "I'll have me that nugget, I'll just
chug-a-lug it,"
And with a gulp made the piece disappear.

The barroom was stunned, but the miner was cool;
From out of his belt came a knife.
He said, "Cough up the nugget, you guzzling fool
If you want to enjoy further life."

The swallower coughed and inhaled but utterly failed
To force a timely regurgence.
So they slapped him in jail without any bail
To wait for the nugget's emergence.

A three day transition from mouth to emission
is a well-known medical fact,
Or so it was said by the local physician
Of a normal intestinal tract.

But three days passed and he failed to expel
And this measurably heightened the tension.
A watch was put on to the prisoner's cell,
Now the focus of growing attention.

I cannot unveil in graphic detail
The nugget's final appearance.
Decorum in poetry must prevail
Lest we offend its cultured adherents.

But it came to pass on the seventh day
When the guard heard the prisoner shout,
"Unlock my cell; I'm on my way.
That damnable nugget came out!"

The authorities gathered to examine the stool,
A procedure I could not recommend;
Not the normal discovery of gold as a rule
But all's well that comes out in the end.

GLENN WASSON
SAN ANDREAS, CA.

CHAPTER 48

THE ECONOMICS OF MINERAL WEALTH

Yes, the world's wealth in located minerals is running out! One of the richest storehouses of minerals is the U.S.A. Yet, advocates termed "good intentioned" want to lock the minerals away forever. They use ploys such as "Wilderness Areas", Environmental rules, and litigation to delay, forestall or block the extractive industries.

WHAT GOOD IS A LOCKED DOOR?

Lead, zinc and copper reserves are already considered inadequate. Millions of public land mineral acreages have already disappeared. When will it stop?

All the more important for the Rockpecker to get off his duff! Stake your claim before another do-gooder closes down the industrial might of the United States. Without raw materials at a reasonable cost, we may as well close the factory doors.

BUT ALL IS NOT BLACK OR WHITE! THE COLOR OF TRUTH IS SOME SHADE OF GRAY.

At heart, we are all environmentalists. Yet, that protective meaning has been withheld from the extractive arts. Inept mining has meant abuse of streams, air quality, acid rains, and fish kills that impact the public mind with ill feelings. Happily, mining abuses are perpetrated by only a few greedlings.

Denuding a mountain side through timber clear-cutting looks like land rape. Every federal timber contract must require reforestation, or in the alternative, the timber industry should have the option to leave one-half of the cutting area intact. We need to consider our future generations.

Where the cut timber is destined for export as a log, then reforestation must be mandated. Additionally, half of the cutting tract standing timber should be reserved to the public for fifty years, measured from the date reforestation is started.

To my mind, worse than clear-cutting is lightning-caused fire that wipes out thousands of forested acres. Why aren't the environmental causes espousing clear-cutting fire lanes along with access trails to act as fire breaks? Isn't it better to allow the timber industry to clear-cut for a beneficial purpose? I don't see anything but good management in protecting national forests with fire breaks. I would extend it even to national parks. Who doesn't remember the despair after the Yosemite fire?

This brings to mind the *Endangered Species* argument. No one wants to erase a life species, but the argument, as applied, can be specious.

Consider the allegedly endangered Fish Darter. All work in the area stopped while the courts took a look. Like the genesis arguments that won the Kramer District borax case in 1929, the expert testimony was in error and later discredited. In the Fish Darter case, other habitats were later found.

As I see it, a finding of endangerment of species should result only in a moratorium, that holds only so long as another habitat is located. At that time the issue is resolved, and work should go forward.

To allow the declaration of large Wilderness Areas simply by declaring that *no economic minerals exist* is an insult to common sense. To declare that mineral prospecting is barred in that area, is simply insurance that no minerals will ever be found. This is merely a locking-out of the public from public lands. This legislation should allow "another look" to verify absence of economic minerals. Where minerals are found, the General Mining Laws should immediately apply, and that section withdrawn from the Wilderness area.

WE CANNOT, AND SHOULD NOT, POST *NO TRESPASSING* SIGNS ON

PUBLIC LANDS.

Still, we must protect against those who care for nothing but maximum profit. Where reasonable controls make a deposit uneconomic, then we have no General Mining Law mineral. It's as simple as that. But I ask you. Is it reasonable to mandate 327 different operating permits before extraction begins? Is it reasonable to file regular, detailed, monitor reports with 25 various agencies? I THINK NOT!

There will always be someone who tries to feel around the edges of legalities to commit abuse for gain. I am reminded of 1848 when gold was first found at Sutter's Mill.

Sutter was a Swiss national who fled debt imprisonment for bankruptcy. In Mexican California, he accepted Mexican citizenship in order to obtain his land grants. It is said that Sutter did not profit directly from the gold finds. That may be partly true. Still, he tried to keep the gold finds quiet, ostensibly to get his mill finished. This was shortly after the end of the California-Mexican War. Sutter had another agenda.

His first objective was to have his extensive land title holdings confirmed by the United States and California. However, he had another secret objective for requiring silence about the gold finds. He quietly needed to treaty with all the Indian chiefs for the gold rights. After he obtained these treaties, he attempted to charge the miners claim rentals. In some cases he was successful.

His next task was to get the United States to approve the treaties. If he could get them ratified, he would control all the known California gold. Fortunately, the United States military governor ruled that the Indians did not hold title to the land, nor were they to be recognized as nations for treaty purposes.

This ruling exploded Sutter's plans. Greed had played for the whole pot and lost. Several years later the California Assembly, of which he was a member, confirmed his land titles.

Environmentalists play on the sensibilities of the public conscience. In many ways their input is valuable. I have great respect for John Muir for his glacial studies of Yosemite. In that he was correct, but being a stubborn Scotsman, he had no room for humility. He had to be acknowledged right at any price.

When the head geologist for the California Department of Mines wrote a study of the Yosemite Valley rock formations, he assigned a genesis, other than glacial erosion. Muir was slighted. He spent the next two years proving him wrong.

Muir experienced his first year in Yosemite as a sheepherder. He is quoted as calling sheep, "locusts with cloven hoofs." I guess that included the Big Horn too !

Muir founded the Sierra Club over a hundred years ago. His forward thinking has benefited the public. Those, who came after him to found other environmental concern groups, have traded on Muir's reputation, only to radicalize the entire movement. Their hidden agenda appears to be the locking away of natural resources; to erase the entire mining industry; to put "no trespassing" signs on all public lands.

With radical groups coming to the forefront, their successes guarantee that the U.S.A. will be a second or third world country in 50 years. If you take the potential scarcity of crustal resources to its unacceptable alternative, dwell on these words:

"War has been a favorite way for great nations to meet their natural resource needs, and if there is another world war, the conflict will most likely be over what industrial states have come to regard as the elements of survival: Oil, of course, and also iron, copper, uranium, cobalt, wheat, and water."

The Lean Years - Richard J. Barnett

CHAPTER 49

AFTERWORD

Whether you are a Rockpecker, Rockhound, Specimen Hunter, Investor, or simply interested in minerals, it is the author's fondest wish, that the language of Geology is now less foreign to your ear. That if, in your mineral research or in examining a prospectus, you should meet the following descriptive words, your mind won't go blank or your eyes glaze over in despair.

GEOLOGY. An elongated body of granodiorite is in the central portion of the district. This body is five miles long in a north-south direction and half to two miles wide. It is intrusive into older metamorphic rocks and itself is cut by various dike rocks. Immediately east and west of the intrusion are dark greenstones classified as metadiabase and metadiabase porphyry (so-called "porphyrites"), and continuing to the northeast are amphibolite schist, serpentine, gabbro and diorite and slate. Just north of the garanodiorite and to the southwest are slates, phyllite, quartzite and schist of the Calaveras Formation (Carboniferous to Permian). A number of intermediate to basic dikes are present also, as well as a few aplite and granite porphyry dikes. Overlying the district to the east and northwest are Tertiary gravels, in turn overlain by andesite.

(Greenstone is a compact chlorite schist) (Amphibolite is a compact hornblende schist)

ORE DEPOSITS. This is the most heavily mineralized and richest fold district in the state with a very large number of productive veins in a relatively small area. The veins fall into two major groups: 1) those of the granodiorite-greenstone area, which have gentle dips, and 2) those of the serpentine-amphibolite area, with steep dips. The veins of the granodiorite area are either in the granodiorite or in the adjacent greenstone, entering the granodiorite at depth. One group of veins strike north and dips gently (about 35 degrees on the average) either east or west. The other group of veins in the granodiorite strike west of northwest and dips gently north. In the serpentine-amphibolite area the veins strike northwest and dip steeply southwest; a few dip northeast. These occur mostly in the amphibolite near or at the serpentine contact.

The veins usually range from one to ten feet in thickness and consist of quartz with some calcite and ankerite. They fill minor thrust faults. Many veins contain several generations of quartz. There are numerous northeast-striking, vertical or steeply-dipping fractures or "crossings" that commonly are boundaries of ore shoots. The ore contains free gold and varying amounts of sulfides, chiefly pyrite. Present in smaller amounts are galena, chalcopyrite, arsenopyrite, sphalerite, and pyrrhotite. Galena is commonly associated with gold. Milling ore averages 0.25 to 0.5 ounce of gold per ton.

NOW, WASN'T THAT EASY?

CHAPTER 50

THE GREAT DEPRESSION AND SURVIVALIST GOLD PANNING

It may be true that more gold was found during the Depression years, than all the gold found during the Gold Rush. Somehow, I have difficulty buying that idea because of the lax record keeping during both periods. Still, many families remained together, as intact families, because of income gleaned from gold panning. It may seem strange to say it, but all the economic hardship felt during the hard 30's by the rest of the nation, was least felt in the gold producing areas. They had it tough too! Survival tough!

Imagine, if you will, a young schoolgirl fainting in the classroom, because it was not her turn to eat breakfast; or the girl school truant who cut school because it was not her turn to wear the only school dress the family owned. This may sound like overstated bull, so let me focus you in closer to some of the realities of those years.

The common beginning cause of the Great Depression is often assigned to the 1929 Stock Market crash, with its rampant stock gambling, and subsequent margin calls that couldn't be met. It was a giant lottery based on margin leverage that dumped shares "at the market", and at a loss, to evaporate paper values. In one swift, leveling process it wiped out the small investor as well as the well-to-do and wealthy. All were suddenly enrolled in the pauper class as the illusion of easy riches collapsed.

Yet, there were many earlier signs of economic hardship in the multitude of so-called vagrants, bums, knights of the road, and pan-handlers that crisscrossed the country, hungry, gaunt, looking for work. Many lay-offs, wage cuts, inventory reductions, cost cuttings took place that individually seemed small and inconsequential, but in total sum sent fathers scurrying far and wide in search of employment of any kind.

Today, are these the same homeless folk we see on our city streets? Are these the same early warning signs we today call "street people" and think of as lazy, druggies, shiftless, and footloose? You know, those people we wish were in someone else's backyard. Or are they an early sign of another national poverty plague of hopeless despair, again unrecognized by society, or the political powers unwilling to take off their rosy glasses?

From 1930 to 1933 the Great Depression worsened progressively. The economic breakdown of society was everywhere. No area of the U.S. was spared from deprivation. Farmers were the first victims of the depression. 600,000 annually left the farm, headed for the cities and unemployment. Consider the plight of the Mid-Western farmer. At a cost of .75 cents to grow a bushel of corn, the commodity price was down to .10 cents a bushel. Unpaid mortgages were soon on the way to foreclosure. The crash of '29 had changed the farmers plight from desperate to critical. The food expansion, encouraged by World War I, led to oversupply and market disruptions, then bankruptcy. Farm owners were paid to take acreage out of production. Tenant farmers, sharecroppers and field hands became surplus to the land. A million farm families were dispossessed. Migration to the cities and California, the promised land, followed. This was the tip of the desperation that prompted survival gold panning. The banks were in secret crisis. Even a modest withdrawal run would force closure of another insolvent bank. In Detroit alone, 900,000 depositors could not reach their frozen accounts. In those days, when a bank failed, the depositors lost their savings. 1400 banks failed in 1932. The nation without spendable funds, or with at least less to spend, led to a steady downtrend in the economy. From 1930 to 1933, 86,000 businesses failed. Price cuts were common as deflation ruled. The cost of living dropped to .35 cents per pound of prime rib roast; eggs at .20 cents the dozen; shoes at 2.98 the pair; men's wool worsted suits $ 7.45 with two pair of pants. In 1932 a new Chevrolet "6" cost $ 445.00. In Canada, a Ford 5-passenger, 4 door, 4 cylinder cost $ 685.00; a tin of Campbell's tomato soup was .08 cents, a loaf of bread .06

cents. In Toronto the rent for a 3 bedroom house was $ 40.00. The house itself could be had for $ 3,000.00.

Those that were employed were the lucky ones. In 1929 the average family income was $ 2,300.00; in 1932 it was down to less than $ 1,600.00. They were lucky, for they had jobs. While the lucky's wages tumbled 40%, living costs only dropped 25%.

These were the dismal economic facts that people sought to escape. But there was no escape. A working society was turned into one of forced idleness. Just as the Depression facts and stories are passed down from generation to generation, so too the tales of the Forty-niners were still fresh in the memory banks of those caught in the economic upheaval. Here was a field of activity that required only a goldpan, shovel, and the barest survival gear. Many rugged individuals turned to the West to try their luck at getting by. Survival was the goal. The old gold fields were revisited. Entire families were kept intact through meager gold finds. But through hard work, they did survive.

Take the story of the unlucky gold panner who was barely making ends meet on Pauley Creek in California. He was a loner. You could tell from the little talking he did, that he didn't have even the barest of goldpanning knowledge. One day he announced that he was leaving. The others working the creek were not surprised. They decided to give him a going-away party at which he ate like a famished man. Much later his old camp was visited by the wife of one of the more successful panners. On her return she reported finding a lot of empty dog food tins in his dump, yet he had no pets.

Twenty years later, a placer gold buyer revisited Pauley Creek. He told me that the only remaining evidence of depression activity was the fallen-down, rusted, sheetmetal roofs that sheltered these stalwarts. Something still more curious happened to me about eight years ago. A man had a placer claim he was offering for sale. In response to my location query, he replied, "Pauley Creek, and it's a good claim!" The claim may have been unimpeachable, but I knew that the depression-worked ground was sub-marginal at best. Any gold still there was mostly illusory.

But who knows, sixty years of run-offs may have replenished the gold. Maybe I'll visit the area. But don't bet on it. There are more fertile avenues to pursue. Incidentally, during our chat, I asked him about the rusted sheetmetal roofs on Pauley Creek. He replied that he had never seen any evidence of roofs or other habitation.

Prior to 1933, the gold price was $20.67 the troy ounce. You could always redeem your paper money for gold or silver. You could get 23.22 fine grains of gold for each dollar. There are 480 grains in one troy ounce.

On 4/19/33 FDR embargoed all export of gold permanently. Any traffic in the precious metals was halted. Gold suddenly became contraband. Gold hoarders by the thousands lined up in a reverse gold rush at the Federal Reserve banks to convert their gold into legal tender before the cut-off date. You need to also know that FDR called a bank holiday. Of 19,000 banks closed down, only 2,400 were deemed solvent enough to reopen at once, the rest needed shoring up or were closed. To illustrate the seriousness of the financial situation, government salaries including those of CONGRESS WERE CUT BY $100,000,000.00. The next day light beer and wine were legalized.

To get more money into circulation, in January, 1934, the Gold Reserve Act gave the president power to reduce the gold content of the dollar. He already had the power to set the value of the dollar at whatever he wanted. The president set gold at $35.00 the ounce, and reduced the gold content of the dollar from 23.22 fine grains to 13.1 fine grains per dollar. He did so at 59.06% of the old weight. Each dollar was now worth .59 cents. It became known as the "Roosevelt Dollar", or the "baloney dollar". Except for outstanding silver certificate dollars, the U.S. was now on a sole paper currency basis. For the survivalist gold panner, the .59 cent dollar gave him 41% more for his meager gold recoveries.

As a political comment on the Great Depression days: Bank and stock market reforms, the freeing of the dollar from gold, had but one single objective. That was, to put the nation's financial resources into the PUBLIC DOMAIN and to take it out of PRIVATELY HELD POCKETS. Bernard Baruch called it "class legislation to benefit the unemployed, the debtor class, and incompetent, unwise people."

What the reform really accomplished, was to free the nation's finances from the clutches of power brokers; the men of power and prestige who ruled the economic and political life of the nation as an invisible government. It democraticized financial resources by leveling the playing field. With the defrocking of the high priests of high finance, the capital market's weaknesses had been exposed; the abuses of the financial system were cleaned up. From now on, the insiders would not play the game at the expense of the suckers. The code of personal greed and social irresponsibility that ruled the financial system was broken.

While the last few lines sound like a cure-all, the recent past tarnishes it a bit. We have junk bond scandals, saving and loan failures, plus a multitude of greedlings nibbling at the edges of legality. At least, gold is now legal to possess. We have the physical gold, such as you see in your pan or in a gold coin. Then there is "paper gold". That is, the gold in your vault that you can sell options against for a small premium at a specified strike price. There is the naked option to sell gold that you don't have, at a future price you hope it doesn't reach; there is the mine owner selling his unmined gold in the forward market, to cover the sale from tomorrow's mining activity. Then again the latest financing tool takes a gold claim, borrows gold to sell for cash today, to finance the opening of the mine from the sale proceeds. Maybe the mine will be a success, and the borrowed gold can be repaid from the new mine output.

At $ 400.00 the ounce, a single pennyweight is worth $20.00 fine. Adjusted for the impurities in your placer gold, you can expect anywhere from $14.00 to $ 18.00 the PWT for your survivalist efforts.

Anyone want to go gold panning? I'm ready now !

CHAPTER 51

28 GOLDPANNING AREAS OPEN FREE TO THE PUBLIC AZ., CA., OR., WA. NO CLAIM — SET ASIDE AREAS

The big question for a novice gold panner boils down to:
" Where can I safely pan gold and really find some ?"

In the western states that question has been answered for you in the form of OPEN TO THE PUBLIC set-aside areas. Usually the charges, if any, are minimal or free. Most often camping is available for a small fee with time stay limits. If you elect to try one of these gold spots, it is wise to get advance information as to whether a reservation is mandated. It may be even wiser to assume a reservation is required, then proceed on that basis. DREDGING REQUIRES A PERMIT & HAS A CLOSED SEASON.

GOLDPANNING REQUIRES NO PERMIT.

National Forests and U.S. Bureau of Land Management areas are generally open for mineral entry.

DO NOT PAN WITHIN:
NATIONAL PARKS
INDIAN RESERVATIONS
WILDERNESS AREAS
MILITARY RESERVATIONS
ON RAILROAD RIGHT—OF—WAY
60 FEET OF THE MEXICO BORDER

ARIZONA — State Trust Lands, in general, are not available for recreational GOLDPANNING use. While the state encourages mineral development on State Land, it is allowed under a system of State permits. IT IS A RESTRICTIVE PROCESS THAT DISCOURAGES RECREATIONAL GOLD PANNING.

EXCEPT FOR THE NEXT 3 AREAS,
IN ARIZONA YOU MUST HAVE A PERMIT FOR RECREATIONAL GOLD PANNING.

1 -- LYNX CREEK off hwy. 69
ARIZONA - 5,000 + elevation.
COUNTY: YAVAPAI -- 6+ miles of Lynx Creek frontage
NEAREST TOWN: PRESCOTT, 6 miles E, R on Lynx lake Rd
NAME OF AREA: LYNX CREEK MINERAL WITHDRAWAL AREA
LAND CONTROL: PRESCOTT NATIONAL FOREST, USFS
TOPOGRAPHICAL MAP: USGS GROOM CREEK 7.5 Min.
MAP LOCATION: T13N, R1W, Sec.5, E2 QUARTER
CAMPING AVAILABLE: YES, Lynx & Hilltop camps
GOLD SEARCH ACTIVITIES ALLOWED:

a. GOLD PANNING (YES)
b. DREDGING (NO)
c. SLUICING (NO)

AREA IS OPEN TO THE PUBLIC.
CONTACT: BRADSHAW RANGER DISTRICT 602/445-7253
HIGHWAY 69
PRESCOTT, ARIZONA 86301

---

2 -- LaPOSA - South of I 10
ARIZONA
COUNTY: La Paz - 2200 ACRES - DESERT LAND
NEAREST TOWN: QUARTZSITE, South of I 10
NAME OF AREA: LA POSA RECREATION SITE
LAND CONTROL: BUREAU OF LAND MANAGEMENT
TOPOGRAPHICAL MAP: QUARTZSITE 15 MIN
MAP LOCATION: T34N R20W MANY SECTIONS GSRM
CAMPING AVAILABLE: LONG TERM O.K.
GOLD SEARCH ACTIVITIES ALLOWED:

a. GOLD PANNING (YES)DRY
b. DREDGING (NO)
c. SLUICING (NO)

AREA IS OPEN TO THE PUBLIC.
CONTACT: BLM YUMA RESOURCE AREA
DISTRICT OFFICE 602/726-6300.

GOOD METAL DETECTOR AREA

---

3 -- LAKE PLEASANT off SR 74
ARIZONA - 1500 elevation
COUNTY: - MARICOPA - 14,000 acres on Castle Hot S. Rd.
NEAREST TOWN: PHOENIX, 20 MI.NW, I17 to SR74
NAME OF AREA: LAKE PLEASANT REGIONAL PARK
LAND CONTROL: MARICOPA COUNTY PARKS & REC. DEPT.
TOPOGRAPHICAL MAP: USGS MINEHAHA 7.5 Min.
MAP LOCATION: HUMBUG CREEK, CASTLE CREEK
CAMPING AVAILABLE: YES, NOMINAL ENTRY FEE
GOLD SEARCH ACTIVITIES ALLOWED:

a. GOLD PANNING (YES)
b. DREDGING (NO)
c. SLUICING (NO)
d. ROCKHOUNDING (NO)

AREA IS OPEN TO THE PUBLIC.
CONTACT: MARICOPA COUNTY PARKS & REC. DEPT.
602/272-8871
NOTE: PANNING ONLY - TRY NORTH END OF LAKE DRAINAGES.

---

4 -- BIG FLAT on SR 299
CALIFORNIA
COUNTY: TRINITY - 40 ACRES
NEAREST TOWN: WEAVERVILLE - 22 miles West
NAME OF AREA: BIG FLAT FREE USE PERMIT AREA
LAND CONTROL: SHASTA-TRINITY NATIONAL FOREST
TOPOGRAPHICAL MAP: HAYFORK 7.5 MIN
MAP LOCATION: T33N R12W SEC 3 MDM
CAMPING AVAILABLE: YES, WITH 14 DAY LIMIT.
GOLD SEARCH ACTIVITIES ALLOWED:

a. GOLD PANNING (YES)
b. DREDGING TO 5" (YES)
c. SLUICING (YES)

AREA IS OPEN TO THE PUBLIC.
CONTACT: DISTRICT RANGER
STAR ROUTE 1, BOX 10
BIG BAR, CA. 96010
916/623-6106

---

5 -- BUTTE CREEK
CALIFORNIA
COUNTY: BUTTE - 300 ACRES
NEAREST TOWN: PARADISE, Skyway N to DeSabla/Mill Rd.
NAME OF AREA: BUTTE CREEK CANYON
LAND CONTROL: BUREAU OF LAND MANAGEMENT
TOPOGRAPHICAL MAP: STIRLING CITY/COHASSET 7.5
MAP LOCATION: T23 24N R3E MDM
CAMPING AVAILABLE: CHECK FIRST
GOLD SEARCH ACTIVITIES ALLOWED:
a. GOLD PANNING (YES)
b. DREDGING (YES)
c. SLUICING (YES)
AREA IS OPEN TO THE PUBLIC.
CONTACT: BLM OUTDOOR PLANNER
355 HEMPSTED DRIVE
REDDING, CA. 96001
916/246-5325
BLM - Small fee & permit required.

---

6 -- AMERICAN RIVER off SR 49
CALIFORNIA
COUNTY: EL DORADO OVER 30,000 ACRES.
NEAREST TOWN: AUBURN North, Rt. off SR 49
NAME OF AREA: AUBURN STATE RECREATION AREA
LAND CONTROL: STATE OF CALIFORNIA
TOPOGRAPHICAL MAP: LOCAL MAP
MAP LOCATION: AMERICAN RIVER N & MIDDLE FORKS
CAMPING AVAILABLE: YES, BUT CHECK FIRST.
GOLD SEARCH ACTIVITIES ALLOWED:
a. GOLD PANNING (YES)
b. DREDGING (YES)
c. SLUICING (YES)

AREA IS OPEN TO THE PUBLIC.
CONTACT: CALIF. DEPT. OF PARKS
P.O. BOX 3266
AUBURN, CA. 95604
916/988-0205

---

7 — SOUTH YUBA TRAIL—Off SR 49

CALIFORNIA
COUNTY: NEVADA - 1500 ACRES
NEAREST TOWN: NEVADA CITY, 12 mi. N-E off SR 49
NAME OF AREA: SOUTH YUBA RECREATION AREA
LAND CONTROL: Bureau of Land Management
TOPOGRAPHICAL MAP: NORTH BLOOMFIELD 7.5 MIN
MAP LOCATION: T17N R9E SEC 13,16,22 MDM
CAMPING AVAILABLE: DESIGNATED AREA 14 DAY MAX.
GOLD SEARCH ACTIVITIES ALLOWED:

a. GOLD PANNING (YES)
b. DREDGING (SOME)
c. SLUICING (YES)

AREA IS OPEN TO THE PUBLIC.
CONTACT: BUREAU OF LAND MANAGEMENT
63 NATOMA ST.
FOLSOM, CA. 95630
916/985-4475

---

8 -- OREGON CREEK on SR 49

CALIFORNIA
COUNTY: SIERRA - 40 ACRES
NEAREST TOWN: NEVADA CITY - N 17 mi.on SR 49
NAME OF AREA: TAHOE NATIONAL FOREST
LAND CONTROL: U.S. FOREST SERVICE
TOPOGRAPHICAL MAP: CAMPTONVILLE 7.5 MIN
MAP LOCATION: T18N R8E SEC 28 MDM
CAMPING AVAILABLE: NO!
GOLD SEARCH ACTIVITIES ALLOWED:

a. GOLD PANNING (YES)
b. DREDGING (NO)
c. SLUICING (YES)

AREA IS OPEN TO THE PUBLIC.
CONTACT: NORTH YUBA RANGER STATION
15924 HIGHWAY 49
CAMPTONVILLE, CA. 95922
916/288-3231

---

9 -- UNION FLAT on SR 49

CALIFORNIA
COUNTY: SIERRA - 40 ACRES
NEAREST TOWN: DOWNIEVILLE, East 5 mi. on SR 49/river
NAME OF AREA: TAHOE NATIONAL FOREST
LAND CONTROL: U.S. FOREST SERVICE
TOPOGRAPHICAL MAP: SIERRA CITY 7.5 MIN
MAP LOCATION: T20N R11E SEC 28,33 MDM
CAMPING AVAILABLE: YES
GOLD SEARCH ACTIVITIES ALLOWED:
a. GOLD PANNING (YES)
b. DREDGING (YES)
c. SLUICING (YES)

AREA IS OPEN TO THE PUBLIC.
CONTACT: NORTH YUBA RANGER STATION
15924 HWY 49
CAMPTONVILLE, CA. 95922
916/288-3231

---

10 -- WILD PLUM off SR 49

CALIFORNIA
COUNTY: SIERRA - 40 ACRES
NEAREST TOWN: DOWNIEVILLE 5+/- mi. on SR49 to W/P Rd
NAME OF AREA: WILD PLUM RECREATION AREA
LAND CONTROL: U.S. FOREST SERVICE
TOPOGRAPHICAL MAP: HAYPRESS VALLEY 7.5 MIN
MAP LOCATION: T20N R12E SEC 26,27 MDM
CAMPING AVAILABLE: YES, 14 DAY LIMIT
GOLD SEARCH ACTIVITIES ALLOWED:
a. GOLD PANNING (YES)
b. DREDGING (YES)
c. SLUICING (YES)

AREA IS OPEN TO THE PUBLIC.
CONTACT: NORTH YUBA RANGER STATION
15924 HWY 49
CAMPTONVILLE, CA. 95922
916/288-3231

---

11 -- RAMSHORN on SR 49

CALIFORNIA

COUNTY: SIERRA - 40 ACRES

NEAREST TOWN: CAMPTONVILLE, Northeast 12+/-mi SR 49

NAME OF AREA: TAHOE NATIONAL FOREST

LAND CONTROL: U.S. FOREST SERVICE

TOPOGRAPHICAL MAP: DOWNIEVILLE 15 MIN.

MAP LOCATION: T19N R9E SEC 1 MDM

CAMPING AVAILABLE: YES - 14 DAY LIMIT

GOLD SEARCH ACTIVITIES ALLOWED:

a. GOLD PANNING (YES)
b. DREDGING (YES)
c. SLUICING (YES)

AREA IS OPEN TO THE PUBLIC.

CONTACT: NORTH YUBA RANGER STATION
15924 HWY 49
CAMPTONVILLE, CA. 95922
916/288-3231

---

12 -- CONVICT FLAT off SR 49

CALIFORNIA

COUNTY: SIERRA - 40 ACRES

NEAREST TOWN: DOWNIEVILLE, West 8+/- mi.

NAME OF AREA: CONVICT FLAT PICNIC AREA

LAND CONTROL: U.S. FOREST SERVICE

TOPOGRAPHICAL MAP: DOWNIEVILLE 15 MIN

MAP LOCATION: T19N R9E SEC 10 MDM

CAMPING AVAILABLE: USE RAMSHORN CAMPGROUND

GOLD SEARCH ACTIVITIES ALLOWED:

a. GOLD PANNING (YES)
b. DREDGING (YES)
c. SLUICING (YES)

AREA IS OPEN TO THE PUBLIC.

CONTACT: NORTH YUBA RANGER STATION
15924 HWY 49
CAMPTONVILLE, CA. 95922
916/288-3231

---

13 -- INDIAN VALLEY on SR 49

CALIFORNIA

COUNTY: SIERRA - 40 ACRES

NEAREST TOWN: CAMPTONVILLE, East 8+/- mi. on SR 49

NAME OF AREA: INDIAN VALLEY RECREATION AREA

LAND CONTROL: U.S. FOREST SERVICE

TOPOGRAPHICAL MAP: DOWNIEVILLE 15 MIN

MAP LOCATION: T19N R9E SEC 17 MDM

CAMPING AVAILABLE: YES - 14 DAY LIMIT

GOLD SEARCH ACTIVITIES ALLOWED:

a. GOLD PANNING (YES)
b. DREDGING (YES)
c. SLUICING (YES)

AREA IS OPEN TO THE PUBLIC.

CONTACT: NORTH YUBA RANGER STATION
15924 HWY 49
CAMPTONVILLE, CA. 95922
916/288-3231

---

14 -SO. FORK STANISLAUS, SR 50

CALIFORNIA

COUNTY: TUOLUMNE - 200 ACRES

NEAREST TOWN: COLUMBIA MAIN St to river

NAME OF AREA: STANISLAUS NATIONAL FOREST

LAND CONTROL: U.S. FOREST SERVICE

TOPOGRAPHICAL MAP: COLUMBIA SE 7.5 MIN

MAP LOCATION: T3N R15E SEC 32 MDM

CAMPING AVAILABLE: FEW NEAR RIVER.

GOLD SEARCH ACTIVITIES ALLOWED:

a. GOLD PANNING (YES)
b. DREDGING (YES)
c. SLUICING (YES)

AREA IS OPEN TO THE PUBLIC.

CONTACT: SUMMIT RANGER DISTRICT
HWY 108 E, STAR ROUTE, BOX 1295
SONORA, CA. 95370
209/965-3434

 # 15 SAN GABRIEL RIVER EAST FORK CALIFORNIA
COUNTY: LOS ANGELES
NEAREST TOWN: AZUZA SR 39 rt on East Fork Rd.
NAME OF AREA: ANGELES NATIONAL FOREST
LAND CONTROL: U.S. FOREST SERVICE
TOPOGRAPHICAL MAP: LOCAL MAPS
MAP LOCATION: EAST FORK ROAD BRIDGE
CAMPING AVAILABLE: YES
GOLD SEARCH ACTIVITIES ALLOWED:

a. GOLD PANNING (YES)
b. DREDGING (YES)
c. SLUICING (YES)

AREA IS OPEN TO THE PUBLIC.
CONTACT: MT. BALDY RANGER DISTRICT
110 N. WABASH AVENUE
GLENDORA, CA. 91740
818/335-1251

#16 –– EAGLE FORKS OREGON
COUNTY: BAKER - 40 ACRES
NEAREST TOWN: NEW BRIDGE 9+/- mi on Eagle Creek Rd.
NAME OF AREA: WALLOWA-WHITMAN NATIONAL FOREST
LAND CONTROL: U.S. FOREST SERVICE
TOPOGRAPHICAL MAP: SPARTA BUTTE 7.5 MIN
MAP LOCATION: T8S R45E SEC S6 WM
CAMPING AVAILABLE: YES
GOLD SEARCH ACTIVITIES ALLOWED:

a. GOLD PANNING (YES)
b. DREDGING (YES)
c. SLUICING (YES)

AREA IS OPEN TO THE PUBLIC.
CONTACT: WALLOWA-WHITMAN NATIONAL FOREST
1550 DEWEY AVENUE, P.O. BOX 907
BAKER, OREGON 97814
503/523-6391

---

17 — McCULLY FORKS off HWY 220

OREGON

COUNTY: BAKER – 40 ACRES

NEAREST TOWN: SUMPTER, West on HWY 220 2 mi.

NAME OF AREA: McCULLY FORKS CAMPGROUND

LAND CONTROL: U.S. FOREST SERVICE

TOPOGRAPHICAL MAP: BOURNE 7.5 MIN

MAP LOCATION: T9S R36E SEC 24 WM

CAMPING AVAILABLE: YES

GOLD SEARCH ACTIVITIES ALLOWED:

a. GOLD PANNING (YES)
b. DREDGING (NO)
c. SLUICING (NO)

AREA IS OPEN TO THE PUBLIC.

CONTACT: WALLOWA-WHITMAN NATIONAL FOREST
1550 DEWEY AVENUE, P.O. BOX 907
BAKER, OREGON 97814
503/523-6391

---

18 — DEER CREEK off HWY 220

OREGON

COUNTY: BAKER – 40 ACRES

NEAREST TOWN: SALISBURY, West on HWY 220; 4+/- N.

NAME OF AREA: DEER CREEK CAMPGROUND

LAND CONTROL: U.S. FOREST SERVICE

TOPOGRAPHICAL MAP: PHILLIPS LAKE 7.5 MIN

MAP LOCATION: T9S R38E SEC 30 WM

CAMPING AVAILABLE: YES

GOLD SEARCH ACTIVITIES ALLOWED:

a. GOLD PANNING (YES)
b. DREDGING (YES)
c. SLUICING (YES)

AREA IS OPEN TO THE PUBLIC.

CONTACT: WALLOWA-WHITMAN NATIONAL FOREST
1550 DEWEY AVENUE, P.O. BOX 907
BAKER, OREGON 97814
503/523-6391

---

19 -- POWDER RIVER on HWY 220

OREGON
COUNTY: BAKER - 40 ACRES
NEAREST TOWN: SALISBURY, West on HWY 220
NAME OF AREA: POWDER RIVER RECREATION AREA
LAND CONTROL: U.S. FOREST SERVICE
TOPOGRAPHICAL MAP: BLUE CANYON 7.5 MIN
MAP LOCATION: T10E R39E SEC 19 WM
CAMPING AVAILABLE: YES
GOLD SEARCH ACTIVITIES ALLOWED:

a. GOLD PANNING (YES)
b. DREDGING (NO)
c. SLUICING (NO)

AREA IS OPEN TO THE PUBLIC.
CONTACT: WALLOWA-WHITMAN NATIONAL FOREST
1550 DEWEY AVENUE, P.O. BOX 907
BAKER, OREGON 97814
503/523-6391

#20-- ANTLERS on County Road 507

OREGON
COUNTY: BAKER - 40 ACRES
NEAREST TOWN: WHITNEY, South on CR 507 2+/- mi.
NAME OF AREA: NORTH FORK BURNS RIVER
LAND CONTROL: U.S. FOREST SERVICE
TOPOGRAPHICAL MAP: WHITNEY 7.5 MIN
MAP LOCATION: T11S R36E SEC 3 WM
CAMPING AVAILABLE: YES, BUT CHECK
GOLD SEARCH ACTIVITIES ALLOWED:

a. GOLD PANNING (YES)
b. DREDGING (YES)
c. SLUICING (YES)

AREA IS OPEN TO THE PUBLIC.
CONTACT: WALLOWA-WHITMAN NATIONAL FOREST
1550 DEWEY AVENUE, P.O. BOX 907
BAKER, OREGON 97814
503/523-6391

---

21 –– QUARTZVILLE CREEK

OREGON

COUNTY: LINN – 1600 ACRES

NEAREST TOWN: SWEET HOME–Northeast to Quartzville Rd

NAME OF AREA: QUARTZVILLE RECREATION CORRIDOR

LAND CONTROL: BUREAU OF LAND MANAGEMENT

TOPOGRAPHICAL MAP: STATE MAP

MAP LOCATION: T11,12S R3,4E SEC Many WM

CAMPING AVAILABLE: YES, 16 DAY LIMIT

GOLD SEARCH ACTIVITIES ALLOWED:

a. GOLD PANNING (YES)
b. DREDGING (YES)
c. SLUICING (YES)

AREA IS OPEN TO THE PUBLIC.

CONTACT: BUREAU OF LAND MANAGEMENT
1717 FABRY ROAD SE, P.O. BOX 3227
SALEM, OREGON 97302
503/399–5646

---

22 –– BRICE CREEK

OREGON

COUNTY: LANE – 600 ACRES

NEAREST TOWN: COTTAGE GROVE to Brice Creek Rd.

NAME OF AREA: UMPQUA NATIONAL FOREST

LAND CONTROL: U.S. FOREST SERVICE

TOPOGRAPHICAL MAP: ROSE HILL 7.5 MIN

MAP LOCATION: T22E R1E SECS 8,9,16,22 WM

CAMPING AVAILABLE: YES

GOLD SEARCH ACTIVITIES ALLOWED:

a. GOLD PANNING (YES)
b. DREDGING (YES)
c. SLUICING (YES)

AREA IS OPEN TO THE PUBLIC.

CONTACT: U.S. FOREST SERVICE
COTTAGE GROVE RANGER DISTRICT, P.O. BOX 38
COTTAGE GROVE, OREGON 97424
503/942–5591

---

23 -- SHARPS CREEK on HWY 2460

OREGON
COUNTY: LANE - 80 ACRES
NEAREST TOWN: COTTAGE GROVE, E rt on HWY 2460 4 mi
NAME OF AREA: SHARPS CREEK RECREATION SITE
LAND CONTROL: BUREAU OF LAND MANAGEMENT
TOPOGRAPHICAL MAP: CULP CREEK 7.5 MIN
MAP LOCATION: T22S R1W SEC 15 WM
CAMPING AVAILABLE: CHECK FIRST
GOLD SEARCH ACTIVITIES ALLOWED:

a. GOLD PANNING (YES)
b. DREDGING (YES)
c. SLUICING (YES)

AREA IS OPEN TO THE PUBLIC.
CONTACT: BLM EUGENE DISTRICT OFFICE
1255 PEARL ST. P.O. BOX 10266
EUGENE, OREGON 97401
503/687-6651

---

24 -- SIXES RIVER off HWY 101

OREGON
COUNTY: CURRY - 160 ACRES
NEAREST TOWN: SIXES, East 14+/- mi on Sixes River Rd
NAME OF AREA: SIXES RIVER RECREATION SITE
LAND CONTROL: BUREAU OF LAND MANAGEMENT
TOPOGRAPHICAL MAP: MT. BUTLER 7.5 MIN
MAP LOCATION: T32N R14W SECS 11,12 WM
CAMPING AVAILABLE: YES, 14 DAY LIMIT
GOLD SEARCH ACTIVITIES ALLOWED:

a. GOLD PANNING (YES)
b. DREDGING (YES)
c. SLUICING (YES)

AREA IS OPEN TO THE PUBLIC.
CONTACT: BLM - COOS BAY DISTRICT
1300 AIRPORT LANE
NORTH BEND, OREGON 97459-2000
503/756-0100

---

25 -- GOLD NUGGET on HWY 234

OREGON
COUNTY: JACKSON - 40 ACRES
NEAREST TOWN: GOLD HILL, North on HWY 234 2 +/- mi
NAME OF AREA: GOLD NUGGET RECREATION AREA
LAND CONTROL: BUREAU OF LAND MANAGEMENT
TOPOGRAPHICAL MAP: GOLD HILL 7.5 MIN
MAP LOCATION: T36S R3W SEC 10, 11 WM
CAMPING AVAILABLE: CHECK FIRST
GOLD SEARCH ACTIVITIES ALLOWED:

a. GOLD PANNING (YES)
b. DREDGING (YES)
c. SLUICING (YES)

AREA IS OPEN TO THE PUBLIC.
CONTACT: BLM MEDFORD DISTRICT OFFICE
3040 BIDDLE ROAD
MEDFORD, OREGON 97501
503/770-2200

---

26 -- TUNNEL RIDGE off HWY 238

OREGON
COUNTY: JACKSON - 80 ACRES
NEAREST TOWN: MEDFORD, West HWY 238 to Applegate Rd.
NAME OF AREA: LITTLE APPLEGATE RIVER
LAND CONTROL: BUREAU OF LAND MANAGEMENT
TOPOGRAPHICAL MAP: STERLING RIDGE 7.5 MIN
MAP LOCATION: T39S R2W SECS 23, 25 WM
CAMPING AVAILABLE: SELF-CONTAINED ONLY
GOLD SEARCH ACTIVITIES ALLOWED:

a. GOLD PANNING (YES)
b. DREDGING (YES)
c. SLUICING (YES)

AREA IS OPEN TO THE PUBLIC.
CONTACT: BLM MEDFORD DISTRICT OFFICE
3040 BIDDLE ROAD
MEDFORD, OREGON 97501
503/770-2200

---

27 — SULLIVAN CREEK off HWY 31 WASHINGTON

COUNTY: PEND O'REILLE
NEAREST TOWN: METALINE FALLS, E on Sullivan Lake Rd.
NAME OF AREA: COLVILLE NATIONAL FOREST
LAND CONTROL: U.S. FOREST SERVICE
TOPOGRAPHICAL MAP: LOCAL
MAP LOCATION: LOCAL
CAMPING AVAILABLE: CHECK FIRST
GOLD SEARCH ACTIVITIES ALLOWED:

a. GOLD PANNING (YES)
b. DREDGING (YES)
c. SLUICING (YES)

$10.00 Recreational Prospecting permit required.
AREA IS OPEN TO THE PUBLIC.
CONTACT: USFS - COLVILLE NATIONAL FOREST
METALINE FALLS, WA. 99153
509/446-2681

---

28 -- SWAUK CREEK on HWY 97 WASHINGTON

COUNTY KITTATAS - 640 ACRES
NEAREST TOWN: LIBERTY, North HWY 97
NAME OF AREA: WENATCHEE NATIONAL FOREST
LAND CONTROL: U.S. FOREST SERVICE
TOPOGRAPHICAL MAP: LIBERTY 15 MIN
MAP LOCATION: T21N R17E SECS many WM
CAMPING AVAILABLE: YES
GOLD SEARCH ACTIVITIES ALLOWED:

a. GOLD PANNING (YES)
b. DREDGING (YES)
c. SLUICING (YES)

AREA IS OPEN TO THE PUBLIC.
CONTACT: USFS WENATCHEE NATIONAL FOREST
301 YAKIMA ST. P.O. BOX 811
WENATCHEE, WA. 98801
500/622-4335

CHAPTER 52

RECREATIONAL GOLD PANNING
32 STATES WITH PLACER GOLD

The fastest growing summer, outdoor, family activity today combines destination camping with the flavor and adventure of the '49er. Youths from eight to eighty are setting up tents near known placer gold locations to test their patience and luck in search of the elusive gold flakes.

I call this activity "GOLDPECKING", because of the appearance of your hand as it holds a tweezer to pick a speck of gold from the pan. To me, it looks like a bird tapping into a pan. A GOLDPECKER ! A recreational gold panner.

From a pure cost standpoint, the only additional expense involves a goldpan and tweezers. You already have the camping gear. The search for placer gold even if fruitless, is a campfire story that can be retold for the rest of your life. When this sport bears fruit, the exhilaration you feel is unlike any other sensation. You are hooked for good, you have gold fever. Believe it!

I think of myself as a Goldpecker. I like the excitement, fresh air, even the solitude. But what I like most is the memory involving the thrill of a new companion finding that first speck of gold. Their faces beam as bright as the shiny gold.

Each Spring, around daylight savings time clock-change, this fever returns to prod me onward. Now-a-days, I'm ready at the drop of a hint. You see, I'm already packed and ready to go. Let me explain: Some years ago I was able to purchase a pigskin rucksack with a tie string outer pocket across the bottom width of the backpack. It's a perfect holder for my fourteen inch gold pan and eight inch Tyler screen. I use a fourteen inch plastic pan with the molded riffle design. Personally, I prefer the maroon or green pan color because it enhances gold visibility.

My Tyler screen is a professional model with one-eighth inch stainless steel mesh soldered to a round, eight inch, brass, frame. I use it for sizing materials to eliminate the larger gangue pieces. You may think that the mesh size is small, but when you think about it, 99.9% of the gold you find will easily pass these openings. Where you are panning an area known for coarse gold, you just inspect the screen more closely before dumping it. It goes without saying that a sorting device makes life easier.

To write this chapter I emptied my rucksack and laid out the contents for inspection. It surprised me to count eight tweezers, but more so, because four were nickel plated steel. I prefer austenitic stainless steel tweezers because they avoid becoming magnetic. Why is this important? Black sands are mostly magnetite, which, as its name implies, is magnetic. Lodestone is a natural magnet. When fishing around in the black sands, steel tip ends become magnetized and your gold speck comes away covered with black sand. The steel tweezers are worthless after a while. For this reason I also carry a demagnetizer.

In the top pocket of my rucksack, which is easiest to reach, I carry a narrow garden trowel and four kitchen spoons of differing sizes. I get most use from the long iced tea spoon. It works well in crevices. The narrow trowel is helpful in reaching tight areas. Seems I use that a lot too.

That top pocket is filled with my most used gear. A sniffer bottle for sucking the tiny fines from the pan edge. A four piece hook and pick for probing cracks; three clear plastic 35mm film roll cases for saving the gold. Those film roll capsules are practical, for when they fall into the stream, they float; also you can cap and decap the container with one hand, very handy when the other hand has a tweezer full of gold. Another item I carry is a 2 inch hand lens to examine gold for angularness. (If the gold is smooth, it traveled far to where found. If sharp and angular, you are near the gold source.) I almost failed to mention an Engineer's Lensatic Compass... just in case.

It may surprise you when I say that the most important

convenience gear in the top pocket is a pair of pruning shears. Kind of unusual, you think? You're right! You see, I prefer to work the inside bend of a gold bearing stream. When the flood water rushes during the rainy season, the main water force strikes the outside bend bank and rebounds along the bottom of the streambed. Any waterborn gold is bumping along the bottom and is struck by the rebound and pushed toward the inside bend. Usually it will catch there, but the rebound force also erodes the inside bank. This means that any vegetation is at the water's edge. It seems my luck always makes that spot full of old and new blackberry bushes. I still carry two scars on my left forearm from them. That taught me the value of pruning shears. The first task I undertake at my selected bend is to clear away the "jaggers". This is a term I learned as a kid to mean anything that can badly scratch you.

I try to pick my bend shortly after I arrive at my camp. I like to make that about mid-day. Of course, both the pathway to the bend and the bend are full of jaggers. By the time I clear the trail and panning spot, three or four hours have passed and it's time for dinner. I count it time well spent, for it is not unusual for me to spend five or six days in the same area. The next morning I'm at work bright and early and unscratched. If the work area is heavily brushed, I carry lopping shears in my camper for larger brush clearing work. One thing I know when I start panning.... no one has been there before me that year. I get first shot at any newly deposited gold.

The final item in that top pocket is a small bottle of halazone water purifying tablets. I'll detail the reason a little later because it's important from a health standpoint.

When I want to go goldpanning, the first thing I grab is my loose-fitting nylon vest with four front pockets, one zippered, two with velcro closures, and one open with two pencil slots. I find the pencil slots convenient to hold my tweezers. Next I strap on my "swat belt" with four movable pockets also with velcro closures. The belt holds my chisel point rock hammer, a nine inch bowie knife, a seventeen inch machete and a plastic water canteen which I always make sure is full.

About the water canteen: This is a critical item, for no matter how clear or pristine the stream water seems, there is the ever-present danger of "giardia", a parasite in the intestines of vertebrates.

Giardiasis is a human intestinal disorder with abdominal discomfort, prolonged intermittent diarrhea caused by drinking untreated water contaminated by feces of infected animals. It is also known as "Beaver Fever". You know darn well that the up-stream deer, bear or beaver left their calling card in the water. Don't drink the water unless you treat it with halazone first.

Now back to the rucksack. In the main compartment I carry a twenty inch screwdriver. I heated the end of it to form a ninety degree, one and one half inch end-hook. I use this quite often to get into narrow, deep crevices to rake out the goldbearing sands. You'll also find an eighteen inch pry bar along with two sizes of rock chisel. Both are used to open up seams and cracks.

A homemade digging tool that I get the most use from is a small shovel made from three inch auto exhaust tail pipe. I cut one end at a forty-five degree angle for penetration and made it six inches long. After slitting the tube, I welded a seven inch Tee handle to the other end to complete the digger. It has served me well in small places. This tool also provides me a bonus in that I can use it as an auger sampler to get a five or six inch deep placer core sample.

After I pack my lunch, mount the rucksack, I step out with my knee boots. I'll be gone till dinnertime.

A word about "rock" chisels. These carry a softer temper than those used to work steel. That temper gives them toughness, whereas a steel chisel is tempered for hardness, which gives it a brittle nature. When working rock, a rock chisel point will stay sharp much longer. A steel-working chisel tends to chip off and damage the edge when working granite. I almost forgot to mention the pair of plastic safety glasses I use to protect against flying rock chips that can knock an eye out.

You should bear in mind that I am describing my gold prospecting kit. After I have found a spot worth working, I use somewhat larger hand equipment that I designed. Let me describe it for you:

You know those round, 15 quart, galvanized, auto oil-change pans that nest nicely into each other. I bought three for an idea I had to increase the amount of placer ground I could work per trip.

In pan #1 I drilled a series of three-eighth's holes in a circular configuration to match the ridges formed into the bottom. About half of the pan bottom is holed. This is my #1 sorter, classifier or material separator.

In pan #2 I cut out the entire bottom leaving about a one inch protruding lip. Inside this lip I pop-riveted a snug fitting, pre-drilled, stainless steel, rigid mesh, sheet screen, with three-thirty-second openings, that had an earlier life as the finish face of a ventilation screen.

Pan #3 is used just as it came from the store. I pop-riveted two handles near the top for easier handling. This, in effect, becomes my goldpan substitute at this point.

Here is how I use these three pans: I nest #1 into #2, then #1 & #2 into #3. Into #1 I enter my placer ground as found. I place a single two inch cobble between #1 and #2 to create a space for the material that enters from #1. I then submerge all three pans and wash the placer ground into #2 until it is fully sized. I remove pan #1. This leaves gangue sizes of over three-eighth's in pan #1 for examination and dumping.

Then I transfer the cobble and with the pan #2 under water, with the material sized to three-eighths or less, I use my hand to swirl and classify those materials. This is the pan I scan closely for larger nuggets after I remove it from the nest. All the material in pan #3 should be in sizes three-thirty-second or less except for the spacing cobble.

With the placer material I wish to work in pan #3, I submerge and swirl the pan contents to let specific gravity work. Any gold and black sand should be at or near the bottom. I have found that excessive start-off material in pan #1, can crowd the final product in #3 to the point where specific gravity will not operate. It is then that you can find unsorted gold throughout the washed placer. However, this is not critical. The whole idea was to ready the ground for the goldpan. If you try to work the #3 pan like a gold pan, the three inch lip causes a certain degree of difficulty when sluffing-off the gangue. I do not recommend it, but it can be done.

Oftentimes, when using pan #3, particularly when I'm nearing the end of my camp-stay, I fill the pan about half full of water and apply a 75 pound pull alnico magnet to the underneath center of the pan. This magnetizes about a five inch section of the pan. With a side to side washing movement, the placer materials move back and forth over the magnetized field to collect in a mass, that I then move to the pan-edge by moving the magnet to that location. I believe that I get most of the adhering fines with the black sands to finish-process at home. Remember, that gold is not magnetic and neither are the fines. If you adopt this method, be sure to work out pan #3 materials in your goldpan. Theoretically, all the gold should still be in the pan. If you dump before working it out, you've just left it for the next Goldpecker to find.

This three-pan method washes larger quantities of placer. It is about as hard as I want to work. You see, I do this for pleasure, not to accumulate or be a gold seller. But your view may differ from mine. But then I'm a Goldpecker !

A word about firearms: In my view they are merely extra weight. The few that carry a sidearm say they want protection from snakes. I say, BULL! They are lucky to hit a barn door at twenty feet. However, snake protection is a valid concern. I solved this problem with a noise-maker I call a "snake-stick". I used an Elm wood rake handle, cut to size with the bottom three inches squared-off to accept a one inch square tube, four inches long.

To the bottom of this .125" sidewall tube, I welded a three inch lag bolt with a half inch diameter stem in which I had drilled a quarter inch hole to accept a cotter key. On the lag bolt stem I placed first a loose-fitting washer, then a twelve inch length of chain with a rounded link at the end, then another washer, loosely secured by the cotter pin. I designed it so that the chain easily rotates around the stem with the washers and chain acting as rattles. If I should meet a snake in the rocks, the chain can coil around an object to deliver a telling blow.

As I use the snake-stick for walking and balance, I know the chain and washer noise alerts any snake in my path. That snake wants about as much trouble as I do. I give it a warning chance to slither safely away. I'm relieved from a sudden unpleasant encounter. In rattlesnake country, I have yet to meet a snake. I prefer it so!

Over my Goldpecking years, I have accumulated some additional gear that gets occasional use. One such piece of equipment I call a rock impact hammer. One lucky day, in a surplus shop, I located a thirty-six inch jackhammer point with a one inch diameter. I had a reverse slide hammer in mind when I designed a pipe tube to house the point. I weld-sealed one end of the tube, then loaded about five inches of lead to form the pipe head slide head. In operation, the point is held against the rock, and the pipe head raised and slammed atop the chisel. What I like about it is that the lead head muffles the sound that can carry noise to bring the curious to investigate.

Another rock attacker I sometimes use (but not often) is an old-time stonemasons trick. The gear is simple and consists of a hand-held star drill you strike with a four pound hammer. By turning the drill after each impact, you run a line of holes about four inches deep along a direction you want the rock to crack. (It takes forever. I usually try to get away with a maximum of three holes.) After you have blown the debris from the holes, you insert two flat metal sliders (called feathers) into each hole. You then insert a chisel opening point between each of the feathers. In a three hole array, you strike each chisel in turn with alternating blows. The rock should crack along the line of tension made by the deeper penetrating chisels. This method is noisy and labor intensive. It's wise to make sure the effort is worth the toil at pay-off time.

Many Goldpeckers carry those white, five gallon, laundry powder, buckets to handle placer material. I favor the black four plus gallon, plastic, paint buckets, only because they easily fit under my camper built-in seats. I use two buckets. One to hold the newly dug placer ground, the other to pre-wash the ground before it gets into my goldpan.

When I'm in an area where serious sniping is possible, I have a series of tools that I have found to be invaluable. Again, I made or adapted these to suit my purpose.

I bought a set of four, 24 inch screwdrivers. You already know about the one with the hook. I cut off the tips of each of the others and silver soldered stainless steel spoons to the end. One, a round spoon end, became a parasol end, so I could scrape with it while blindly fumbling. Another has a plain teaspoon end; the third, an elongated finger-type spoon tip. Along with my hook end, I can clean out cracks that are wider than one inch. Of course, the clean-out material is then normally panned... but much more profitably.

In conjunction with the long spoons, I use a large set of hook tools and three sizes of pry bars. Oftentimes, gravel is wedged in a crack to interfere with, or stop, clean-out. Various sized pry bars make short work of the obstruction.

After I collected these crevice tools, I had a problem transporting them to where I wanted to use them. I came up with the bright idea of fashioning a golf bag style carrier. I took a thin-wall four inch plastic pipe, and with the proper solvent, welded an end cap to form the bag bottom. I mounted another end cap by lanyard to act as the removable, but not losable top cap. After determining the proper balance point, I mounted the carrier straps from an old golf bag. It works fine. Try it!

Sure, I'm a Goldpecker, but I'm a mineral prospector too! In my travels I search out outcrops for interesting mineralization. I pick up float that hefts "heavy", and particularly rocks of unusual color for later inspection. I always make a location note of where I found the specimen. If you are going to rely on your memory as to where the rock came from, you might as well drop it. That's the stuff of lost mine stories.

About getting lost and using your compass: Areas of high mineralization will play havoc with your compass needle. There are iron hills in northern Michigan which make a compass totally unreliable. In fact you may think you are heading north, when in fact you're walking south. Fortunately, these situations are few and far between. But it does bring up a piece of equipment that I would like to have. I'm referring to a GPS system.

Imagine you've tracked into unfamiliar territory. A high altitude snowstorm strikes. Darkness is close behind. How do you find your way back to camp safely?

Through the magic of modern technology, a sophisticated, handheld, global positioning device tells you precisely where you are, where you're going, and where you've been. The ads say that wherever in the world you are, day or night, in good weather or bad, you'll never be lost or disoriented again. Sounds great doesn't it?

This locating feat is accomplished by 24 satellites, orbiting 12,000 miles above the earth twice a day. Each satellite constantly transmits its position in space and the precise time using atomic clocks. These satellites form the basis for a Global Positioning System. (GPS)

The device in your hand receives information from the satellites to determine your exact position, speed and direction. It lets you return to your campsite quickly, with no guesswork. Think of it as an electronic fire on a tree to guide you back. You can use your topographic map to program 15 legs of any route, then it automatically tracks your course, tells you what direction you're heading, how fast you're going, how much farther you need to go.... All at the touch of a button.

This unit, which is not much larger than a hand-held telephone, sells for about $700.00. I would love to possess this unit.

34 STATES WITH PLACER GOLD

NO MATTER WHERE YOU LIVE IN THE CONTINENTAL UNITED STATES, YOU ARE NEVER MORE THAN A FEW HUNDRED MILES FROM A GOLDPECKING SITE. PICK A CONVENIENT PLACER AREA THAT APPEALS TO YOU AND GO!

IF YOU ARE NEAR OR HEADING WEST, REVIEW CHAPTERS 45 AND 51.

ALABAMA - At least seven counties with lode and placer gold.

GOLDPECKER TIP # 173 - TALLADEGA NATIONAL FOREST - Upper reaches of Chulafinnee Creek and southwest lying water courses. Always seek permission to prospect. Some areas within the Forest are privately owned.

ALASKA - Gold available in very extensive areas that are best defined as regions.

GOLDPECKER TIP # 174 - American Creek near EAGLE VILLAGE has a campground to the south. THIS IS WILD BEAR COUNTRY! USGS TOPO MAP: EAGLE, ALASKA.

ARIZONA - At least 14 counties with lode and placer gold.

GOLDPECKER TIP # 175 - ELEPHANT BACK MOUNTAIN AREA south of Quartzite on HWY 95. Rough four wheel drive country. Try Colorado Wash area for metal detecting and drywashing. This is BLM land. Remember, Arizona requires a prospecting permit for State Lands search.

CALIFORNIA - At least 37 counties with lode and placer gold.

GOLDPECKER TIP # 176 - STANISLAUS NATIONAL FOREST, USGS TOPO: COLUMIA SE., TUOLUMNE COUNTY. Try north of American Camp Station. Respect existing claims. This is a good sniping area.

COLORADO - At least 40 counties with lode and placer gold.

GOLDPECKER TIP # 177 - BLM LAND, MESA COUNTY, USGS TOPO: Juniata Arch. Try Delores River on HWY. 141, Blue Creek area. ELEVATION IS 5,000'. Take warm clothing.

GEORGIA - At least 11 counties with lode and placer gold.

GOLDPECKER TIP # 178 - Try LUMPKIN COUNTY, DAHLONEGA HWY 52 EAST TO ROCK HOUSE ROAD, then right on HIDDEN VALLEY ROAD. This private campground in the North Georgia hills rests in an area of gold-bearing streams. Fee-paying campers may pan free.

IDAHO - At least 21 counties with lode and placer gold.

GOLDPECKER TIP # 179 - NEZ PERCE NATIONAL FOREST, USGS TOPO: GOLDEN. Check for campground nearby. HWY 14 near Elk City, South Fork of Clearwater River, north of Fall Creek. This area has operating placer mines. Be a good Goldpecker!

INDIANA - At least 3 counties with placer gold.

GOLDPECKER TIP # 180 - South of Indianopolis in Morgan and Brown Counties try any streams in the area. This is placer gold deposited by the Ice Age glaciers. Assume you are on private lands. Ask for prospecting permission.

IOWA - These erratic placer deposits are glacier related.

GOLDPECKER TIP # 181 - Try Northeastern county streams.

MAINE – At least 8 counties with placer gold.

GOLDPECKER TIP # 182 – Try Oxford County, Bryon Township, Swift River and its tributaries near Byron.

MARYLAND – At least 4 counties with lode and placer gold.

GOLDPECKER TIP # 183 – Try Howard County, Simpsonville area streams. Potomac river upstream from Georgetown for miles. Generally prior permission is required for any panning activity.

MICHIGAN – At least 19 counties with lode and placer gold.

GOLDPECKER TIP # 184 – Try St. Joseph County, Burr Oak area. Northwest of Sturgis streams. US 41 streams around Ishpeming area. These are mostly glacial gold deposits.

MINNESOTA – At least 2 counties with lode and placer gold.

GOLDPECKER TIP # 185 – Try Scott County, Jordan East of Bloomington streams. Also try the Zumbro River Mazeppa area.

MISSOURI – At least 3 counties with placer gold.

GOLDPECKER TIP # 186 – Try Macon County, Elmer, South of Kirksville. Sand Creek and tributaries. Also the Chariton River and tributaries. This is Canadian glacial drift gold.

MONTANA – At least 24 counties with lode and placer gold.

GOLDPECKER TIP # 187 – Try Missoula County, Greenough, East of Missoula. Elk and Union Creeks. Regional stream have been productive in the past. Jefferson County, USGS TOPO: Delmoe Lake. BLM land. HWY 90 off Lake Delmoe Rd. West and South streams. This is rugged country lying at the 5,000 foot level.

NEVADA – At least 17 counties with lode and placer gold.

GOLDPECKER TIP # 188 – Try Nye County, Beatty area, Northwest of Las Vegas. Amargosa River. Clark County USGS TOPO:

Shenandoah Peak. BLM land. HWY 161 to Goodsprings try the Wilson Pass area streams. This is at 5,000 foot elevation.

NEW HAMPSHIRE - At least 2 counties with lode and placer gold.

GOLDPECKER TIP # 189 - Try Coos County, Milan area streams, North of Belin. Grafton County, try the Baker River near Marren. Written permission is required from landowners before panning on private property. NOTE THE WORDS "WRITTEN PERMISSION".

NEW MEXICO - At least 15 counties with lode and placer gold.

GOLDPECKER TIP # 190 - CARSON NATIONAL FOREST, Rio Arriba County USGS TOPO: East Galivan Canyon. US HWY 64 to Hopewell Lake Campground south to Placer Creek. This is rugged 4x4 country at the 9500 foot elevation. Water is seasonal. Also try Grant County, Northeast of Silver City, Bear Creek and tributaries.

NEW YORK - Minor amounts of gold have been reported from the Catskill and Adirondack Mountain stream drainage areas. This is most likely glacier gold deposited during the Ice Ages. The State Parks associated with these mountains require prior District Ranger permission. (Just let him laugh, what does he know?) Remember, private property requires prior permission too!

NORTH CAROLINA - At least 40 counties with lode and placer gold.

GOLDPECKER TIP # 191 - NANTAHALA NATIONAL FOREST, Cherokee County. US HWY 19 south of Rhodo try Harris Creek. Some of the surrounding land is private and requires permission. Check with District Ranger to be sure you are on Forest lands. Also try Union County, Waxhaw area, South of Charlotte all area streams.

OREGON - At least 19 counties with lode and placer gold.

GOLDPECKER TIP # 192 - SISKIYOU NATIONAL FOREST, Josephine County, USGS TOPO: Chrome Ridge. Grants Pass to Briggs Valley, to Big Pine Campground area. Try Briggs Creek and tributaries.

Also try Jackson County, Ashland area, Bear, Ashland, Wagner Creeks.

PENNSYLVANIA – 1 county. Gold is byproduct of iron ore.

GOLDPECKER TIP # 193 – Try Lebanon County, Cornwall area South of Lebanon streams. This is glacier deposited gold.

SOUTH CAROLINA – At least 18 counties with lode and placer gold.

GOLDPECKER TIP # 194 – SUMPTER NATIONAL FOREST, Oconee County. Try area streams outside Oconee State Park. Also try Lancaster County streams around Tradesville.

SOUTH DAKOTA – At least 4 counties with lode and placer gold.

GOLDPECKER TIP # 195 – BLACK HILLS NATIONAL FOREST, Pennington County. USGS TOPO: Rochford. HWY 385, Rochford area. Rapid Creek and tributaries. Also try Custer County, Fairburn area South of Rapid City streams.

TENNESSEE – At least 3 counties with lode and placer gold.

GOLDPECKER TIP # 196 – CHEROKEE NATIONAL FOREST, Try Coker Creek and tributaries. Campground at Coker Creek Village. Also try Blount County, South of Marysville all streams. GET PRIVATE PROPERTY ENTRY PERMISSION.

TEXAS – At least 13 counties with lode and placer gold.

GOLDPECKER TIP # 197 – HWY 71 OR 16 north of Llano. Try Pecan Creek and tributaries. Also try Culberson County, Van Horn area all streams. GET PRIVATE PROPERTY ENTRY PERMISSION.

UTAH – At least 15 counties with lode and placer gold.

GOLDPECKER TIP # 198 – FISH LAKE NATIONAL FOREST, Beaver County. USGS TOPO: Beaver. US 91 north of Manderfield, try Wildcat and Indian Creeks and their tributaries. Also try Wasatch County, Heber City area, Snake Creek.

VIRGINIA - At least 18 counties with lode and placer gold.

GOLDPECKER TIP # 200 - BUCKINHAM/APPOMATOX STATE FOREST, Buckingham County, try Slate River and tributaries near Argonia. Also try Orange County, Fredericksburg area all streams.

WASHINGTON - At least 27 counties with lode and placer gold.

GOLDPECKER TIP # 201 - CITY OF EVERETT LAND, Snohomich County, USGS TOPO: Monroe, north of Sultan, try Sultan river and tributaries. Also try Douglas County, Stevenson's Ferry areas streams.

WEST VIRGINIA - At least 1 county with lode and placer gold.

GOLDPECKER TIP # 202 - Try Tucker County, Parsons area streams.

WISCONSIN - At least 1 county with placer gold.

GOLDPECKER TIP # 203 - Try Pierce County, Rock Elm Township, Plum Creek area streams.

WYOMING - At least 17 counties with lode and placer gold.

GOLDPECKER TIP # 204 - MEDICINE BOW NATIONAL FOREST, Albany County, USGS TOPO: Keystone/Albany, HWY 230 west of Laramie to Keystone, try Douglas and Lewis Creeks.

GOLDPECKER TIP # 205 - SOME OF THE LISTED PLACER DEPOSITS MAY BE ALREADY MINERAL-CLAIMED OR REST ON FEE LANDS THAT REQUIRE THE OWNER'S PERMISSION TO GOLDPAN. DON'T TRESPASS ! IF PERMISSION IS GRANTED, RESPECT THE OWNER'S PRIVACY AND PROPERTY.

DON'T MESS IT UP FOR THE NEXT GOLDPECKER!

ALWAYS FILL YOUR PROSPECT HOLES

CHAPTER 53

GOLDPANNING SCHOOL

The classic gold country anecdote has the motor home pulling up to a gold-bearing stream with an eager gold panner leaping out to try his first-time hand at the recreational art. He slips on a pair of knee-high boots to walk into the stream to dip his pan into the nearest loose gravel. Standing erect he swishes the pan contents, ever watching the pan action. After a minute or two, he discards the contents into the stream with the comment, "There's no gold in this stream". Another recreational gold panning know-it-all has been lost to a beautiful loner sport through self-induced ignorance.

If he makes any other comment it will be that, "All the gold is gone, why bother?" If you notice his words, he blames the stream's lack of gold, or if gold was ever there, it no longer is. Yet, that same stream is furnishing recreational gold to the many who have bothered to learn more than pan swishing. The fact of the matter boils down to an annual gold replenishment as winter rains and spring floods transport and redeposit new stream materials..... in natural gold catchment basins. Knowing where to seek alluvial gold is a first step toward fun and success.

If you are one, or know someone, who was disappointed and quit after your first goldpanning effort, then you need to meet Ralph Shock. In point of fact, there are many Ralph Shocks under different names, who will take you to their claim, and for a fee teach you panning fundamentals. (You keep all the gold you find.)

Let's go visit the real "Ralph Shock", the man who operates from Jamestown, California.

California's Golden Highway, SR 49 crosses Woods Creek just before reaching Jamestown. For its size, Woods Creek was the richest stream in all of California. While the discovery date of 1848 is questionable, this creek produced a 75 pound mass of gold and quartz, which was mostly gold. Nothing but a memory today, the town of Woods Crossing was born and died here.

As SR 49 enters Jamestown, a side road branches off to become the main street of the downtown area; the sign says elevation 1405 feet, population 1200. On the left is a rusted, metal roofed, stable filled with gold mining memorabilia. No, it's not a museum, it's Ralph Shock's place of business. The building sign says: "GOLD PROSPECTING EXPEDITIONS". (It may now read: "GOLD PROSPECTORS INTERNATIONAL") We enter the old livery stable built in 1852. The address is 18172 Main Street. The phone number is: 209/984-GOLD. As you enter, the inside is nothing like a livery stable, as your eyes greet two life-size, mute, wooden Indians.

The gray-bearded, smiling man behind the counter is the one we seek. His balding-gray is always covered by a western hat; his striking feature (to a gold panner) is the massive nugget dangling from a neck chain. If asked, he will tell you he found the nugget at Parrot's Ferry on the Stanislaus River while working a twelve inch deep crevice. I think that thrill still stays with him, because one of his favorite gold-finding techniques is sniping or crevicing. Here you learn both classroom and field methods that work to make recreational gold panning pleasurable and exciting.

Obviously, Ralph can't do it all alone, so he calls on a staff of instructors to help out as demand requires. After fifteen years, he has the routine smoothly in place. The most popular offering is the special trip leaving at ten in the morning and returning at three in the afternoon. Guests take their own car and lunch, then caravan to the designated claim for the day's adventure. Two and three day trips are also available.

Is the gold all gone? October 12, 1984 saw a construction crew trenching Main Street for a new sewer line. A local prospector picked up five gold nuggets from the excavated material. When the word of his find spread, all the gold pans in town were sold out. The gold-bearing dirt was being carried away faster than the work crew could fill the trench.

Is the gold all gone? Commemorating the date January 4, 1985 is a constantly running video that captured an unusual event. A couple entered Shock's Jamestown shop carrying over eleven pounds of gold nuggets in a shopping bag. Eight of the nuggets weighed from eight to twentyfour ounces each. The total gold value was $140,000. Neither the identity of the couple, nor where they found the gold was revealed. It is known that it is local gold. Every gold nugget has a fingerprint of combined elements, with the percentage of each contained element identifying the gold district.

TOUCHING ALL THE BASES

Bring old clothes and a pair of tennis shoes, all other equipment is supplied. You can be taught the basic techniques:

1. How to find the "pay streak" in a stream.
2. Gold panning.
3. Sniping (Crevice searching).
4. Using a sluice box.
5. Dredging.
6. Pocket hunting.
7. Electronic prospecting (metal detecting).
8. High-banking.

If you know your way around gold prospecting, and want to spend your vacation working a claim, you can rent a gold mining claim for a week or month and keep all the gold you find. Want to dredge? Seasonal operating permits are available here.

If you feel adventurous, you can take a dredging trip, a white water raft panning fling, helicopter flights to remote prospecting areas, or four wheel drive expeditions. Sorry, but the mule trips have been discontinued. (For safety's sake, their actions were too unpredictable... And they can bite!)

CHAPTER 54

GOLD PROSPECTING CLUBS

In States such as Arizona, where there are only three claim withdrawal set-aside areas for recreational gold panning, the onus of obtaining a prospecting permit before panning in the 247 known placer gold areas, effectively discourages fun panning. To overcome the need for an individual permit, prospecting clubs have been formed to welcome you.

Joining a club provides excellent educational and social benefits. They are a reasonable cost way to learn about gold panning and places to pan. The worry about claim jumping and trespassing is eliminated, for the clubs have their own claims for member use. The outstanding benefit, to my mind, is that you will pan a promising prospect. Nothing gets the juices flowing more than the glint of gold in your pan.

INTERESTED IN ARIZONA PROSPECTING CLUBS?

THESE LISTED CLUBS HAVE OPEN ENROLLMENT

Goldpanning, Dredging. Metal Detecting
Mesa Gold Diggers
464 E. 7th Place
Mesa, AZ. 85203

Placer Gold Prospecting
Arizona Prospector's Association, Inc.
106 S. 43rd Place
Phoenix, AZ. 85044

Recreational Prospecting - Club Owns 200 Claims
Phoenix Roadrunner Prospector's Club
P.O. Box 56804
Phoenix, AZ. 85079

Recreational Prospecting
Desert Gold Diggers
P.O. Box 30401
Tucson, AZ. 85751

THERE ARE MANY PROSPECTING CLUBS AVAILABLE TO YOU. SOME REQUIRE INVESTMENT AS IN A PARTNERSHIP SHARE; SOME ARE RESTRICTED TO RESIDENTS OF ARIZONA, OR TO EMPLOYEES OF A PARTICULAR WORK GROUP.

For a more complete list of Arizona Prospecting Clubs and Rock-hound Clubs, CONTACT the Arizona Department of Mines.
Ask for EARTH SCIENCE CLUBS, 1993, Circular No 45, May, 1993.

THERE ARE PROSPECTING CLUBS IN ALL THE STATES. SOMETIMES THEY ARE KNOWN AS:

GEM & MINERAL CLUBS.

NATIONAL GOLD PROSPECTING CLUBS

WHERE YOU DO NOT WANT TO RESTRICT YOUR PANNING TO ANY SPECIFIC STATE OR LOCATION, YOU SHOULD CONSIDER THE G.P.A.A. AS PROVIDING A GOOD NATIONAL FOCUS FOR RECREATIONAL GOLD PANNING, DREDGING OR SLUICING.

GOLD PROSPECTOR'S ASSOCIATION OF AMERICA
P.O. Box 891509
Temecula, CA. 92590
909/699-4749

For the current $25.00 annual membership fee you receive many benefits... if you elect to take advantage of the offerings.

1. GOLD PROSPECTOR MAGAZINE
   Published every other month.
2. PICK & SHOVEL GAZETTE
   Published every third month.
3. Access to GOLD CLAIMS in 37 States & Canada

G.P.A.A. publishes an annual GOLD PROSPECTOR'S MINING GUIDE which lists all the claims available to members. The 1993 issue is one inch thick with directions and maps. This is the most valuable benefit of membership, apart for group outings and normal social events. The Mining Guide sells for $ 74.50. New or withdrawn listings are detailed in the Pick & Shovel Gazette to keep your guide current.

From time to time the G.P.A.A. offers a membership special for $ 67.50. Currently you get a one year membership, the Mining Guide, a gold pan, sucker bottle, club patch and decal, magazine and gazette subscription, ball cap, plus a gold panning video. This membership deal is called the "Buzzard Special". It may still be available by contacting the club's headquarters.

This club is a prospector's club with a special, exclusive emphasis on gold prospecting. It is an easy way to get in on some serious gold panning whether you are novice or seasoned player. Wherever you live in the continental United States, there is probably an available claim within easy driving distance.

LEARN

ENJOY

JOIN A CLUB !

CHAPTER 55

COMMERCIAL PLACER FLOW CHART

GOLDPANNING PROCESS COMPARISON

Let's pretend we own a sand and gravel plant. Our mineral analysis tells us that our materials carry modest gold values. We decide to set up our recovery processing system to include placer gold as a by-product. We know that the recovered gold values will not pay our overhead, but will exceed any extra costs we incur plus a small profit.

Our sand and gravel market is the construction industry. Since we must wash our product before it is saleable, it makes good sense to design a sluice-type arrangement to trap the gold particles.

LET'S FOLLOW THE MATERIALS THROUGH OUR RECOVERY SYSTEM, THEN COMPARE IT TO THE GOLD PANNING PROCESS.

First, the material is dug and transported to a screening, sizing device. This could be a drum-shaped, rotary, sorting, sizing screen known as a "trommel", or a series of various spaced bars called a "grizzly". As our placer ground travels through the trommel or grizzly it is washed by a water spray, then directed to the sized collection areas.

COMPARE: When panning we dig, classify by screening or sorting out the larger pieces in a washing process. We search through the classified pieces to make sure no nuggets larger than our screening are discarded.

THE CONCENTRATION PROCESS

Our commercial flow takes the pieces that fell through the classifier, and runs them over a sluice-riffle-carpet system. This catches the larger gold values on the top of the carpet at the riffle; the flakes and gold dust known as "fines" settle into the carpet fibers. At the same time the continuous water flow over the sluice recovery system moves the gravel and most of the sand, including black sands, down into the sand collection conveyer and onto the storage piles.

COMPARE: With the goldpan we need to hand pick or wash away the larger pieces of gravel, sediment and other foreign materials, until only the black sands and heavier gold remain in the pan.

After a long run of sand and gravel, our commercial plant will shut down to take up the carpet and riffles. We hang the carpet to air dry, after which we will shake the DRY carpet free of all captured particles. This will usually be black sand and free gold. Our shaking unit is a large rotary bin much like a household clothes dryer. The particles are collected in a gravity trap for further processing.

COMPARE: With the goldpan, you can now separate your concentrates into gold values by time consuming water separation, or you can save the concentrates for later investigation. MOST PANNERS PICK OUT THE LARGER VISIBLE GOLD, THEN SAVE THE BLACK SANDS FOR LATER REWORK.

Our plant employs a vibrating concentration table. This is a water-fed slanted device that moves the lighter black sands ahead of the gold in a band of black sands first, then a gold/black sand mix in the second wave, and in the last wave our gold-only. The second wave containing the gold/black sand mix is removed from the concentration table and saved. It is called the "cons", short for "concentrates".

The cons must be dried before we can process it in our magnetic separator. Those black sands are mostly magnetite and hematite. Though both are iron, magnetite is highly magnetic; hematite is weakly magnetic. Some of the other black products may be weakly magnetic if they contain iron, but our system is set up to disregard them. We also know that some of the black sands may be gold with a surface coloring of black or brown. Gold, being non-magnetic, will not react to magnetism. We can now separate out the gold and other non-magnetics particles.

COMPARE: When we return to home base, we take a six foot long, 5 inch plastic, corrugated sewer pipe; that we have sliced in half diameter and nailed to a plank. The plank is placed on a rise of four inches per foot; a hose attachment at the top end gurgles a slight water flow down the incline; preferably a catch basin with a return water system completes the cycle. With the water flowing, deposit one TEASPOON OF CONS at the top so that the gold washes into the corrugated traps. SAVE THE SURVIVING CONS FOR IMPORTANT PROCESSING OF INVISIBLE MICRON GOLD.

Our magnetic separator consists of a conveyor belt with three A.C. magnets mounted underneath. Overhanging the belt we have two D.C. magnets. As the cons move along the beltway, the D.C. magnets charge the slightly magnetic hematite black sands; the A.C. magnets cause the now-magnetic sands to be charged with opposite polarities. Eddy currents cause the iron particles to spin in a hopping, slight airborne progression to move to the sides of the moving belt where at belt-end they drop into a sand catcher. This goes to the tailing dump.

Whatever black sand still in the center of the belt is attracted to and picked up by the second D.C. magnet. The material remaining in the center of the belt are our cons of gold and small amounts of nonmagnetic material. We discard the tail, take the cons and walk it over to the gravity separator, where it will be worked a spoonful at a time. This is our primary cons.

The secondary cons picked up by the D.C. magnet is collected by cutting the current from the suspended magnet to make it drop onto the belt. It is now ready for amalgamation on its way to a "dore" bar.

Now back to the primary cons! This is the first or richest separator concentrate composed mainly of gold, but intermingled, with non-magnetic, non-gold particles. We must get rid of the non-gold impurities.

HOW TO DO THIS? The most efficient method is to reclassify the primary cons as to particle size known as "mesh size".

Mesh sizes are described by the number of openings per square inch. A size 10 mesh means 10 holes per square inch. Mesh sizes of 20 to 100 or more are used to obtain each grouping of particles. The reason for this step is a practical one. It's simply easier to deal with similar size particles when extracting the tiny gold.

Each classified mesh size cons is saved in its own container, and worked on a slanted square pan with a hand-held water jet. One teaspoon at a time is jetted. The gangue is washed down into a catch basin leaving the free gold on the side of the pan. This process continues until all the cons are worked. Since lighter gold is also washed down with the gangue, the contents of the catch basin are also ready for amalgamation with the secondary cons. Our gold dore bar or button will be a combination of the gold in cons #1 and #2.

Our gold/black sands cons are put in a mini-grinder, stamp mill device called an amalgamator. This centuries-old process uses mercury or quicksilver to alloy itself with the gold, silver, copper and zinc. The grinding, mixing of cons and mercury product is called amalgam. Iron and other impurities will not alloy with mercury.... BUT NEITHER WILL IT ALLOY WITH GOLD THAT HAS AN IRON OR MANGANESE SURFACE BUILD-UP. (This is known as black gold.) Therefore, any surface scale must be cleaned from the gold with an amalgamator that mauls, rotates and grinds off the impurities. It is at the point of "clean gold" that the mercury is added. Mixing continues for a thorough amalgamation of the gold. While not technically true, this gives a practical 100% recovery of gold values.

COMPARE: The old timers mixed the mercury and pan-cons in their frying pan. (Those copper gold pans you see are primarily made for amalgamation purposes.) Sometimes, if they had access to nitric acid they would first clean the cons before mercury addition. Why a frying pan? THEY WANTED TO DIE FROM THE MERCURY FUMES THAT ESCAPED INTO THE CABIN AS THEY HEATED OFF THE MERCURY OVER THE COOK STOVE. If the old timer played it safe, he used his shovel as a pan to vaporize the mercury over a camp fire. There is, however, another step before this frying pan/shovel vaporization was done. SUFFICE IT TO SAY MERCURY VAPORS ARE DEADLY TOXIC. THIS IS AN UNSAFE OVER-ALL PRACTICE AND SHOULD NOT BE USED. (But see later.)

If you should find mercury droplets in your cons, before you add your mercury, chances are the floured droplet is loaded with gold. Flouring means the quicksilver became separated into many small drops that in turn became stiff from gold pick-up. The mercury loses its fluidity. Someone upstream threw it out with the gangue.

After the amalgamation step, the material is placed in a gold pan and the lighter gangue washed off. When the mercury is visible in the bottom of the pan, it is poured off into a pyrex measuring cup. The material at this point still has a small amount of gangue. The gangue is removed by alternate pouring with clean water from cup to cup, until only mercury and gold amalgam remains.

To separate the mercury from the amalgam, a prime, thin skinned chamois is soaked in clean water, spread out over an empty pyrex cup, with the skin pushed down into the cup to form a catch basin. Pour the mercury material into the basin and gather up the chamois to form a bag of mercury. Twist the wet bag around the contents over a container and watch the mercury exude from the pores of the chamois skin. The gold amalgam particles are too big to pass through the pores. They are the remainder product left in the chamois that feels like a hard lump. Carefully, open the chamois and allow the gold amalgam to fall on a sheet of parchment paper. Using your fingers, the amalgam is balled and placed in a sealed glass jar. The reason you ball the amalgam is to fuse the gold together so as to form a sponge when retorted.

FOLLOW THESE CAUTIONS

1. MERCURY IS EXTREMELY TOXIC.

2. MERCURY VAPORIZES AT ROOM TEMPERATURE. BREATHING THE FUMES IS DEADLY.

3. USE A MINIMUM OF TOUCH. WASH YOUR HANDS FREQUENTLY.

4. NEVER HANDLE MERCURY WITH A SMALL CUT OR BREAK IN THE SKIN OF YOUR HANDS.

5. WHENEVER POSSIBLE, CONDUCT ALL MERCURY OPERATIONS IN THE OPEN.... NEVER IN A CLOSED OR SEALED ROOM!

Our ball of amalgam is now half gold and half mercury. How can we get rid of the mercury so that only the gold is left?

A process known as distilling is used to utilize the low boiling point of mercury. Mercury forms a gas at about 700° F. Gold melts at about 1900° F. By using a retort-still the quick-silver vapors in a gaseous state are driven from the amalgam. The mercury gas is cooled and recovered for reuse as liquid mercury.

CAUTION: THESE RETORTS ARE MADE OF STEEL OR CAST IRON. IF YOU USE A GLASS RETORT, BE PREPARED TO JOIN THE DEAD MINERS, WHO USED THE CABIN COOK STOVE AS A HEAT SOURCE, TO VAPORIZE THE MERCURY. DISTILLING AMALGAM BUTTONS IS A DANGEROUS TASK. TAKE NO RISKS!

COMPARE: IN THE CHAPTER 59 WE PRESENT, WITH PERMISSION, HOWARD F. STEINER'S, COPYRIGHT 1981 BOOKLET: "MAKING GOLD DUST INTO GOLD BEADS". SUBTITLED -- "HOW TO: AMALGAMATE "FINES" AND MAKE A "MINI" RETORT" - WITH THE TIP: "HOW TO MAKE A GOLD NUG-GET."

After the mercury distillation, the gold remaining in the retort is known as "sponge gold". It is then melted in a furnace at over 2000° F along with soda ash and borax to serve as a flux which covers the gold in the melting process.

COMPARE: THE STEINER PROCESS DETAILED IN THE CHAPTER 59 USES LITHARGE INSTEAD OF BORAX AND SODA ASH. LITHARGE IS LEAD OXIDE. RED LEAD OR LEAD FILINGS ALSO WORK WELL.

Next the molten gold is poured into a cast iron conical gold mold. After solidification, the flux on top is broken away. The gold cone is washed in hot soapy water to remove all trace of flux. The gold button is between 70% and 90% pure gold. The impurity is usually silver. When these buttons are cast into bars, they are known as "dore bars". Further refining is needed to produce .999 fine gold. (Pure gold is 1000 fine.)

Remember, we are dealing with a commercial operation where the by-product is gold. You may correctly think that the effort is excessive. It is!

Let me take you back a few years when ownership of gold in any form was illegal. To deal in gold you needed a license. However, in the operation just described, it was legal to have on hand up to 200 ounces of sponge button gold. Free gold or virgin gold could be hoarded in any quantity. GOLD IS NOW UNCONTROLLED.

A WORD ABOUT SELLING YOUR FREE GOLD:

If your gold is 70 fine, then it will bring only 70% of the world price. Nuggets in the one to three pennyweight size (DWT or PWT) have open offers of spot plus $50.00. HOWEVER, NUGGETS IN THIS SIZE RANGE ARE NORMALLY VALUED AT SPOT PLUS $175.00.

JEWELRY GRADE NUGGETS WILL BRING WHATEVER THE TRAFFIC WILL BEAR.

CHAPTER 56

FINDING THE PAYSTREAK

CONVERSION TABLE FOR MEASURING GOLD

WEIGHT	: Grains	: Pennyweight :	Troy Ounce :	Avoir. Ounce :	Grams
1 GRAIN =	1 =	0.0417 =	0.0021 =	0.0023 =	0.0648
1 PENNYWEIGHT =	24 =	1.0000 =	0.5000 =	0.0549 =	1.5552
1 TROY OUNCE =	480 =	20.0000 =	1.0000 =	1.0971 =	31.1040
1 AVOIR. OUNCE =	437.5 =	18.2292 =	0.9115 =	1.0000 =	28.3500

A WORD ABOUT SAFETY

TAKE PLENTY OF WATER: Do not count on drinking stream water, it may be unfit to drink..

IF YOU GET LOST: Make sure you let a third party know your plans. A search may be necessary!

TAKE AN AREA MAP: 7.5 min. topographic maps are good field maps. Learn to read them!

BE A NOISEMAKER: It alerts unfriendly wildlife to move on.

STAY OUT OF UNDERGROUND MINE WORKINGS: What looks sound to you, probably is not.

BEWARE OF DRY GULCHES: Be alert to area storm clouds.

Always have an escape route ready in case of flash floods.

RESPECT PRIVATE PROPERTY AND POSTED PROPERTY. DO NOT TRESPASS !

GOLD TRIVIA

A cubic foot of gold weighs over 1000 pounds. At $500. the ounce, it is worth over nine million dollars on world markets.

Placer gold purity runs from 70% to 90% PURE GOLD. Usually, it runs about 86% gold with silver making up difference. From the ratio of contained impurities, it is possible to generally area localize the source.

There is no exact definition of "nugget". If you can pick it up with your fingers, it's a nugget.

Gold is six times heavier than gravel, but UNDER WATER the gold ratio is closer to ten times heavier than gravel.

The density of gold will cause it to settle as deep as it can in the streambed. Bedrock or false bedrock is the end of the line.

The iron ores of magnetite and hematite, usually referred to as "black sands", behave in your pan like gold because of their high specific gravity. While not 100% true, it is fairly safe to say, "NO BLACK SANDS, NO GOLD". However, all black sand does not contain gold.

When panning for gold, the last washing residue in your pan will be black sands. It is in the black sand mix that you will find most of your gold.

While it is a lifetime thrill to see a nugget in your pan, retain all black sands for further processing. Properly processed, the black sands CONTAIN INVISIBLE, PROFITABLE, MICRON GOLD. The amalgamation – retort process will recover it!

Flour gold can float, particularly when in contact with oils. Before you use a new gold pan or finishing pan, be sure to use a little dish detergent and steel wool to clean away the oils left from the manufacturing process. If you favor the newer plastic pans, use the steel wool to abrade the inside of the pan. You do this to roughen the inside to thereby slow down the slip and zip of the panned materials. If you see water beading in your pan, you did not get all the oils.

PAYSTREAKS

Placer gold accumulations are called "paystreaks". If the gold has had time to be fully sorted, the richest portion of a placer rests on bedrock. True bedrock must be distinguished from false bedrock which is usually a hardpan layer of clay above the true streambed.

Favorable gold accumulation places are any natural pocket that can trap the yellow metal. Placer gold has a water-worn pebble appearance caused from abrading on the rough stream floor during transport. If you find sharp, angular, nuggets you are near its lodc source.

OBJECTIVE: WE ARE LOOKING FOR THOSE AREAS WHERE GOLD ACCUMULATED OVER THE LAST 100 YEARS, A DEPOSIT SEGREGATED BY HIGH WATER, FLOODS OR RUNOFF INTO A RICH ALLUVIAL FIND.

Concentrations of black sands, bits of iron, buckshot, old square nails are generally associated gold. THE IDEA IS TO FIND A PLACER THAT IS RICHER THAN OTHERS SO AS TO SPEND OUR WORK EFFORT MORE PRODUCTIVELY.

PREDICTION: If you are fondling your first gold pan adventure, all that follows will flit out of your mind. You will be mesmerized by an attitude that you know something unknown to anyone else; that Lady Luck will guide you to a big find. You'll dip your pan into some stream spot that strikes your fancy; swirl the pan contents; dump the pan and shout, "There's no gold left."

Hopefully, you will remember this prediction. Finding a pay streak is more than a matter of luck. It is a matter of gold lore, pure and simple. These tips were passed down from the generations who had to find gold to survive. Of course, there's a lot of geology involved too!

INSIDE BENDS: The first bit of geology involves knowing the mechanics of how a stream operates to move gold. Let's pretend we are standing in the middle of a stream that is running at full pace. We are facing upstream, the water is about nine feet deep at the center. Like Moses, we stop the stream so we can view a cross section of the water flow.

Let's imagine that in cross section view the bottom three feet is colored pink from bank, stream bed to bank. That pink section is the SLOWEST running part of the watercourse. Why? Well, believe it or not, it is being slowed down by friction with the banks and bedrock plus any obstructions in the streambed. The pink section is where our gold is being hopped, tumbled, pushed and abraded until it is trapped. Yes, the gold is in the slowest flow.

Now suppose that our straight running stream is approaching a bend in the streambed. If the bottom third is the slowest, the upper two thirds above the pink water is faster. As the water reaches the bend, the faster upper water tries to continue straight ahead against the OUTSIDE BEND. That FAST WATER hits the bend, rebounds to push the lower, slower, pink water with its gold across the stream bed and onto the INSIDE OF THE BEND. The fastest water displaced the lower slow moving water which crowded the inner bank in a twisting pattern. The rolling, sweeping motion slackened the current to drop what it could no longer carry. REMEMBER, GOLD BEING HEAVIEST IS RIDING THE LOWER WATER CURRENTS IN ITS PATH TOWARD DEPOSITION. THAT IS, THE INNER BEND.

WIDENING STREAM COURSE: As the stream fans out, the swiftness of the water diminishes and slackens. This reduced push-pull-tumble power causes some of the stream burden, including gold, to drop at the nearest settling site.

ROCK FORMATIONS AND BOULDERS: Island-like obstacles in the stream, such as a large boulder, when met by the waterflow, develop eddys BEHIND on the DOWNSTREAM side of the obstacle. The calmer water behind the obstruction causes eddys that encourage the dropping of heavier sediments. Again, this is due to a sucking, slackening water flow. THIS IS A FAVORED SPOT FOR DREDGING. WHEN THE WATER IS LOW OR GONE, THIS CAN MAKE YOUR PANNING DAY. BE SURE TO GET DOWN TO BEDROCK.

STREAMCOURSE PROJECTIONS: As the running stream strikes a barrier projecting from the bank, the water diverts to the open side to swirl to the opposite bank. Generally the bedrock on the open side is the same composition as the barrier. The heavier materials carried by the stream will not move with the swerving stream, but will slide and slip around the barrier, hugging it as close as possible. The main force of the stream has lessened its power over the gold; a slowing in the drag aims it for the barrier side of the bank, where it seeks the first series of natural riffles. Start your panning about ten feet below the stream barrier sampling toward the bank and toward the barrier. Any pay dirt will lie within that area.

SMALL WASH ENTERING MAIN STREAM: You can try this experiment yourself with two garden hoses. Turn one on to represent the main stream, then vary the flow of the other hose as you direct the flow into the main stream hose. Notice that the stronger stream will throw the water toward the stream with the softer flow. Remember this point. When you see a dry wash entering the stream, the pay dirt will be found at some point close to where the streams meet. If you conclude that the wash had the lighter flow, then the place to begin will be the first series of riffles below the point on the bank where the streams met.

It is always important to check your gold recoveries as to size. If you are finding larger pieces, work downstream for the lighter gold which took longer to fall. Conversely, if you are finding gold specks, work upstream for the coarser pieces. On the theory that the entering stream receded first, work up the wash bank. At full flow and ebb, gold was being deposited on the side of the wash bank as the water level lowered.

STREAMBED POCKETS: Logic tells us that the hole acts like a pocket to catch and hold the heavier gold. Many prospectors have cleaned the hole down to bedrock, only to zero-out for the effort. Sure, maybe someone was there ahead of you, but also consider this: The formation of the hole in bedrock may be accounted for by the presence of a softer rock material eroding from stream action. The hole caught rocks which swirled around like a drill enlarging the pocket. Those same rocks also acted as a grinder to wear away any gold present. Your probable recovery will be gold dust.

Now assume the force of water kept the hole clear of sediment because of an inclined entry point. Didn't the gold also wash further down stream? Look for bedrock that rises in the stream below the cavity. That will be the catch basin.

Think of the pocket when empty of sediment as a whirlpool location. Estimate or discern the exit point of the eddy, then begin your search at the point outlined in the preceding paragraph.

My own conclusion about working out stream holes boils down to wasted effort. But, who knows, the hole you clean to bedrock may be a really good pay strike. Remember, I'm only trying to lay down guidelines of search. Small variations in incline of bedrock, velocity or water, undercurrent forces that are now unknown because the watercourse is dry, will vary expected results.

The best test is to take nothing for granted; to spend your time in thoughtful prospecting. If you are working in known gold country streams, the placer is within reach. The only unknown is where... and are you the first to discover it?

FOLLOWING THE WASH: Recall that the wash entered the main stream with gold bearing gravel, thin, but still auriferous. Try this! Follow the wash up-grade. Take samples every 50 or 75 feet along the way. You are trying to determine whether the gold came down the watercourse from a higher elevation, or was washed in with the run-off. Pay particular attention for evidence of a fault in the terrain caused by a slip. It's the slip you want to find. Look for evidence of a line of springs. Could the slip have been the original water course path centuries ago?

The clue that you are really searching for is the presence of rounded gravel in the hillside. A strip of layered gravel may well be a portion of the original streambed. Any evidence of rounded gravel should be sampled. You may be on virgin ground with good gold concentrations.

SANDBARS: Direct your attention to where the larger rocks are deposited, then search the area below the rock. Clear the overburden testing along the depth for color. If you find color, move closer to the rock. If you find coarse gold, don't forget to work further downstream to get the smaller pieces that carried a little further. Remember, go to bedrock. That is the deepest possible down-point for the gold concentration.

Consider the bar again. No rocks are surface visible. It's an inside bend, perhaps with an obstacle extending into the waterway. How best to handle the prospecting? Visually check the bar; count-pace it off for length and width. Dig down to bedrock at midpoint, quarterpoint and two end points, sampling along the way.

Assuming you find color at the quarter section, trench the area from stream to bank. This start is a trade-off: easier digging for muddy water filling the trench. The real problem lies in the gold diving deeper as you work your shovel. You need a water run-off; start your tail ditch low enough to assure ditch drainage. Digging in the direction of your test hole down to bedrock, is designed to cut and locate the paystreak. Watch for a paystreak above the bedrock too. The overburden may have been laid down from some later, heavier storm. When you get to bed-rock, pay close attention to cleaning out cracks and crevices. If there's a nugget in your future, that's where you'll find it. If your selected bar is good, your trenches and cuts will all be drained by a single ditch. Once you find the pay stripe, follow your efforts in the line of deposited gold.

BENDS IN THE WATER COURSE: When water rounds a bend, the lighter material swings to the outside bend and the heavies to the inside bend. The heavies start around the inside bend with diminishing water power because of the spread-ing-out effect of stream widening. The heavy materials deposit at concentration points, traps and barriers which gave birth to the bar. The greater the down-slope of the stream, the more of the heavies will move to the inside curve. This is a pulling drag effect on the heavies, rather than a water-pushing force.

CHAPTER 57

HOW TO PAN GOLD

Goldpanning is like learning to parallel park your car between two parked cars. No one does it well the first few times without a system. Even with a system, the more you practice, the more proficient you become.

CHOOSING YOUR PAN: It is wise for the beginner to select one of the newer plastic pans with the catch-lips molded into one side. The purpose of the lips is to catch the black sands and nuggets as you ship water from the pan. When actually panning, the lips face the pour-off side.

FOUR BASIC PRINCIPLES

1. SEDIMENT REMOVAL: Silt in all its varieties operates to glue particles together. Water is the solvent that separates the particles through a process called washing. As you rotate the pan contents, slurry off the brownish water. Replenish it as often as necessary until the pan water is clear enough to see the bottom contents. There is no short-cut for this vital step to successful panning.

2. SPECIFIC GRAVITY: Water separation of materials with differing densities depends on those materials being continually submerged. THE CARDINAL RULE IS TO KEEP THE PAN CONTENTS UNDER WATER AT ALL TIMES.

3. STRATIFICATION: Water separation of the materials in the pan will result in layers of material. The gold and black sands, being the heavies, will settle to the bottom of the pan. THIS ONLY WORKS WITH SUFFICIENT AGITATION OF THE PAN CONTENTS UNDER WATER.

4. **CLASSIFICATION:** This is a sizing, sorting process. The lighter materials in the pan will sluff-wash-off to leave the heavies at the bottom. THIS ONLY WORKS WHEN THE PAN CONTENTS ARE UNDER WATER; SUFFICIENTLY AGITATED TO SETTLE THE HEAVIES; WITH THE PAN AT THE PROPER ANGLE, TO DIP-WASH-OFF THE LIGHTER TOP-MOST LAYERS.

EQUIPMENT NEEDED

1. **GOLD PAN:** Any kind of a pan will do, but it is preferable to use a plastic pan with built-in riffle lips.

2. **TWEEZERS:** Used to pick out the small gold pieces that are too small for picking with the fingers. (pickers)

SPEED-UP EQUIPMENT

3. **SNIFFER BOTTLE:** This is a plastic bottle with a plastic straw through the cork. It is a sucking device used to vacuum the fine gold from the pan. It is fast, and saves a lot of time. (Sometimes called a snuffer bottle.)

4. **CLASSIFIER SCREEN:** These are made in multiple mesh sizes to fit over your pan or a five gallon plastic pail. A good all around choice is the quarter inch mesh size. Throw your placer material into the pail; wash it until the water is clear. Place classifier over your pan and screen about a one-third pan. Check the classifier for nuggets before you dump it.

5. **FINISHING PAN:** This is a smooth-sided pan used to work the black sands for gold. It is usually a six inch pan, oftentimes of blue color. Its size makes it easy to handle.

FIVE BASIC PANNING PROCEDURES

1. Find a comfortable sitting site near the water. Prepare and remove enough material from the streambed so that the pan will

sit flat, submerged under water. With your classifier over the pan, both under water, place your gold-bearing gravels into your classifier to wash off all the silt. Repeat until water is see-through clear. If you only have a pan, use your hand to remove excess large pieces.

2. Lift the pan to a comfortable position, but with pan contents under water, rotate and swirl the materials in a circular pattern. It is best to maintain the same circle path to stratify the heavies.

3. With pan in stream water, tip the pan to sluff off the top-most materials. Repeat the circular stratification and sluffing until most of the lighter materials are washed out. Sluffing-off is stream water entering one side of the pan with the water and "lights" escaping with the outward flow. Think of it as a "U" flow. Continue sluffing until you see mostly black sands. Re-stratify as often as needed.

4. Inspect the pan contents from time to time CONTINUING with a swirling pan-tapping motion. The idea is to get the gold into the pan's leading edge. You should see mostly black sand mixed with gold. Try tapping the pan to keep the gold at the pan edge while moving the black sands toward you. This is done with a GENTLE HOPPING of the pan slightly tilted away from you with BARELY enough water in the pan to keep the contents in fluid condition.

5. Next, pick out the pickers. Swirl the contents in the pan in a fanning-out pattern display. You want the gold fines settling at the top edge near the bottom rim of the pan. Use your sniffer bottle to suck up the fines. It's O.K. if you get some black sands. Repeat as often as necessary. If you think you are finished, dump the remaining black sands into a saver can. THIS IS LOADED WITH MICRON SIZE GOLD AND CALLED THE "CONS".

NOTE

ALL THE PROCEDURES INVOLVE THE CONTENTS REMAINING UNDER WATER SO THAT SPECIFIC GRAVITY CAN OPERATE. THE SNIFFER BOTTLE SHOULD BE ABOUT ONE-THIRD FULL OF WATER. WHEN USING ITS SUCKING ACTION, THE VACUUM TIP MUST BE UNDER WATER. (TO GET A VACUUM ACTION.)

OTHER GOLD RECOVERY METHODS

THE SLOW-POKE CONTEMPLATIVE WAY

It is essential to learn good gold panning techniques that work for you. Most other hand methods end with the gold pan separating the gold from the concentrates. That's how important it is to polish your panning skills. Let me illustrate a method I use that works for me. However, you need to bear in mind that the panning process to me is the end in itself. My personal objective is not to gather and amass gold for the sake of gold, but rather for therapeutic recreation. Let me explain:

When you are working a placer that is providing you with the thrill of seeing the gold as you work the pan, each little shine is a moment of surprise. I can sit and work a single pan for a good hour. Remember, I said that gold itself was not the end for me. I just use it to keep score. I certainly would be bored and bummed-out if I went blank for an hour, a day, or heaven forbid, a trip. IT IS A BETTER PRACTICE TO WORK YOUR PAN FOR A SIMPLE SEPARATION PROCESS, THEN WORK THE FINES WHEN YOU GET HOME. I'm suggesting what works for me. It's therapeutic because it causes me to relax; to lose what I call the "jitts"; to obtain a state of "calm". If you suffer from high blood pressure, my method of panning can lower your pressure readings 30 points on each side after three days of effort. Most of this book was put in mental order during the thought patterns operating while gold panning. You see, there's more to it than just looking at wet dirt sloshing in a pan. After a while, your gold pan may turn into your special security blanket too! For your enjoyment, I hope so!

THE ROCKER BOX METHOD

This tried and true method allows you to work more gravel for a day's labor. It is most effective when partnering. That way one man can dip and rock, while the other digs and feeds the box with placer material. The fines are then PANNED. There is the problem of getting the box to the stream, so make sure you buy or make a lightweight unit.

THE SLUICE BOX METHOD

This is a sliding-board way of washing gravel. It has the merit of working much more gravel than by rocker box. The stream current powers the gravel down a slope (usually one inch drop per foot of sluice). The gravel is washed and the freed silt flows out the sluice end. The gold and black sands are trapped by built-in riffles, then PANNED. When not properly set up, a sluice will lose most of your fine gold. Using a sluice is a practical way of working placer material... if the water supply is adequate.

THE HAND DREDGE METHOD

For the serious gold seeker this a popular option that requires a sizable investment in equipment. If you can visualize a motor-powered suction device vacuuming the bedrock, you have a dredge in operation. This is a practical gold recovery method if you can devote a goodly amount of time to reaching bedrock. If you enjoy working in a wetsuit; have a lot of time , and want to work weeks just to get to bedrock, you'll love dredging.... and the payoff too! This method's shortcoming runs most of the fines back into the stream. After the nuggets are picked out of the dredge sluice, the concentrates are PANNED.

Dredging requires a STATE permit to operate. The permit is easy to obtain with only a small permit fee charge. Most areas limit the size of the nozzle to six inches to qualify for hand-held status. Maximum nozzle size in some streams is three inches. Certain streams are closed to dredges. Some streams have closed seasons to protect the fish spawn.

IF YOU PLAN TO DREDGE, IT MAKES SENSE TO CHECK OUT THE NOZZLE SIZE LIMITS AND OPEN SEASON BEFORE YOU LUG THAT HEAVY EQUIPMENT TO A PLACE THAT MAY BE CLOSED TO YOU OR YOUR EQUIPMENT.

CHAPTER 58

HOW TO HANDLE "FINES"

"Fines" are the black sand-gold mix that is found at the bottom of the gold pan after all other materials are washed out. It is a mix of magnetite and heavy minerals along with visible and invisible gold particles that run-out at 100,000 or more gold particles the ounce. TECHNICALLY, FINES ARE ONLY THE GOLD PORTION. FINES ARE ALSO KNOWN AS "HEAVIES" OR "CONS". ("Cons" is short for "concentrates".

You have just returned from a goldpanning expedition. You have that comfortable, pleased feeling that the time was well spent. You are ogling your first nugget. Even your wife has gold fever now. Remember how she fought you about going along? You have several ounces of material in your sniffer bottle, but it looks mostly like black sand flecked with tiny gold specks. IT IS A GOOD BET THAT NEARLY HALF THAT WEIGHT IS IN GOLD.

HOW SHOULD YOU PROCEED TO SEPARATE THE GOLD FROM THE BLACK SANDS?

The old time method of separating sands from gold involved drying out the mixed material in a spreading-out on a piece of stiff material. Then by using a magnet under the flooring piece, to remove as much of the black sand as was magnetic. By working a teaspoonful at a time of the remaining mix, that amount can be spread into a pan, plate or other flat container. Here's where the task gets tedious and boring. (For me.)

Next get a flat-type toothpick, wet the small end with spittle to act as adhesive, and pick out each fleck of gold. This is best done in strong light such as sunlight, and with a good pair of eyes. REMEMBER, THAT YOUR NON-MAGNETIC BLACK SAND CONTAINS MICRON SIZE, INVISIBLE GOLD. PUT IT IN A SEPARATE CONTAINER. SAVE IT FOR LATER AMALGAMATION.

TRY THIS MORE EFFICIENT WAY

THE BACKYARD MINI-SLUICE

This home-made device works efficiently and is worth the small cost to make. You will need a 3 foot piece of scrap lumber about 6 inches wide. To this board you will use roofing nails to secure a 3 foot piece of plastic, corrugated, non-perforated, 4 inch, sewer pipe which you have cut lengthwise in half.

You will also need some bricks for height adjustment purposes plus 2 five gallon plastic buckets. Take one of the buckets and insert a plastic shut-off valve at the bottom. The sewer pipe is available from hardware or home improvement stores. The plastic shut-off is easily obtained from garden shops or nurseries.

THE MINI-SLUICE IN OPERATION

Set up your board with the pipe corrugations acting as riffles so that one end of the board rests on the edge of bucket #1. Raise the other end of the board high enough to approach an angle of 30 to 45 degrees. Use your bricks to support the high end for stability. Next place bucket #2 with the shut-off valve spout extension overhanging the top end of the riffle-board. Use a hose to fill bucket #2 with water.

Reviewing the operation: Open the shut-off just enough to dribble water into the plastic trough. Notice the water gently runs down the riffles and is saved by bucket #1. When bucket #1 is full, simply pour it back into bucket #2. You now have a water return system for separating the gold from the fines.

THE SEPARATION OPERATION

Place the fines from your sniffer bottle and/or other fines collection container into a pie pan or other low-lip vessel. You will need a teaspoon to handle the fines. An iced tea spoon gives you a longer reach advantage. You also need an empty sniffer bottle to vacuum the riffles.

1. Set your water spout shut-off valve to provide a small dribble or rivulet of water flow down the tube. Take a teaspoonful of fines and gently rinse the spoon clean of fines. Don't just dump the spoon of fines in the water stream. THIS IS A SLOW, GENTLE OPERATION. CHECK TO SEE THAT THE FLOW IS STRONG ENOUGH TO WASH THE FINES WELL, BUT NOT OVERLY FAST IN DOWNWASH. RECHECK FOR THE CORRECT RAMP SLOPE TOO.

2. After a few spoons have been washed, check the condition of the riffles. If they are close to full, use your sniffer bottle to get the visible gold. The top few riffles should be nearly free of black sand with the gold particles clearly visible. If not, readjust the ramp angle or water flow to obtain that result. Repeat until you have worked your fines out.

3. As each bucket #1 fills, pour the water back into bucket #2. Then salvage the fines found in bucket #1. These are usually washed again. DON'T WORRY ABOUT GETTING ALL THE GOLD FROM THE FINES, YOU WILL RECOVER IT ALONG WITH THE MICRON-SIZED GOLD IN THE LATER AMALGAMATION PROCESS. SAVE THE FINES FOR THIS PROCESS.

4. The gold you recover from your fines is the "dust' that is valued at the gold spot price. It is usually worth enough to recoup your trip expenses.

PRELUDE TO AMALGAMATION

HOW TO HANDLE FINES

In the last chapter you learned how to separate fines to obtain the VISIBLE gold. You know (or have heard) that the residue of your non-magnetic fines carry INVISIBLE OR MICRON SIZED GOLD. Since you cannot see the gold, you take that idea on faith. You are a believer. You have heard that amalgamation of your fines can clean out the remaining gold. You decide to give it a try.

AMALGAMATION IS A PROCESS FOR RECOVERING FINE GOLD THAT INVOLVES THESE STEPS:

1. THE FINES ARE MIXED WITH AND BALLED IN MERCURY.

2. THE MERCURY BALL IS SQUEEZED THROUGH A CHAMOIS TO REMOVE THE EXCESS MERCURY.

3. THE MERCURY/GOLD IN THE BALL IS RETORTED TO DRIVE OFF THE MERCURY AND LEAVE THE GOLD IN A FORM KNOWN AS SPONGE GOLD.

4. TO GET THE SPONGE GOLD INTO BEAD FORM, THE SPONGE IS HEATED IN A CUPEL WITH OXIDE OF LEAD. THIS MELTS THE GOLD INTO A BEAD OR GOLD BUTTON.

5. WHILE THE GOLD BUTTON IS MOLTEN, YOU CAN FORM A FAKE NUGGET BY POURING THE GOLD ONTO A LAYER OF WATER-DAMPENED SAW-DUST.

MERCURY CAUTION

MERCURY CAN BE ABSORBED THROUGH THE PORES OF THE SKIN. SAFETY REQUIRES USING RUBBER GLOVES.

MERCURY VAPORS ARE INVISIBLE, ODORLESS AND TOXIC. DO YOUR RETORTING IN THE OPEN AIR OR IN A WELL VENTILATED WORK SPACE.

HOW SERIOUS?

THIS WARNING MUST BE TAKEN SERIOUSLY FOR UNDER CALIFORNIA'S PROPOSITION 65 ALL PRODUCTS CONTAINING MERCURY MUST BE LABELED AS BEING HAZARDOUS TO YOUR HEALTH.

To give you an idea of how seriously the manufacturers are treating mercury products in California:

You know the product "tincture of iodine" that stings when you brush it on a cut? It contains the skull and crossbones poison label, and has for years. That product is still freely available over the counter at your local drug store.

You know the product "Methiolate" (TM Eli Lilly) that you brush on the same cut, but with no sting? Well, it contains 0.1% THIMEROSAL as an antiseptic. Methiolate Tincture is 0.1% Thimerosal in 1000 parts of alcohol. Thimerosal is a mercury product. For a while the old stock of Methiolate carried a warning sticker that read: "This product contains a chemical known to the State Of California to cause Birth Defects or other Reproductive Harm."

METHIOLATE IS NO LONGER AVAILABLE IN CALIFORNIA.

THIS WARNING IS INCLUDED HEREIN TO PROTECT THE WRITER AND MR. HOWARD F. STEINER, THE AUTHOR OF THE CHAPTER ON AMALGAMATION, FROM "FAILURE TO WARN" CLAIMS. BE CAREFUL ! USE COMMON SENSE.

AMALGAMATION CHAPTER COURTESY:
HOWARD F. STEINER, Copyright 1981

CHAPTER 59

MAKING GOLD DUST
INTO
GOLD BEADS

HOW TO:

Amalgamate "Fines"

and

Make a "Mini" Retort

These instructions are written for those who pan gold for a hobby, or just for fun. I hope that they may gain greater delight from their field trips because of it.

HOWARD F. STEINER

Copyright, 1981

CONTENTS

MATERIALS REQUIRED

1. Mercury (Hg) -- A thimbleful is sufficient.

2. Litharge (Oxide of lead) -- A teaspoonful is plenty.

3. The "Mini" Retort -- See sketch.

4. Soldering torch -- I use a Propane gas torch for removing the mercury from the gold. A can of Sterno or a cookstove alcohol heater will work too.

5. White blackboard chalk.

6. Teflon pipethread tape or plumbers putty to seal the cap on the retort.

7. Cupel -- small size. Get six to start with. The cupel, pronounced "koo'-pel", is a molded, bone ash cylinder, about one inch in diameter and one inch high. It has a conical depression in one end. It is used to filter and absorb molten lead from gold.

SOME WORDS ABOUT PANNING AND EQUIPMENT.

This material is not intended for the people who work professionally in the mining industry. It is intended solely for those people who enjoy wandering along the creeks and ravines with their gold pan, hoping to find some gold to show their friends. Most of these hobbyists are looking for nuggets. They do not realize the value of the "fines" they lose or ignore. Buy $10.00 worth of gold flakes or fines, --and see how little you get!

There are books available telling where to go or how to pan. Also, there are many kinds of special equipment offered to "prospectors". All have their adherents. A few years ago, a friend went down to South America and visited a placer operation. It was a family affair. Men carried the gravels from bank to stream; their women and children panned with pans woven from local reeds and grasses. Gold was saved in quills from hawk wings with the ends sealed in wax. Placer gold was taken from streams long before there were iron tools! It took skill then, just as it does today.

Just last summer I came upon two couples "panning" on a stream in the California Mother Lode Country. I inquired about their success and asked if they knew how to pan. (Their motor home had an out-of-state license.) They assured me that they knew how to pan, but they had concluded that the talk about gold in the West was "baloney". They had panned in Montana, Colorado and in California, and had yet to find a piece of gold! I watched as they waded about, scooping up sand here and there, sloshing it around in the pan, then tossing it out. A mile away, others were panning, and coming back year after year! It's not just the equipment or the book, --there's more to it than these things. One can never acquire too much skill!

RECOVERING THE "FINES"

Presuming the panner has done a good job, there will be very fine gold with the sands in the pan, plus mica or iron. Use a magnifying glass and you will see that the tiny yellow specks are really gold nuggets.

There are many inventions on the market for removing fine gold from either black or white sand. I do not question the claims of the manufacturers of these devices, but most of them are rather expensive or bulky. From ancient times, fine gold was recovered using mercury, - popularly called "quicksilver". About the year 50 A.D., Pliny wrote about gold mining. He described how the gold miners squeezed the excess mercury from the gold by using soft leather as a filter. The same practice is still followed, using a chamois or chammy.

To the sands in the pan, add a ball of mercury about the size of a pea. Tilt the pan and roll the bead back and forth over the sands and gold specks. Some specks of gold may be floating on the water. Add a few drops of kitchen detergent to the water. It will cut the oils from the gold, which will then sink. Some gold will adhere to the surface of the mercury ball, but not all of it! This is because there is some film separating the contact. Add a teaspoonful of table salt to the water. The chlorine in the salt will cut the scum, and the metallic gold will then cling to the mercury. It may be necessary to rub mercury into the "salty" sands with the thumb. (Ed. Note: Wear surgical gloves.) If the ball breaks up, it will come together again. If there is much gold, the mercury will become mushy and will have to be "pushed" out from the sands. Pour out all the water you can, then with a clean sponge or paper towel, mop up the rest of the water. The sands will stay put, but the mercury with its gold can easily be separated. The ball is "Amalgam".

"MINI" MERCURY RETORT WITH A BRASS COMPRESSION FITTING.

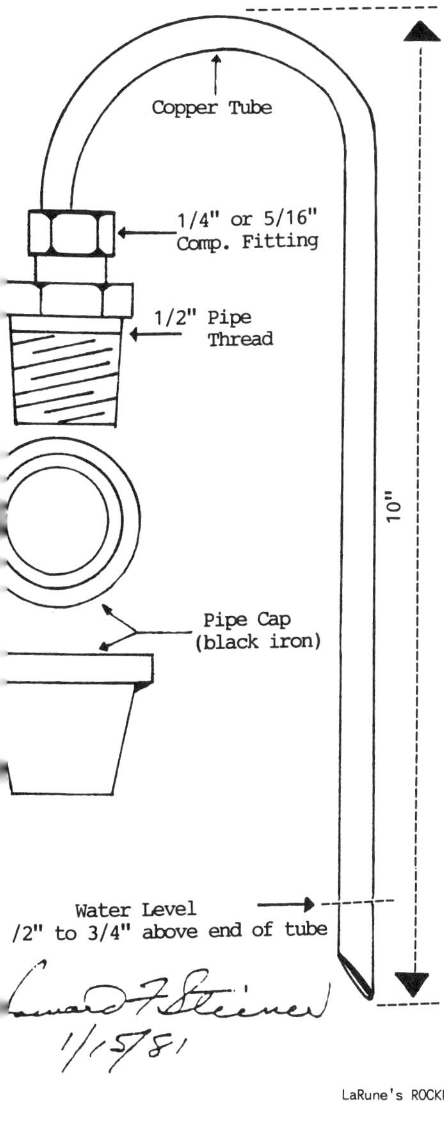

One end of the brass compression fitting should be threaded for 1/2 inch pipe. However, a 3/4 inch fitting, and copper tube of 1/4 or 5/16 inch will work as well.

A shorter tube leg to 7 inches can be used, but tube overheating and sputtering of water is more likely to occur.

RETORT OPERATION

A tall, one pound, coffee can does very well. With a nail, punch two holes spaced about 1/2 inch apart, at the top of the can. With a wire, make a loop to hold the tube upright in the can. Place water in the can, covering the tube end, NOT OVER 3/4 INCHES.

Coat pipe threads with white chalk or teflon thread tape. Coat the interior of the cap with chalk.

Place your amalgam into the cap and screw the cap finger-tight firmly to the fitting. MAKE SURE YOU HAVE AT LEAST 4 FULL THREAD TURNS.

APPLY HEAT TO CAP AND FITTING.

The mercury can be boiled off the gold by heating. This requires special equipment. Perhaps the miner needs the mercury, but does not want to take the time to heat it. He can regain most of it by a simple filtering method. If a chamois is not at hand, a nylon stocking, a paper towel, or even a Kleenex (TM Kimberly-Clark) will do!

Make a pouch in the filter and put the amalgam in it. Then place the pouch inside a large-mouthed fruit jar or clean can. Twist, squeeze, and "milk" the pouch. The mercury will spurt through the filter, but the gold will not. Open the filter, and you ll find beads of mercury in the creases of the filter. Scrape these back down into the bottom of the pouch, and move the position of the pouch so that it will not tear. Then continue to strip the filter. When the mercury has been removed, there will be a silvery sponge of hard particles in the pouch.

The mercury in the receptacle may not be clean enough to reuse. Place it in a pan with water, add salt, and pan the mercury. Withdraw the water, and pour out the mercury. It will now be bright.

NOW FOR THE "RETORT".

The amalgam should be transferred from the pouch to a clean receptacle. ALWAYS WORK OVER A PAN in case of accidental handling.

Before going further, it may be helpful to recall some of the characteristics of mercury:

Mercury is a metal. It becomes solid when temperature falls below -65 degrees F. It will boil, like water, at above 640 F., and return to a liquid when below 640 F.

Vapors are invisible and have no odor. The vapors can be inhaled; will condense as liquid mercury in the lungs, and are very poisonous!

Handle mercury only in a well-ventilated area; keep it away from children, and do NOT handle it around food for humans or pets.

Mercury is heavy and easily spilled. ALWAYS HANDLE IT OVER A CATCH BASIN.

IF THESE PRECAUTIONS ARE FOLLOWED, MERCURY IS SAFE TO HANDLE AND VERY USEFUL.

THE "MINI" RETORT

THE "RETORT" IS AN APPARATUS FOR DISTILLING MERCURY FROM THE AMALGAM.

A retort is composed of three main parts:

1. A receptacle for the amalgam.
2. The heating torch.
3. The discharge tube for the vapors.

Of course, add the water to cool the vapors and condense them back into liquid mercury. This retort is designed for the small operator. Cost is a major consideration. (See accompanying construction diagram.) [ED-NOTE: Mercury clings to metal cans. A wide-mount fruit jar is O.K., but risks breakage.]

MELTING AMALGAM INTO GOLD IS DANGEROUS BECAUSE OF THE POISONOUS FUMES GENERATED BY THE PROCESS. The operator should be cautious and SAFETY MINDED when handling the equipment. Parts for making the retort can be purchased in most plumbing stores.

You will need a brass compression fitting attached to a bent copper tube. Braze the fitting and tube to assure a tight joint.

The container holding the cooling water need not be some special item. A tin can or a glass jar will do. If the condensed mercury is to be saved for reuse, it should be clean.

Hang the retort on the container and place sufficient water

in it to cover the tube end about one half inch deep. If necessary, add some weight (rocks) to the container to hold it upright. Put a slight coating of chalk over the interior of the cap to prevent gold from adhering to it. Coat the plug threads with a generous amount of chalk, teflon pipe tape or plumber's pipe sealer. A good joint seal is essential to prevent vapor fumes from escaping into your work area. MERCURY VAPORS CAN BE INJURIOUS IF INHALED. WITH PROPER PRECAUTIONS, MERCURY IS SAFE TO HANDLE.

With a good thread on the cap and fitting, a finger-tight joint will be adequate. The brass under heat, expands more than the iron cap and thus improves the joint seal. You may need a wire loop made from a coat hanger to stabilize the upper section of the retort.

Next place the amalgam into the cap. Then screw it to the fitting as tight as you can WITH YOUR FINGERS.

When these preparations are complete, insert the charged cap and tube through the wire loop and add water to the can. It is now ready to be heated.

Slowly apply heat with the torch. It would be wise to try "boiling away" a small bead of mercury before trying the amalgam.

WHAT TO EXPECT

First, bubbles will come from the tube. This is the hot air being expelled from the retort. Then there will be a time of no action in the water. Just like water, mercury does not evaporate suddenly. BE PATIENT AND DO NOT OVERHEAT. If you see the iron cap at red heat... it is 900 degrees and too hot. Keep the mercury below 850 degrees. If the retort is overheated, the tube will become overheated and boil the water. This will create a vacuum in the tube. The water will be sucked up the tube, turn to steam and cause minor explosive forces, which drive the gold and mercury into the water, which turns milky. This breaks up the mercury, which will be impossible to recover.

Play the flame of the torch MOSTLY on the bottom of the cap, only occasionally on the rest of the retort OUTSIDE THE CAN. As the retort gets hotter the water will become calm. You should begin to see small beads of mercury appear beneath the tube mouth. When the vaporization is complete, tap the tube portion that rests inside the can to dislodge any mercury clinging inside the tube. NEXT WAIT FOR THINGS TO COOL DOWN.

You may need to use pliers to unscrew the cap. If you find a gray-yellow lump or granules in the cap, the job is complete. If a silvery deposit remains, the mercury has not been completely removed. You will need to recap and reheat.

When you are trying to unscrew the cap, place a safety pan under the retort. If the cap slips and drops, you can still save the contents. If the mercury has boiled away, scrape the contents from the cap into a clean container with a knife or a small screwdriver. If your scraping breaks a fused lump into fines again, don't worry. The gold is captured and free from the sands. It is now ready to be melted into a bead.

MAKING THE GOLD BEAD

It is best to know some of the properties of the metals we are working with.

METAL	MELTING POINT	METAL	MELTING POINT
GOLD		SILVER	
Pure Gold .985	1913 F	"Fine"	1762 F
Half Gold	1620 F	Sterling	1640 F
10 Karat Gold	1575 F		
Boiling Point	4860 F	----------------------	3212 F
LEAD	585 F	BOILING POINT	2817 F

The blue flame of a FULL bottle of Butane gas with a soldering torch reaches approximately 2100 degrees F. This will barely melt pure gold, even if all drafts are kept from the metal. Most natural gold is not pure or .985 "fine". Impure natural gold, when air is excluded, is quite easily melted. There are no lead hazards either because the boiling point of lead is never reached.

To exclude air when melting "fines", I use common litharge, which is lead oxide. Red lead, or lead filings will work, but I find litharge easier to handle. Others may use borax or saltpeter (potassium nitrate) in lieu of lead. Borax is easily blown out of the cupel; saltpeter percolates into the cupel too quickly for me.

As the lead becomes molten it can be flushed around the cupel's depression to wash any stray grains of gold to the bottom to unite with other grains that will make the bead.

You will need a stand for the cupel in order to evenly heat the bottom and sides. Make one from a wire coat hanger.

Apply heat from the torch gently and evenly to the cupel to dry out any moisture. In an assay furnace, the cupel gets uniform heat until it is withdrawn. Here we cannot do this. The assay furnace delivers a much higher temperature than our torch will produce. The effect will be apparent as we start the melting process.

PLACE A PAN UNDER THE CUPEL. THEY HAVE BEEN KNOWN TO CRACK.

Put the sponge gold into the bottom of the cupel and cover the gold with litharge. Do not use more than twice as much litharge as there is gold. The lead is absorbed by the cupel. An excess of litharge will leave a pool of lead with the gold. This means you will need to reheat in a NEW cupel. (The first one has absorbed all it can.)

As you evenly heat the cupel it will begin to glow. Alternate the flame from the cupel to the litharge which will turn yellow, then black. Soon beads of lead will begin to sweat out of the litharge and form a pool. Continue to heat the litharge. The pool will grow and become more fluid. Keep heating! Remember the flame cannot boil the lead. Your gold, not being pure, will soon melt.

In due course a small speck of yellow will appear in the pool of lead. As the lead gets hotter, it can be flushed around the gold. As the lead begins to sink into the bone ash, the gold bead will appear larger and more prominent.

Quite suddenly the lead will disappear leaving only the gold bead visible. Heat the gold button until it stays constant in size, then remove the torch. There may be some dross floating over the bead which can be brushed off when the bead cools.

If you find the bead WHITE IN COLOR, you used more lead than the cupel could absorb. If only the top surface of the bead is WHITE, it could be silver which is lighter than gold. In the melt it would rise to the surface of the gold.

When the torch is removed from the bead, the bead will almost immediately "flash", indicating surface solidification. Do not touch the bead yet, for it is not totally solid. Wait for the cupel to cool. It may take longer than you think. Expect to find the cupel cracked, and a break probable as you dig the bead from the depression. Again, remember to use a pan under the work as you pick the bead out of the ash with your knife. I have made a bead from as few as 6 grains in a demonstration. The average is closer to 85 grains.

To remove any silver or similar metal from the bead, place the bead in a glass and cover it with a weak solution of nitric acid to dissolve the foreign material. If copper is present, the bead will be "reddish".

Prolonged heating of the retort when working with larger amounts of amalgam can overheat the copper tube, especially if a tube shorter that 12 inches is used. Overheating results in mercury "spitting" into the cooling water. This causes mercury to divide into fine particles which cannot be recovered. You can avoid this by using a lower heat source. You must allow extra time and provide a draft-free heating area. (Tube spitting can be minimized with a heat sink on the down-leg.) The flame of a low-heat source should be kept about 2 inches from the cap.

Always tap the tube before removing it from the can. When unscrewing the cap, hold the cap over a safety pan. You may drop it.

Gold may discolor when alloyed with contaminating metals such as iron or copper. To improve the color, place the gold button in a glass or porcelain container with dilute nitric acid. Leave it for 24 hours. Sometimes the discoloration can be removed in this way. Nitric acid alone will not dissolve gold in that length of time.

Now hand the bead to anyone. The first question will be, "How much is it worth"?

WEIGHTS AND MEASURES

Perhaps someday there will be a universal scale of weights and measures. Today, however, we use many different scales, terms, and formulas. Because of similar terms or pronunciation of different words, it is easy to confuse descriptions. We need to brief ourselves on these terms.

Precious metals are weighed by the troy scale of weights. I once thought this term came from the ancient city of Troy. However, the French gold and silversmiths of Troyes, France developed this scale about three hundred years ago.

The basic unit of measurement is the ounce. It is subdivided into pennyweights, then into grains

24 grains (Gr) equal 1 pennyweight
20 pennyweights equal 1 troy ounce
480 grains equal 1 troy ounce.

The term grain is derived from the ancient use of the Carob seed in balance scales. The pennyweight, from the silver penny in use in Europe, is abbreviated DWT or PWT. More commonly DWT.

Gems are priced by their weight and beauty in the Metric scale of weight. The term "carat" is 200 milligrams in weight. Purity is no factor in the gem-trade weight-word "carat". In gems, purity is thought of as "free from inclusions or defects".

The precious metals, such as gold and platinum, are sold by weight and purity. Purity is designated in "karats", sounded the same as "carat". The "karat" is simply a fraction of the number 24. One karat is called "one marc". 24/24ths is pure gold. This would be stamped "24K". 12K is one half gold with the other half an alloy addition. An error can be made when using balance weights. The gram weight is often stamped "GR" instead of "GM". One gram is equal to 15.4324 grains.

In the United States, all jewelry or articles sold as gold must be stamped as to purity with its "K" identification. It is easily falsified. Also, certain markings are not well understood. "Solid G.F." means that only the coating is gold. In such jewelry, less than 4% by weight is the gold content.

In some countries, the purity of the metal is marked in the percentage of 1000, with .925 being considered pure" for gold, and .925 "pure" for silver. Hallmarks on tableware and other articles are beyond the scope of this investigation.

TESTING FOR PURITY

Because natural gold is seldom pure, the buyer of gold must determine its purity. He cannot afford to pay the price of gold for the alloys which will be removed by the refiner. Therefore, he becomes familiar with gold in his area, and learns to make positive purity tests himself.

There are several ways he can do this. The simplest method

involves the use of a touchstone. This is a flat, dark rock with some abrasiveness. Gold rubbed across its surface leaves a gold streak which is color-compared with the color of one of 24 gold-tipped needles. Each needle represents one karat with its distinctive karat-color. By matching the needle karat color with the streak, he can quickly determine the karat purity of the gold and its karat rating.

Since the gold buyer does not pay for gold's alloy content, let's follow through a sale of a bead to market. The gold price for that day would be taken from the London spot market gold fixing price made public that morning.

We wish to sell our bead weighing 85 grains. The price of gold on the date of sale is $350.00 the ounce. (480 grains) If we divide 480 into the price of $350.00, each grain of pure gold is worth $0.73. Since we have 85 grains @ .73 cents each grain, our 85 grains are worth $62.05 if pure. If our bead tested 12 karat in purity, this would be 50% gold or 12/24th of $62.05. Our buyer would pay us $31.03 for the bead. However, most placer gold runs 85% and higher gold content. In the real world our bead would sell at least for $52.74.

ON SELLING GOLD ITEMS

From time to time large ads appear in newspapers announcing that there will be a gold buyer in town for only a day or two. He will be dealing from an office in some prominent place. He will buy any gold and silver items, watches, spectacles, or heirlooms. HOW CAN HE PAY MORE THAN LOCAL DEALERS? Generally, he can't. He hopes to pick up some items that have a greater value as antiques, or have a better resale value than the offered precious metal value. When interest rates are high, many sellers release items for sale that have been held for years. If later they find they have been gypped, where is the recourse?

Gold buyers are not regulated. A license to deal in gold is no longer required. Business reputation is about the only clue you have as to fair dealing. If you are considering the sale of jewelry, find the local dealers of prominence. Get several appraisals, preferably three. If the dealers you have contacted all quote within a short price range, take the best offer and run. You have been fairly dealt.

The gold buyer is a gambler. After he purchases the gold, the price can quickly change either up or down. Like any other merchant, he has an overhead cost. For the gold buyer to sell his gold he needs a certified purity stamp from an assayer or refiner who charge fees for their services. Insurance, storage, brokerage, and shipping costs increase costs of doing business. The gold buyer gets a small, cost off-setting, bonus from the silver or platinum values that may be alloyed in the gold. All in all, he is generally as honest as your grocer.

MAKING A FAKE NUGGET

Many purists turn their nose up at non-natural nuggets.. and properly so! Each natural nugget has its own character and configuration. Like snowflakes, no two nuggets are alike, though they may be similar. Considering the complications in marketing the gold, the panner may want to melt several beads together. This can be quickly accomplished by using a heat-resistant dish or crucible and saltpeter as a flux. THE MOLTEN GOLD CAN THEN BE POURED INTO DAMPENED SAWDUST. This will result in a beautiful artificial nugget with jewelry value.

If troy weights are not available, go to a gun shop or find a cartridge reloader. Powder and bullets are weighed in grains.

Mining magazines carry ads of chemical firms who can supply your needs. Expensive? Try golf or fishing or most any other relaxation sport for equipment costs. GOLD PANNING IS A BARGAIN!

I WISH YOU FUN AND REAL GOLD BEADS TO SHOW FROM YOUR FIELD TRIPS.

MELTING TEMPERATURES

FINE GOLD (.999 Pure)	1975 F
18 K	1620 F
10 K	1575 F
PLATINUM	3193 F
PALLADIUM	2829 F
NICKEL	2619 F
SILVER, "fine" .925 Pure	1762 F
STERLING -92.5% SILVER, 7.5% COPPER	1640 F
1981 F	
LEAD	585 F

COLORS OF GOLD IN JEWELRY

GOLD COLOR	ALLOYS ADDED
YELLOW	SILVER WITH COPPER
RED	25% COPPER
DEAD LEAF	30% COPPER
GREEN	25% SILVER
WATER GREEN	40% SILVER
BLUE	25% IRON
WHITE	NICKEL, ZINC OR PALLADIUM

Note: This chapter courtesy:
HOWARD F. STEINER, CALIFORNIA - Copyright 1981

BIONOTES: HOWARD F. STEINER was born in 1898 and is still active in recreational gold activities. His latest project is a new type gold pan shaped in a VEE. He calls it a VEE PAN. As a young man representing the Caterpillar Tractor factory, he was sent to Nevada. He met a gold panner in Tonapah who induced him to invest in a mine. He lost $1,000. in that venture. Gold was then $20.00 the ounce. He determined to learn more about gold.
HE IS WELL KNOWN FOR HIS MINI RETORT METHOD DEMONSTRATIONS.

CHAPTER 60

GEMSTONE SCREENING

YOU SHOULD BE SO LUCKY!

Rarely, but sometimes, your goldpan will show bright, little, rolled pebbles colored light to dark green, blue-green, red, red-brown, or colorless. If you are working a sluice, they will clog up your riffles. You will be working on a gravel bar in the eighteen inch area above the hardpan.

Why lucky? You may have hit a favorable sapphire or garnet gemstone deposition spot. Sapphire is the mineral corundum, second in hardness only to diamond. The blood-red variety is called "ruby". Large rubies are the rarest, most valuable of the gemstone class, even more costly than diamonds. Star sapphire is corundum with inclusions of another mineral oriented in parallel fiber bundles, that when a cut as a stone, dance in the light.

As a general class of gemstones, sapphire is corundum of all colors except red, which is the ruby. The garnet group comprises six minerals with the same basic structure, but with the addition of different atoms of other elements. This leads to a variety of colors, forms and occurrence. Spinel occurs in a wide variety of colors. The ruby spinel has the color of a fine ruby. Red spinel is often mistaken for the corundum ruby. A case in point is the Royal Crown of England with its massive frontal ruby which in later years was determined to be spinel. I have often wondered if a successful thief substituted the spinel for the real thing.

In a water environment, such as working the gold pan or sluice, specific gravity operates to stratify materials of differing specific gravity.

Quartz = 2.6; Garnet = 3.5; Spinel = 3.8; Corundum = 4.0.
You can see that the gemstones would settle lower in your pan.

SCREENING GEMSTONES CAN BE A UNIQUE, EXCITING, PROFITABLE TASK.

HOW TO SCREEN GEMSTONES

The first secret in preparing to dry screen for gemstones is to scrape the hardpan clean.

You will need about a 10 inch frame with a half inch mesh screen plus a 12 inch frame with a one eighths inch mesh screen. Fill the coarse screen box with gravel and shake over the fine mesh screen to collect it. This gets rid of the larger gravel and acts as a classifier.

Next, take and shake the fine screen material letting the fines drop to the ground. Any gemstones falling through the mesh are too small to keep. The remaining gravel is high-grade. Pour it into a bucket until you have enough to proceed to the washing phase. This washing step is vital to gem recovery.

Washing removes the mud and silt coating from the pebbles; those that are sapphire crystals will gleam when wet and clean. You can see them in the gravel. But there is another trick!

Load your fine mesh screen with the materials and immerse it in water while swirling and bouncing the screen. The object is to remove the clinging silt, and let specific gravity operate to settle the heavier gemstones.

The next underwater step is to center the sapphires in the screen. This is accomplished by swirling, bouncing, tipping the screen, dipping it from side to side, all the while turning the screen in a rotation pattern. After some practice and skill, the sapphires are centered at the bottom of the screen.

Here's the next trick! Question: How can you get the sapphires laying at the center-bottom of the screen into view for picking? Answer: Remove the screen from the water and let it drain. Make sure you have a piece of sheeting spread flattened on the ground. Now, quickly flip the screen upside-down onto the sheet. The centered pebbles will be in bright contrast to a black layer of hematite/magnetite overlying the gravel.

THE ONLY REMAINING QUESTION ASKS: IS IT GEM GRADE MATERIAL?

59 GEMSTONE CHART

You may have noticed that the word "stone" in the sense of "rock" has not been used in any of the previous chapters. This is for the reason that the word "stone" has no real meaning. However, when it comes to gems, the term "gemstone" is common.

To be counted a gemstone, the mineral must be durable, rare and beautiful. Beauty is a matter of light reflection and refraction in the way it plays the rays to create fire and color effects when cut and polished. Durability combines the qualities of hardness, toughness and chemical resistance into a lasting asset. Rarity adds to the value of a stone. Size, weight, color, clarity and current market demand all enhance value. The weight of a gemstone is measured in "carats". A carat is one-fifth of a gram.

REFLECTION AND REFRACTION

The internal atomic structure of a mineral governs how light behaves. Values for different gem minerals can be measured with a refractometer. Those measurement readings can be used as a means of identification.

Where optical properties are the same in all directions, the mineral has an isometric crystal system. Each ray of light is slowed down and bent or refracted to the same degree. Isometric minerals are said to be singly refractive.

In all the other crystal systems the entering light is split into two rays. Each ray is refracted by a different amount. These crystals are described as doubly refractive.

Refractive index [RI] indicates the number of degrees a light ray is bent. Isometric minerals have only one RI. Doubly refractive minerals have a range of RI.

If you subtract the minimum RI degree from the maximum RI

degree, that difference is called "birefrigence" [BI]. A high BI reading reflects light rays in a cut gemstone off the back facets when viewed through the front facet.

A colorless gemstone is one that absorbs no rays of the light spectrum. A crystal-absorbed color ray is the viewed gemstone color. The ability of a gemstone to split white light into colors of the spectrum is called "dispersion" which can be measured by refractive index.

Luster is the amount of light reflected from the SURFACE of the mineral. It is variously described as "sparkling", "vitreous" ranging down to "earthy" and "dull".

SINCE VERY FEW OF NATURE-MADE GEMS ARE PERFECT, DEFECTS SUCH AS INCLUSIONS WERE USED TO DISTINGUISH THEM FROM SYNTHETICS. THIS TEST IS NO LONGER VALID. MAN-MADE GEMS ARE AVAILABLE THAT DUPLICATE NATURAL GEMSTONES. IN MANY CASES THE SYNTHETICS ARE BETTER AND ALWAYS LESS COSTLY.

ALL GEMS AND PRECIOUS STONES CAN BE MAN-MADE. NO EXCEPTIONS !

NEED MORE BACKGROUND ON PHYSICAL TESTS ? SEE PART III.

FOR DETAILED INFORMATION ON A SPECIFIC MINERAL SEE:

PART III - CHART F

ARRANGED ALPHABETICALLY MAIN GEM GROUPINGS Page 1

COPYRIGHT 1993 - T. D. LaRUNE

GEMSTONE	MINERAL	HARDNESS RANGE	SPECIFIC GRAVITY RANGE	REFRACTIVE RANGE	SOURCE - OCCURRENCE
01					
ACHROITE - The colorless variety of tourmaline.					
02					
AMAZONITE	MICROCLINE	6.0 - 6.5	2.56	1.53	IN PEGMATITES & GNEISS
COLORS: Bright green, Blue-green -- Easily mistaken for jadite & nephrite.					
03					
AMETHYST	QUARTZ COARSE CRYSTAL	7.0	2.65	1.54 - 1.55	HOLLOW GEODE ROCK CAVITIES
COLORS: Pale violet to dark purple.					
04					
QUARTZ	QUARTZ COARSE CRYSTAL	7.0	2.65	1.54 - 1.55	PEGMATITES & HYDROTHERMAL VEIN
[CITRINE - Yellow] [ROSE - Shades of Pink] [MILKY - White] (also known as BULL or SUGAR QUARTZ) [CAT'S EYE - Green-gray, yellow, brown] [HAWKS EYE - Blue] [TIGER EYE - Gold-brown] [BLUE - Blue] [AVENTURINE - Green, brown-red, blue-white] [MORION - Black] [SMOKY - Brown] [ROCK CRYSTAL - Colorless]					
05					
ALMANDINE	BLOOD RED GARNET	7.5	3.9 - 4.2	1.76 - 1.81	METAMORPHIC MICA SCHISTS
06					
ANDALUSITE	ANDALUSITE	7.0 - 7.5	3.1 - 3.2	1.63 - 1.64	IN METAMORPHIC ROCKS
COLORS: Pink-red, Green-brown, Brownish-pink, Gray-green, Olive green. CHIASTOLITE IS A VARIETY.					
07					
ANDRADITE	GARNET	6.5	3.85	1.89	CONTACT METAMORPHIC LIMESTONE
COLORS: Brown, Red, Black, Yellow, Emerald Green.					
08					
AVENTURINE	FELDSPAR	6.0 - 6.5	2.64	1.54	IN VEINS OF WHITE QUARTZ
COLORS: Reddish-brown, Reddish-orange reflective tinges in colorless background. Known as Sunstone.					
09					
APATITE	APATITE	5.0	3.1 - 3.3	1.63 - 1.64	IN ALL TYPES OF EXTRUSIVE ROCKS
COLORS: Yellow, Green, Brown, Red, Blue, Violet, Colorless.					
10					
AQUAMARINE	BERYL	7.5	2.69	1.57	PEGMATITES & ALLUVIALS
COLORS: Blue-green, Sky-blue.					
11					
AZURITE	AZURITE	3.5 - 3.7	3.7 - 3.9	1.73 - 1.84	FOUND WITH MALACHITE
COLORS: Shades of blue.					

MAIN GEM GROUPINGS Page 2

GEMSTONE	MINERAL	HARDNESS RANGE	SPECIFIC GRAVITY RANGE	REFRACTIVE RANGE	SOURCE - OCCURRENCE
12					
BENITOITE	BENITOITE	6.5	3.6 - 3.7	1.73 - 1.80	BLUE SCHIST WITH SERPENTINE
COLORS: Pale to dark sapphire blue with zoning. VERY RARE.					
13					
CHALCEDONY	QUARTZ MICRO CRYSTAL	7.0	2.60	1.54	PEGMATITES & ALLUVIAL
VARIETIES	QUARTZ MICRO CRYSTAL	7.0	2.60	1.54	PEGMATITES & ALLUVIAL
[AGATE - Banded] [ONYX - Black & white layers] [SARDONYX - Red & white layers] [PRASE - Dull green] [SARD - Brown to yellow, red-brown] [CARNELIAN - Red to orange] [JASPER - Red, red-brown, yellow-brown] [CHRYSOPRASE - Apple-green, green-yellow] [PLASMA - Dark dull green] [HELIOTROPE - Green with jasper red spots]					
14					
CHESSYLITE - Another name for azurite.					
15					
CHRYSOBERYL	CHRYSOBERYL	8.5	3.68 - 3.78	1.74 - 1.76	METAMORPHIC MICA SCHISTS ALLUVIALS
COLORS: Colorless, Green, Yellow, Gray, Brown, Greenish-yellow.					
16					
CORDIERITE	CORDIERITE	7.0 - 7.5	2.5 - 2.7	1.53 - 1.55	PEGMATITES, SCHISTS, GNEISSES
COLORS: Blue-violet, Gray-blue, Smoky yellowish, Green-gray. Also known as IOLITE & DICHROITE.					
17					
DIAMOND	DIAMOND	10.0	3.52	2.42	PERIDOTITE ROCKS & ALLUVIALS
COLORS: Colorless, yellow, brown, pink, blue, green.					Found in many USA states.
18					
DICHROITE - Another name for cordierite.					
19					
DIOPSIDE	DIOPSIDE	5.0	3.2 - 3.3	1.67 - 1.70	CONTACT META-MARBLES, ALLUVIAL
COLORS: Pale green, Blue, Whitish, Yellowish, Brown, Dark green, Purple is known as VIOLAN.					
20					
DIOPTASE	DIOPTASE	5.0	3.2 - 3.4	1.64 - 1.71	DRUSE CAVITIES & ALLUVIALS
COLORS: Emerald green with blueish or blackish tint. Also known as Emerald of the Poor.					
21					
DRAVITE - The brown variety of tourmaline.					
22					
EMERALD	BERYL	7.5	2.71	1.57	PEGMATITES & ALLUVIALS
COLORS: Green, yellowish-green.					

MAIN GEM GROUPINGS Page 3

No.	GEMSTONE	MINERAL	HARDNESS RANGE	SPECIFIC GRAVITY RANGE	REFRACTIVE RANGE	SOURCE - OCCURRENCE
23	ENSTATITE	ENSTATITE	5.5	3.1 - 3.3	1.66 - 1.67	HIGH GRADE METAMORPHIC ROCKS ALLUVIALS
	COLORS: Green, Yellowish, Brownish-green, Olive green.					
24	FLUORITE	FLUORITE	4.0	3.1 - 3.2	1.43	PEGMATITES & LIMESTONES
	COLORS: Yellow, Brown, Green, Blue, Violet, Pink.					
25	GROSSULAR	GARNET	7.0 - 7.5	3.65	1.74	CONTACT METAMORPHIC LIMESTONE
	COLORS: Yellow-brown, Orange-red, Colorless, Pale to emerald green.					
26	HESSONITE - The orange-red variety of grossular garnet.					
27	IOLITE - Another name for cordierite.					
28	INDICOLITE - The blue variety of tourmaline.					
29	JADITE	JADITE	6.5 - 7.0	3.3 - 3.5	1.66	METAMORPHIC ROCKS & ALLUVIAL
	COLORS: Emerald green, White, Pink, Brown, Red, Orange, Mauve, Blue, Violet, Black, Green Shades.					
30	KORNERUPINE	BOROSILICATE	6.5	3.3	1.66 - 1.68	ALLUVIAL
	COLORS: Green, Brownish-green, Brownish-yellow.					
31	LABRADORITE	FELDSPAR	6.0 - 6.5	2.70	1.57	IN IGNEOUS ROCKS
	COLORS: Gray background with flashes of blue, green, red, yellow, purple.					
32	LAPIS LAZULI	LAZURITE	5.5	2.7 - 2.9	1.50	CONTACT METAMORPHIC LIMESTONE ALLUVIALS
	COLORS: Greenish-blue, Light blue, spotted with calcite and pyrite.					
33	MALACHITE	MALACHITE	4.0	3.3	1.85	FOUND INTERGROWN WITH AZURITE
	COLORS: Pale green, Emerald green, Blackish-green.					
34	MOONSTONE	FELDSPAR	VARIETY OF PLAGIOCLASE WITH BLUE SHILLER EFFECT.			
35	OPAL	QUARTZ	6.0	2.10	1.45	ROCK CAVITY FILLINGS
	COLORS: White in spectrum colors; Black, Blue, Green, Gray; FIRE - Yellow, orange, red; WATER - Clear.					

MAIN GEM GROUPINGS Page 4

No.	GEMSTONE	MINERAL	HARDNESS RANGE	SPECIFIC GRAVITY RANGE	REFRACTIVE RANGE	SOURCE - OCCURRENCE
36	NEPHRITE	NEPHRITE	6.5	2.9 - 3.1	1.62	TALC-SERPENTINES & ALLUVIALS
	COLORS: Dark green, Creamy white, Sand, Yellow-green.					
37	ORTHOCLASE	FELDSPAR	6.0 - 6.5	2.56	1.53	DRUSE CAVITIES
	COLORS: Pale Yellow, Pink, Blue, Gray, White, Colorless					
38	PERIDOT	OLIVINE	6.5 - 7.0	3.34	1.64 - 1.69	VOLCANIC ROCKS & ALLUVIALS
	COLORS: Shades of Green, Gray-white, Yellow, Brown, Black.					
39	PYROPE	RED GARNET	7.0 - 7.5	3.7 - 3.9	1.73 - 1.76	PERIDOTITES & ALLUVIALS
40	RHODOCHROSITE	RHODOCHROSITE	4.0	3.5 - 3.6	1.60 - 1.80	VEINS, PEGMATITES, SEDIMENTS
	COLORS: Pale pink, sometimes Gray or Brown banded.					
41	RHONDONITE	RHODONITE	6.0	3.5 - 3.7	1.72	LIMESTONE SKARNS
	COLORS: Deep pink with Black marks that tend toward brown.					
42	RUBELLITE - The red and pink variety of tourmaline.					
43	RUBY	RED CORUNDUM	9.0	4.0	1.76 - 1.78	MICA SCHISTS & ALLUVIALS
44	SAPPHIRE	CORUNDUM	9.0	4.0	1.76 - 1.78	PEGMATITES & ALLUVIALS
	COLORS: Colorless, black, purple, violet, dark blue, light blue, green, yellow, orange.					
45	SCAPOLITE	SCAPOLITE	6.0	2.6 - 2.7	1.54 - 1.58	PEGMATITES & LIMESTONE SKARNS
	COLORS: Yellow, White, Pink, Violet, Blue, Lavender, Colorless.					
46	SCHORL - The yellow, green and black variety of tourmaline.					
46	SIBERITE - The violet variety of tourmaline.					
47	SILLIMANITE	SILLIMANITE	6.0 - 7.5	3.1 - 3.4	1.64 - 1.71	PEGMATITES & META-ROCKS
	COLORS: Pale green, Dark green, Blue, Gray, Brown. Also known as FIBROLITE.					ALLUVIALS

MAIN GEM GROUPINGS Page 5

No.	GEMSTONE	MINERAL	HARDNESS RANGE	SPECIFIC GRAVITY RANGE	REFRACTIVE RANGE	SOURCE - OCCURRENCE
48	SODALITE	SODALITE	5.5 - 6.0	2.1 - 2.4	1.48	METASOMATIZED LIMESTONES
	COLORS: Bright blue, White or Gray with green tints.					
49	SPECTROLITE - A variety of labradorite that comes from Finland and is similar to that from Labrador.					
50	SPESSARTINE	GARNET	7.0	4.16	1.80	GRANITE PEGMATITES & ALLUVIAL
	COLORS: Orange-pink, Orange-red, Red-brown, Brownish-yellow.					
51	SPHENE	TITANITE	5.5	3.5 - 3.6	1.88 - 2.05	CONTACT METAMORPHIC LIMESTONE ALLUVIALS
	COLORS: Yellow, Brown, Green.					
52	SPINEL	SPINEL	8.0	3.60 - 3.70	1.71 - 1.75	CONTACT METAMORPHIC ROCKS ALLUVIALS
	COLORS: Red, Blue, Purple, Mauve, Violet, Dark Green, Brown, Black.					
53	SPODUMENE	SPODUMENE	7.0	3.1 - 3.2	1.66 - 1.68	PEGMATITES, ALLUVIALS
	COLORS: Lilac-pink is called kuntzite; Bright green is called hiddenite.					
54	SUNSTONE - Another name for Aventurine oligoclase feldspar.					
55	TOPAZ	TOPAZ	8.0	3.50 - 3.60	1.62 - 1.64	PEGMATITES & ALLUVIALS
	COLORS: Yellow, Blue, Green, Violet, Reddish-brown, Colorless					
56	TOURMALINE	TOURMALINE	7.5	3.06	1.62 - 1.64	PEGMATITES & SCHISTS
	COLORS: Red, Pink, Blue, Brown, Violet, Yellow, Green, Black, Colorless, also Pink center with Green surround.					
57	TURQUOISE	TURQUOISE	6.0	2.6 - 2.8	1.62	ENCRUSTATIONS OR ROCK VEINS
	COLORS: Sky-blue, Green.					
58	ZIRCON	ZIRCON	6.5 - 7.5	4.5 - 4.7	1.92 - 2.02	IGNEOUS ROCKS & ALLUVIALS
	COLORS: Yellow, Red, Orange, Brown, Colorless, Sky-blue, Yellow-green, Bright green, Dark green.					
59	ZOISITE	ZOISITE	6.5	3.3 - 3.4	1.69 - 1.70	METAMORPHIC ROCKS, VEINS
	COLORS: Blue, Pink, Green, Yellow, Brown. The pink variety is called thulite.					

CHAPTER 61

TERTIARY CHANNELS OF GOLD

The fondest wish of a dedicated goldpanner or Rockpecker is the discovery of a hitherto unknown "TERTIARY CHANNEL" in an area that can be claim-filed. The term is descriptive, in that "channel" refers to a streampath, watercourse or even a lake that in the long-ago carried or received the products of erosion, but is no longer visible as such. "Tertiary" refers to a geological period during which time catastrophic events occurred at various intervals to erase any watercourse evidence from view. What were these devastating powers that could cover, alter or disguise an ancient streampath?

First, a bit of background on the available geologic thought. Call it Earth history theory.

"Tertiary" is a term used to define one of the geological time scales in the Earth's developmental history called the TERTIARY PERIOD. In terms of time, the period had a duration of 65 million years which ended about two million years ago. The Tertiary Period along with the Quartenary Period embrace a time frame known as the CENOZOIC ERA. Not all geologists agree with this time scheme, but they do agree that the Cenozoic Era, based on fossil evidence, was the AGE OF MAMMALS.

During the Tertiary important landmass reactions took place which elevated the continents to produce the Alps and Himalayan Ranges from ocean sediments. Marine fossil evidence can be traced in Tertiary rocks from the mountain tops to the sea. The Arabian Peninsula was split from the continent of Africa to create the Red Sea. These dramatic movements and upheavals are known as the "Alpine Revolution" which ended the MIDDLE TERTIARY PERIOD leaving its mark as one of the Earth's greatest, severest surficial disturbance.

At the beginning of the Tertiary 65 million years ago, North America had achieved its present shape and size, but was still attached by Greenland to the European landmass. Later events rifted Greenland from Europe to make the opening for the North Atlantic Ocean, and still later for the Greenland separation from North America.

The Tertiary extensively altered the Pacific Coast of North America and its Coast Ranges. Massive lava flows, cinder, and pumice outpourings were widespread into Washington, Oregon and Idaho. The basalts in the Columbia River area are 10,000 feet thick, and have been measured as containing at least 23 separate lava flows piled one atop the other, extending through an area one half million square miles.

The Pacific Coast, Sierra Nevada placer goldfields were formed over a wide area of the western slopes. Roughly, 100 miles North and South by 70 miles East and West, generally describes the area. These are called *HIGH LEVEL* Tertiary and Quartenary gravels. They are the erosion product of many thousands of feet of quartz veins. The gravels containing the gold eroded to allow the free gold to travel.

Many active, explosive volcanoes blanketed the American West with deposits of ash, cinder and pumice hundreds of feet in thickness. The process of erosion has since worked to expose the pre-eruption topography. This was the Tertiary phase that covered streambeds with lava, cinders, ashes and pumice to force the watercourse to carve new pathways through the landscape. The re-routing of waterways to hide the ancient pathways had another contributing factor at work too. That was a general uplift of a large part of North America. Growth of the Rocky Mountains had raised them over 3000 feet. Subsidence between the Sierra Nevada and Wasatch Ranges created water-catch basins for future Ice Age lake formations. The uplift, in effect, rearranged the surface so that existing stream channels and flood plains had to find entirely new travel lanes and run-off pathways. The tertiary channels had been erased from view under the impact of eruption debris. The uplift, with its folding and faulting of stratified layers, changed streamcourse and flow direction. The new water courses were no longer a clue to the past.

We are indebted to the Tertiary Period for the products of its activity. That is, most of the metal-bearing ores of Western North America and South America were deposited. Gold, silver, lead, copper, zinc and mercury are widespread and associated mainly with igneous intrusive rocks. Almost everywhere these deposits have been eroded. Native metals such as gold, silver, tin, and platinum were released to form placer deposits. Non-metallic minerals such as clay, gypsum, salt, phosphate rock and building stone are mined from Tertiary rocks or Tertiary sedimentation.

At this stage in our review of Tertiary Channels we have them hidden from view under tons of volcanic debris. Next we will examine events that occurred in the last two million years of the Tertiary Period. This was the beginning of severe climate changes which culminated in the Ice Age. This is what produced the water necessary to cause erosional forces to release the native metals from their intrusive igneous rocks.

The North American ice sheet blanketed nearly all of Canada and the United States as far South as the Ohio and Missouri Rivers. Large independent mountain glaciers formed in the Rocky Mountain and Sierra Nevada Ranges. Great areas of land, equal to a continent, were alternately laid bare and inundated by each freeze-melt interval. Accordingly, this choice living territory alternated between man, marine and land plants and animals, as new living space became available or was lost. Far from the actual ice, tropical regions experienced precipitation changes and stream volume variations. Each advance of the glaciers forced total depopulation of habitat species throughout millions of square miles of the Earth's surface. In terms of geological time, the end of the Tertiary, known as the Pliocene, and the beginning of the Quartenary, known as the Pleistocene Epochs, were the Ice Ages. It was the freeze-thaw cycles that produced the water that fed the run-off into watercourses. Weathering and erosion worked their magic to reduce and relieve once deep-seated rocks of their mineral content including the resistant precious metals.

During the tectonic events of the Tertiary, volcanic ash and other extrusive materials buried the placers. Earthquakes activity upset the direction and flow of the gold-bearing streams. The Tertiary gravels are said to lie in the remains of old valleys, high on the valley slopes. The author believes this is only partially true. Since the volcanic cap overlying the hidden Tertiary streams is more resistant to erosion, it is more probable that the fossil placers now lie at, or near, the *top* of the valley slopes. This would be especially true if the capping is of basalt. (lava)

During the Quartenary Period, new streams cut the old Tertiary channels to reconcentrate much of the Tertiary gold that became the basis for the 1849 Gold Rush. The richest gold lying in the channels of the East-West streams has been recovered. Many smaller channel deposits remain for your discovery.

Altered streamcourses flowed over and through the ancient, now dry, streambeds to wash their golden particle content into new streamcourses onward toward the sea. Where an ancient Tertiary channel was cut, it would not be long before a Rockpecker was following that golden trail to its source.

It has been said that many of these Tertiary channels lie as yet undiscovered, and that the unfound gold exceeds that which has already been located.

Even if you locate a Tertiary channel, there is no assurance that it will contain fossil gold. Consider for a moment the fact that deposition in Tertiary time was no different than it is now. There had to be catchment factors favorable for gold to deposit, such as a bend in the Tertiary stream. In real world terms, locating the gold in a Tertiary channel is very different from a non-Tertiary stream search. It becomes much more difficult because of the overburden ccient streambed. How can you really distinguish the old streamcourse from other ground? There is a clue you can follow if you are back-tracking gold finds. If the gold source is Tertiary, then you may have located the stream bend or other place of deposition. But chances are, that if you find the source-gold, it will be locked in a cemented conglomerate or sandstone. It will require hard-rock drift mining methods to recover.

Based on the erosion deposits in the lower valleys, it has been calculated that the Sierra Nevada's were erosionally reduced 10,000 feet in height. After being erosion-freed and transported by ancient river systems, the gold available in favorable settling sites was locked in place by the volcanic materials that choked and covered the streams. Where pumice was the covering material, *white channels* were formed when the pumice turned to rock. This was only possible if the pumice was later covered by a lava cap that provided a more erosion resistant shielding.

ROCKPECKER TIP # 206 – Where your prospecting travels carry you to a formation that lies capped with pumice, cinders, scoria or lava flow, search beneath the pumice strata. The overlying material may hide an undiscovered Tertiary Channel holding its golden treasure in a matrix of sandstone, mudstone or claystone for your discovery.

ROCKPECKER TIP # 207 – Where you suspect a high-lying Tertiary channel, but can find little gold particle evidence, your metal detector can be of help in locating black sands. Higher concentrations of black sands, together with rounded quartz pebbles are clues to an ancient river channel. Pay particular attention to the mix of rounded gravels with other types of materials. Take frequent samples to bedrock. If you locate gold particles in your sample, look closely to see if water-worn alluvial-smoothing is apparent. You would normally expect to see both black sands and abraded gold particles in any alluvial deposit. WATCH FOR WHITE OR BLUE ROUNDED QUARTZ GRAVELS. THE BLUE COLOR WILL MOST LIKELY BE SURFACE COLORING OVER THE WHITE QUARTZ.

CHAPTER 62

THE LOST UNCLE JOE MINE

"Take these keys, unlock the box, and bring the contents to the hospital." These were the instructions he gave me during my visit.

I always called him "Uncle Joe". In point of fact, he was the brother of my aunt's husband, which gave no blood relative status to me, or does that make him my uncle twice removed? It really doesn't matter. We knew each other, but not well. But then, no one knew Uncle Joe very well. He was a loner; kept to himself; a free spirit who never married. His penchant for privacy irritated my aunt. He never told anyone where he had been after a two or three month absence. As for never holding a steady job, he lived fairly well; made a point of keeping his small bungalow in good repair. Because his well-being worried my aunt, who transferred the agony to my mother, I was sent, like a spy, to discover where he was spending those months of absence. As I thought about my espionage mission, I concluded that I would not satisfy anyone's curiosity, including my own. But what can you expect from a lad of ten?

As the years passed, I saw him less frequently. I never pried into his affairs, so as to have nothing to report. Then the message to see him came from the hospital.

He needed a shave, but looked to me quite healthy for a man over eighty. He handed me two keys with the statement, "Bring the box contents as soon as you can. I may not have too long to live". He uttered a deep cough; I saw blood on the tissue.

I hastened to the hiding spot, opened the home-made safe he had expertly hidden in a closet wall, reached to fetch the contents. It contained a heavy sealed box weighing several pounds. By this time my curiosity level was at a breaking point. Without tampering with the box, the next day I turned it over to him at the hospital. Carefully he examined the exterior of box, then slid it under the bed sheet. I headed toward the door, so that he could enjoy his privacy.

"Hey, don't leave, I want to give you something!". Over the years the only thing he ever gave me, were apples from his backyard tree. His hand reached into the now-opened box handing me a gold nugget that covered my entire palm. "I want you to use the proceeds from the sale of this nugget to have this monument placed on my grave." His hand held a crude drawing of a kneeling prospector holding a gold pan. "Anything left over is yours to keep. For your trouble this smaller nugget is yours to remember me by." The second nugget was a little less than half the size of the first. Too stunned to question, I stared at the piece of gold that was to be mine. "It weighs nearly four troy ounces, it's all yours."

"Aren't you going to ask where I found the nuggets?", he coughed again. " Well, I brought you here to tell you about my mine, more specifically to give you location directions. I want you to have it; to enjoy it as I have."

My heart rate must have instantly doubled; my eyes bugged out; I felt my jaw slacken. If ever I were at a loss, the glint of sunlight on these shiny freaks of nature struck me dumb, causing my insides to churn with excitement.

Our eyes met and locked into a silent stare. "Yes, I'll tell you where! Perhaps it will serve you as well as it has me." Was that a sneaky grin of pleasure I noted on his lips?

He sucked in a gasp of air, then let out another cough. He began his story:

"I was a young twenty at the start of the Great Depression.

With no jobs available and prospects ever diminishing, I decided to head for the gold fields to try gold panning to make ends meet. Survival was my only objective; Coloma on the south fork of the American River was the only place I remembered from 49'er folklore. I was determined to survive, no matter what!

I settled in my first camp on the American a few miles upstream from Coloma a few weeks later. I was a green pilgrim who lucked out by meeting and teaming up with an old timer. He taught me the basics of panning along with tricks of the trade. In a short time I was adept at "making beans". I would outwait the damned Depression.

One day the old timer announced that he intended to prospect a few miles upstream to "look over the ground". He thought he saw a quartz ledge high up near the top of a cliff. It might be too steep for him to reach, but he wanted "to give it a try".

"Don't look for me for two or three days," he said as he packed his rucksack. When four days passed with no return, I prepared to search for him. I was miles upstream a few days later. I could find no evidence of any apparent mishap. Was he O.K.? On the seventh day of search, I turned back toward camp. Perhaps he was waiting there for me to return. In an area of high cliffs, the stream swiftly narrowed. I lost my footing to slip rearend backwards into the stream. In the course of recovering from my fall, I saw a movement high on the south cliff wall. It was just a peripheral eye catch; it could have been wind in the brush.

As I steadied upright, a mountain lion sat high on a game trail just watching me flounder around. My eye followed the game trail down to the stream line, then back up again. That goat trail disappeared in some places behind brush. That's when I found the old timer. Midway up the cliff, he was hanging, choked by rucksack straps that had caught on the brush as he tumbled downward. From my vantage point, he appeared to be dead. Higher up on the game trail, I saw the white ledge that had lured my partner to his death. The next day, I buried my pard. I experienced a sense of loneliness that I had never felt before or since. I had depended on that man.

During the next days I moped listlessly thinking of my next move. With not enough gold to survive for long, I decided to spend the remaining good weather panning. I wondered what drove my pard to try that neck-snapping climb. He knew he was too old for that kind of exertion. At that moment I reached a decision; I would check out his killing quartz ledge.

It took me a whole day to climb up to the quartz vein, only to find out it wasn't quartz or a vein of anything. It turned out to be a four inch white strata line overlying some blue and white rounded gravel locked in conglomerate rock.

I had found what appeared to be the remnants of a fossil placer capped with a white pumice-like rock, with a narrow game trail near the base of the gravel. But I didn't know that at the time. I thought, what a waste of life for this bit of nothing.

I decided to follow the game trail upstream to look for a highrise cliff exit. This precipice wasn't going to claim me too. The trail just dead-ended.

A short distance along that highrise trail, I saw a small cave-like depression dug in the conglomerate wall. As I peered into the opening, I could see a man's skeleton on the cave floor. A shiver ran over me as I noticed two sets of rattler tails near him. Was he double-hit by those pit vipers, only to expire from the venom? Double-strike poison is just too much to overcome.

A search of the seven foot deep tunnel led to my discovery of a pick, shovel, chisel, hammer and a leather poke with those two nuggets covered by another ten ounces of smaller nuggets. From the contents of the sack, he just didn't bother collecting gold dust.... I didn't bother either

Because the discovery was so well hidden and difficult to reach, I erased the pumice line from view to protect the find. I also decided to avoid filing a formal claim and to keep the gold source secret. For the past fifty years, that gold has given me a good living for a couple months work each year. I determined not to be greedy. To take only enough gold to live modestly. I didn't want to fabricate any sudden wealth explanations. However, I did stash enough for my old age years.

I haven't visited my secret mine in more than ten years. The gold you hold is all that's left, but that's O.K., because I won't be needing any more.

" 'Cause they are tricky, write down these directions".

I wrote furiously, and in detail, as he dictated. He was correct when he said "tricky". I knew that without the notes, I could never find his mine. I was pleased with his confidence in me. The only string on the deal involved the burial marker. I guess that in death he wanted everyone to know he was a gold prospector. I agreed to honor his request as I left with the box, notes and nuggets. As the hospital room door was closing, I heard another coughing seizure. I determined to put the box and contents back where he kept them. He looked too healthy to die.

When his will was read, I was named his sole heir. He gave me the house with its contents all free and clear. I arranged for his custom grave marker making sure it was well placed. It did generate a lot of questions and comments. He would have been pleased.

I pondered my next step. I knew I would be off prospecting as soon as the weather permitted. I had a secret gold mine.

About a week after Uncle Joe's funeral, his house burned to the ground along with my discovery notes which were unreadably charred. I should have made a back-up copy. I prepared to set down the location directions again, searching my memory banks for each clue. I was confident that the copy was accurate.

As soon as spring arrived, I was on my search path. It was a. hopeless trip. Uncle Joe or I had forgotten some important detail. All collateral evidence was erased during the ten years of Joe's absence. All I have to show for the effort is this four ounce nugget. I call it my "Uncle Joe". It's also all I have to back up my story.

Uncle Joe thought he had the remnants of a tertiary stream bend that may have been responsible, in great part, for the 1848 Coloma gold discovery. I like to think so too.

But then again, my mother may have been right about Uncle Joe. She consistently called him a lying, practical joker. Was I his last victim? Still, I had that four ounce nugget, or was it merely a gold plated natural copper nugget? I really don't want to know. REST IN PEACE UNCLE JOE!

AUTHOR'S NOTE: The author believes that there are no lost mines, only lost mine stories; imaginative fantasies that thrill the reader with a pleasurable excitement. The "Lost Uncle Joe Mine" story is presented here to illustrate that point.

Copyright 1993 - T. D. LaRune

LaRune's ROCKPECKER Notes

PART THREE

LaRUNE'S

277 MINERAL RECOGNITION NOTES

PHYSICAL IDENTIFICATION TESTS

PART FOUR

LaRUNE'S

COOKBOOK CHEMISTRY NOTES

CHEMICAL IDENTIFICATION TESTS

Copyright 1993 by T. D. LaRune

All rights reserved

Memo

to: MINERAL PROSPECTORS, ROCKHOUNDS, STUDENTS

from: T. D. LaRune

" ROCKPECKER ™ v.1 DOS PROFESSIONAL "

DESCRIBE, IDENTIFY & CHEMICAL TEST UNKNOWN ORE

PAGE 412

Computerized Versions

of PART THREE & PART 4 ARE AVAILABLE

NOW FOR PC's

DOS 3.3+, 400K+ RAM,

XT\486 COMPUTERS.

277 MINERALS ARE DETAILED.

Prospector can identify unknown mineral / ore;

then PRE-ASSAY chemical test to confirm finding.

WRITE TO: Skill//Quest Co.

675 Fairview Drive, Suite 246,

Carson City, Nevada 89701 U.S.A.

PART THREE

LaRUNE'S

MINERAL IDENTIFICATION

NOTES

PHYSICAL TESTING DATA

FOR 277 COMMON MINERALS

From the book

LaRune's ROCKPECKER Notes

A Mineral Prospector's Primer

COPYRIGHT NOTICE:

COPYRIGHT 1990 BY T.D. LaRUNE.
All rights reserved

No part of this publication may be copied or distributed, transmitted, transcribed, stored in a retrieval system, or translated into any human or computer language, in any form or by any means, electronic, mechanical, magnetic, manual, or otherwise, or disclosed to third parties without express written permission of: T. D. LaRune, P.O. Box 620995, San Diego, California 92162 U.S.A.

CHAPTER 63

THE LaRUNE METHOD OF MINERAL IDENTIFICATION USING THE PHYSICAL TESTS OF POWDER COLOR STREAK MOHS HARDNESS SPECIFIC GRAVITY LUSTER

TABLE OF CONTENTS

MINERAL CHARTS

STREAK --- HARDNESS SPECIFIC GRAVITY --- LUSTER

DETAILED EXPLANATIONS

LaRUNE'S IDENTIFICATION METHOD

The LaRune mineral identification method takes the physical properties of *STREAK, HARDNESS, SPECIFIC GRAVITY, LUSTER* to eliminate the minerals that do NOT meet those TESTS. Of the minerals that DO FIT the tests, only a few will show as our TESTED PROBABLE MINERAL. We may wind up with three possibles, because we are using a RANGE of hardness and specific gravity. For example: Our STREAK TEST shows the powder as a shade of PINK. CHART C GIVES US 3 POSSIBLES. Of the three, a LUSTER finding of VITREOUS, TRANSLUCENT eliminates 133 HAUERITE. Had our LUSTER finding been SEMI-METALLIC, OPAQUE, we would have identified # 133 HAUERITE. A hardness test reading of 2.0 eliminates 125 RHODOCHROSITE, and identifies 009 ALURGITE. This is a simplistic example because I picked a streak with only 3 possibles. If you had 25 or 100 alternatives to eliminate, you would need to display precision in your tests to maximize PROBABLE certainty.

It is too much to expect precision measurement until a familiarity with the method is developed. MUCH IS PRACTICE AND SIMPLE LOGIC.

After completing the streak, hardness, specific gravity and luster tests, the remaining probables are noted and identified by their chemical name. (CHART A, column -F-) If you have three probables, two of which have manganese and iron content, the third has copper and iron. A *chemical test* (See Part 4) for iron would tag all three. A chemical test for copper shows the element positive or indicated. You still are not sure yet. Suppose that the chemical name identifies the three minerals as, a carbonate, a silicate, and the third, a sulfide. Your copper probable is the sulfide. Running the sulfide test confirms your test mineral as # 068 CHALCOPYRITE, COPPER IRON SULFIDE.

STREAK TEST

CHART C, COLUMN – G –

COLOR STREAK – POWDER TECHNOLOGY

The most obvious physical property of a mineral is its color. If you are trying to identify your mineral by color alone, chances are you will be fooled. The inclusion of various elements in the atomic structure of a mineral, (CHART F, COLUMN -G- and -T-) can make it appear in a wide range of colors.

ROCKPECKER TIP # 208 – If relied upon as the sole test, the overall color of a mineral, with a few exceptions, is at best a guarantee of ambiguity.

The *STREAK* of a mineral is the powder color of the mineral as it abrades and shows on a *STREAK PLATE*, which is nothing more than the unglazed side of a ceramic tile. A ceramic tile has a hardness range of 6.0 – 6.5. A mineral with a hardness over that range will not powder on a streak plate.

ROCKPECKER TIP # 209 – A two inch plumber's cast iron end cap, along with a one inch steel rod, long enough to fit the hand, makes a good mortar and pestle for grinding your mineral to powder. A hammer head can be substituted for the steel rod. (You will need this equipment to do the chemical tests in Part IV.)

Place the powder on a sheet of pure white paper to discern your powder color. THINK OF THE POWDER COLOR AS A COLOR RANGE OF SHADES OF BROWN OR BLUE ETC. Rather than try to distinguish the color as blue-gray or brownish-black, test STREAK for blue, then gray; or brown, then black. It is a process of elimination.

Dark minerals of a metallic nature often have a powder streak that differs from the overall mineral color. For example: Mineral # 135 - HEMATITE is commonly black, yet its powder streak will show red with sometimes a brownish cast.

ROCKPECKER TIP # 210 - In employing the streak test make sure of two things. #1 - That the streak is all mineral and not a portion of ADHERING country rock which may disguise the streak. #2 - Before you run your streak, make sure you are working on a fresh break. Surface weathering can deplete the specimen of some of the more soluble elements and give a false color streak.

The streak test is down-played by many mineralogists, probably because of the many minerals that streak white. However, for the Rockpecker who needs to know in the field, the streak test can be helpful. For example: You find a BLACK mineral that streaks BLACK. Of the 277 minerals shown in CHART -C-, the CLUE CODE COLUMN -C- shows 18 with a BLACK STREAK. You have, with this single test, eliminated 259 minerals from consideration.

IN THE STREAK CHART C, COLUMN -G-, YOU ARE ONLY DEALING WITH THIRTEEN COLORS, NOT THE SHADES OF COLORS THAT CAN RUN IN THE THOUSANDS. THE STREAK COLORS TO CONCENTRATE ON ARE VARIATIONS OF:

1 - BLACK, 2 - BLUE, 3 - BROWN, 4 - COLORLESS, 5 - GRAY,
6 - GREEN, 7 - ORANGE, 8 - PINK, 9 - RED, 10 - SCARLET,
11 - SILVER, 12 - WHITE, 13 - YELLOW.

I suggest to you that the *color streak clue code* has an important preliminary recognition value. By performing the other physical tests, you further reduce the possibles to pinpoint your tested-for mineral. Your next test should be:

HARDNESS

HARDNESS is defined as the resistance to abrasion, general wearing away, or scratching. The operative word is "resistance". A scratch is a small furrow plowed into the surface of a solid which breaks the atomic bonds. Resance to scratching depends on the strength of the atomic bonds. Recall, the streak plate of ceramic tile had a hardness range of 6.00 - 6.50. Harder minerals need other methods for powdering.

The MOHS hardness scale was drawn up by and named for Frederich Mohs. (1773 - 1839) Based on the resistance to scratching, MOHS scale of hardness is only a COMPARATIVE scale. A rough scale, that even today, is a generally accepted guide. More important to a Rockpecker it is convenient, and can be used in the field.

MOHS scale is set up so that the lower *hardness number* assigned could be scratched by a mineral of a *higher hardness number*. There is available a set of pencil-type tools with hardness point made of the minerals in the scale. Jewelers are the primary users, but that does not mean you should overlook these convenient tools. The set I have, consists of three aluminum pencils with six sides to keep them from rolling. Each end of a pencil, which is 3.5 inches long, has a hardness mineral firmly inset the end. That gives me 6 test points. My hardness points are #5 & 6, #7 & 8, # 8.5 & 9. All are fitted in a small leather pouch. You will find them convenient to use.

When you field-use the hardness test, it is important that you test a DRY mineral on a fresh mineral break or cleavage, rather than on a softer weather-worn surface. The MOHS CHART B, COLUMN MOHS SHOWS HARDNESS NUMBERS AS A RANGE OF HARDNESS. One reason for this involves crystal system and bonding. KYANITE (149/216) has a hardness of 5.0 parallel to its crystal length, but a 7.0 hardness across the length. This boils down to scratching with the atomic chain or across the chain. We chart the mineral as 6.0-7.0. Without multiple tests, you could be following the CLUE-CODE CHART to error.

ROCKPECKER TIP # 211 – Hardness is an extremely useful testing property, but care should be taken to test hardness on a smooth, fresh surface. Your mineral specimen may have surface alterations, grains or splinters that appear to submit to the test. Be sure your scratch is not merely displacing surface fragments.

ROCKPECKER TIP # 212 – Always take two hardness readings, one on each different surface. Some minerals can give you a different read on each surface. Always take the hardest reading for accuracy.

Where you are working with a known mineral hardness, you can confirm the hardness by using a needle number below the known hardness to make your scratch. If there is no furrow, go to the next higher hardness needle. If that needle does not produce a visible mark either, you know the material is at least the hardness of the last needle test. This test is most often used by jewelers to distinguish a genuine ruby from spinel or red tourmaline, though now-a-days the better jewelers use an index of refraction instrument to avoid a scratch mark on the gem.

ROCKPECKER TIP # 213 – The simplest method to determine hardness is to test with a higher hardness and work down in decreasing hardness until the material cannot be scratched anymore. The hardness of the examined material then equals the hardness of the test needle that was not able to scratch the material.

YOU NEED NOT OWN A SET OF HARDNESS PENCILS TO CONDUCT SCRATCH TESTS, OTHER COMMON MATERIALS, AS SHOWN BELOW CAN BE EFFECTIVELY EMPLOYED.

MOHS arranged his scale in a sequence-of-hardness numbers ranging from 1 to 10. He assigned a 10 to the diamond; a 1 to talc.

The following hardness scale was devised by MOHS: TO THE RIGHT OF EACH HARDNESS NUMBER IS SHOWN A COMMON MATERIAL USEFUL TO YOUR HARDNESS TEST.

MOHS SCALE OF HARDNESS	*CAN BE SCRATCHED BY:*
# 1.0 TALC	GREASY OR SOAPY FEEL
# 2.0 GYPSUM	# 2.5 ------ FINGERNAIL
# 3.0 CALCITE	# 3.0-4.0 -- COPPER PENNY
# 4.0 FLUORITE	# 4.0 -- COPPER PENNY
# 5.0 APATITE	# 5.5 -- STEEL KNIFE BLADE OR WINDOW GLASS.
# 6.0 ORTHOCLASE (Feldspar)	# 6.5 -- STEEL FILE OR STREAK PLATE.
# 7.0 QUARTZ	# 7.0 -- QUARTZ
# 8.0 TOPAZ	# 8.0 -- TOPAZ
# 9.0 CORUNDUM	# 9.0 -- CORUNDUM
# 10.0 DIAMOND	# 10.0 -- DIAMOND

ROCKPECKER TIP # 214 - Minerals under 2.5 will leave a mark on paper; those under 5.5 can be scratched by a steel knife; those over 5.5 will scratch window glass. Corundum will scratch all materials of 9.0 or lower.

ROCKPECKER TIP # 215 - A fair test guide rule breaks all minerals into two groups; those that can be scratched by a knife blade and those that cannot. THIS IS USUALLY A GOOD TEST-STARTING POINT.

Remember, these are only relative tests. A mineral with a hardness of 9.0 is not three times harder than one with a hardness test of 3.0.

FINDING SPECIFIC GRAVITY — THE SIMPLE WAY.

FIRST SOME BASICS:

1. Specific Gravity is a ratio measurement. SPECIFIC GRAVITY OF A MINERAL IS THE RELATIVE NUMBER INDICATING HOW MANY TIMES HEAVIER THE MINERAL IS THAN THE SAME VOLUME OF WATER. It is the ratio of the mass of a body (mineral) as compared to a mass of an equal volume of water at a specified temperature.

THEREFORE, IF WE WEIGH OUR MINERAL IN AIR TO GET A MINERAL WEIGHT OF ....14 GRAMS.

THEN WEIGH OUR MINERAL IN WATER. DRAW OFF THE WATER THAT THE MASS OF THE MINERAL DISPLACED, FOR A DISPLACED WATER WEIGHT OF .... 6 GRAMS.

NOW WE HAVE A RATIO OF WATER WEIGHT DISPLACED OF 6 GRAMS,
AND
A MASS WEIGHT OF MINERAL IN AIR OF..............14 GRAMS.

IF WE DIVIDE 14 BY 6 WE GET A SPECIFIC GRAVITY OF 2.33 FOR OUR MINERAL.

BUT WE HAVE A PROCEDURAL PROBLEM. THAT IS, HOW TO WEIGH THE MASS OF WATER DISPLACED BY THE MINERAL.

THE ANSWER LIES IN KNOWING HOW MUCH WATER TO DRAW OFF. WE DON'T PHYSICALLY NEED TO WEIGH THE DRAWN OFF WATER TO GET OUR ANSWER.
WE CAN GET OUR ANSWER A LOT EASIER.

IF BEFORE WE PUT OUR MINERAL INTO THE SEE-THRU WATER CONTAINER, WE MARK THE WATER DEPTH WITH A PENCIL MARK. PUT OUR MINERAL INTO THE WATER AND WEIGH IT. THEN WITH THE MINERAL STILL IN THE CONTAINER, WITHDRAW WATER DOWN TO OUR PENCIL MARK, WE CAN REWEIGH THE MINERAL IN THE REDUCED WATER TO CALCULATE HOW MUCH WAS WITHDRAWN. NOW SUBTRACT TO FIND THE WATER DRAWN OFF.

SO MUCH FOR BASICS:

LaRUNE's SPECIFIC GRAVITY METHOD

EQUIPMENT NEEDED

1. A scale to measure the weight of the mineral in question. As long as all the weights are taken in the same weighing units, it does not matter whether the scale unit is measured in ounces or grams. Gram scales are more suited to this purpose.

2. A glass container or wide mouth beaker.

3. A pipette or a turkey baster.

THIS METHOD OF DETERMINING SPECIFIC GRAVITY IS ACCURATE TO THE SECOND DECIMAL PLACE.

TEST ACTION:

1. WEIGH YOUR DRY MINERAL IN AIR. (Example: 14 grams.)

2. TAKE YOUR WIDE MOUTH GLASS CONTAINER AND FILL IT WITH ENOUGH WATER TO COVER YOUR TEST MINERAL. (But do not insert the mineral yet.)

3. MARK THE WATER LEVEL IN THE GLASS WITH A THIN PENCIL LINE. (After the mineral is weighed in water, this line will tell you how much water to draw off.)

(Make sure the specimen is completely covered with water.)

5. WEIGH THE GLASS, WATER AND SPECIMEN. NOTE THE WEIGHT. (Example: The weight is 152 grams.)

6. WITH THE SPECIMEN STILL IN THE GLASS, DRAW OFF ENOUGH WATER TO REACH THE MARKED LINE ON THE GLASS. (A turkey baster works fine.)

7. WEIGH THE GLASS, REDUCED WATER AND MINERAL AGAIN. NOTE THE WEIGHT. (Example: We now read a weight of 146 grams. Remember, we reduced the water level, so it should weigh less.)

8. NOW HERE COMES THE MATH: Take the weight in # 5 of 152 grams and SUBTRACT the weight found in # 7 of 146 grams. THE DIFFERENCE of 6 is THE WEIGHT OF THE REMOVED WATER. IT IS ALSO THE WEIGHT OF THE MINERAL IN WATER.

9. WE NOW HAVE OUR RATIO OF 6 FOR THE WATER REMOVED, (This is the weight of the mineral in water) TO 14 FOR THE WEIGHT OF THE MINERAL IN AIR.

10. DIVIDE THE WEIGHT OF THE DISPLACED WATER (6) INTO THE WEIGHT OF THE MINERAL IN AIR (14) TO GET THE SPECIFIC GRAVITY OF OUR SPECIMEN. (14 divided by 6 equals 2.33. Our mineral has a specific gravity of 2.33.)

There are many methods that can be used to accurately determine SPECIFIC GRAVITY. The LaRune method requires the least cost outlay for equipment. He now uses a Health o meter[R] brand, digital, office, ounce\gram postal scale. This scale reads from 1 to 1,000 grams; ounces to 2 lbs., 3 oz. The scale lists for under $50.00. It can be found on sale for $ 35.00.

ROCKPECKER TIP # 216 – Make sure you are dealing with a mineral as distinguished from a rock. The mineral will be an entire piece of material, of any size, of IDENTICAL COMPOSITION; the rock will be a MIX OF MANY MINERALS as an entire whole. FOR A ROCK, SPECIFIC GRAVITY WILL BE VIRTUALLY MEANINGLESS AS AN INFORMATIONAL FACT.

ROCKPECKER TIP # 217 – To insure against unintentional errors, run specific gravity tests twice for each mineral.

ROCKPECKER TIP # 218 – IF YOU SEEM TO BE GETTING A SPECIFIC GRAVITY READING LOWER THAN YOU EXPECTED, THERE COULD BY AN AIR POCKET OR BUBBLE WITHIN THE TESTED SPECIMEN. TRY ANOTHER SPECIMEN FOR THE NEXT TEST. If you retest the same specimen, positively make sure that it is dry, or that contained cavities are not holding water. The bubble adds buoyancy, the water cavity adds weight.

LUSTER —

TRANSPARENCY & TRANSLUCENCY

LUSTER is the way a mineral reflects light. The words *shining, glistening, glimmering, sheening, and splendent* more closely express the idea of luster. EXCEPT IN THE CASE OF METALS, LUSTER DOES NOT AFFECT THE COLOR OF A MINERAL. Light also reflects overall color of a mineral. Except for cinnabar - red, malachite - green, graphite - black, and a few others, COLOR of a mineral is so variable that it serves no diagnostic purpose, whereas LUSTER IS AN IMPORTANT AID IN THE IDENTIFICATION OF MINERALS.

The most common type of luster is VITREOUS or GLASSY. Minerals with a vitreous luster are usually also TRANSPARENT or TRANSLUCENT. Where the LaRune charts show "tp" meaning transparent, read that term as simply letting light pass through the mineral in various degrees of clearness. Where the charts use "tl" to indicate translucent, read that term to mean light is trapped inside the mineral in various degrees.

MINERALS ARE GENERALLY CLASSIFIED AS TO WHETHER THEY HAVE A METALLIC OR NONMETALLIC LUSTER.

Metallic lusters are divided into a bright sheen polish for METALLIC, and a duller sheen polish for SUB-METALLIC. ALL ARE OPAQUE "opq", meaning no light passes through the mineral.

LUSTER	MINERAL EXAMPLE
METALLIC [MET]	PYRITES & GALENA
SUB-METALLIC [S-MET]	BLACK SPHALERITE

NON-METALLIC LUSTERS

VITREOUS [VIT] (The luster of glass)	QUARTZ
SPARKLING or ADMANTINE [SRK] (Like glass but with flash)	DIAMOND & CORUNDUM
RESINOUS [RSN] (Like an amber cast)	REALGAR & MONAZITE
GREASY [GSY] (Like an oily film)	SERPENTINE, SOME QUARTZ
PEARLY [PRL] (Like mother of pearl)	MICA, TALC & STILBITE
SILKY [SKY] (Like sheeny rayon fibers)	ASBESTOS
EARTHY [ERY] (Like crumbly soil, no real luster)	LIMONITE
DULL [DUL] (No real reflectivity or luster)	LEUCITE

MINERAL INDEX & TEST CODES CHART A

CHART A HEADINGS

COLUMN — EXPLANATION

A. - MINERAL I.D. NUMBER = This number identifies the same specific mineral throughout ALL CHARTS.

B. - MINERAL NAME & FIELD NAME = Identifies the mineral name. Note: Some minerals are better known by their field name.
Example: Silver Glance for Argentite.
(The charts use the U.S.A. mineral name.)

C. - CLUE CODE = This code number gets you in COLOR STREAK, Chart C for quick pick-up of streak, hardness, specific gravity, luster & mineral color as it relates to the mineral with that clue code.

D. - ORE % = Once you identify your mineral you want to know if it has a market value. Ore % lets you know if your market is for an ore, gem, or collector's item.

F. - CHEMICAL NAME = To avoid chemical formulas, the chemical name is written out.
Example: " Anglesite " is a 68% lead ore.
It has the chemical name " LEAD SULFATE."
RUN LEAD & SULFATE TESTS TO CONFIRM.

G. - ELEMENT SCREENING TEST CODES = This gets you into PART IV, PRE-ASSAY CHEMICAL TESTS. Once you have identified your mineral with the physical tests, you can then confirm the ELEMENT(S) CONTAINED IN THE MINERAL.
(either "positively" or "as indicated")
This nails the I.D. down.

ROCKPECKER CHART TIP: FOLLOW YOUR MINERAL EASILY THROUGH THE CHARTS. If it appears on page 5 of 12 in chart "A", find it on the same page number in chart "B" & chart "D". (But not in "C".)

MINERAL INDEX & TEST CODES

Copyright 1993

CHART A

-A- I.D. #	-B- MINERAL NAME (& Field Name)		-C- CLUE CODE	-D- ORE OF %	-F- CHEMICAL NAME	-G- MINERAL ELEMENT TESTS QUALITATIVE ANALYSIS (P= Positive) (I = Indicated)
001	ACMITE [*Aegirine ^]	* [P]	032	Coll.	SODIUM IRON SILICATE	1-A, 2-E, 1--A-P
002	ACTINOLITE [Nephrite ^]	*	196	Gem	HYDROUS CALCIUM MAGNESIUM IRON SILICATE	3-B-P, 10-C-P
003	ADAMITE		151	Col.	HYDROUS ZINC ARSENATE	1-G-M, 1-T-P, 3-A-P, 11-D-P
004	ADULARIA [*Feldspar ^]	*	211	Gem	POTASSIUM ALUMINUM SILICATE	1-B
005	ALBITE [*Feldspar ^]	*	214	IND.	SODIUM ALUMINUM SILICATE	1-A
006	ALLANITE [*Orthite ^]	* (REO)	061	Ce.5%	RARE EARTH SILICATE	[*[-?-]
007	ALMANDINE [*Garnet ^]	*	224	Gem	IRON ALUMINUM SILICATE	2-E, 10-A-P
008	ALUNITE [*Alum Stone ^]	*	159	Coll.	HYDROUS POTASSIUM ALUMINUM SULFATE	1-B, 12-A-P
009	ALURGITE	*	106	Coll.	HYDROUS POTASSIUM ALUMINUM SILICATE	1-B
010	AMBLYGONITE		206	Li.4.7%	HYDROUS LITHIUM ALUMINUM PHOSPHATE	1-L, 1-T
011	ANALCITE [*Analcime ^]	*	185	Coll.	HYDRATED SODIUM ALUMINUM SILICATE	1-A
012	ANATASE		249	Ti.	TITANIUM OXIDE	2-J
013	ANDALUSITE		186	Gem	ALUMINUM SILICATE	*[-?-]
014	ANDRADITE [*Garnet ^]	*	223	Gem	CALCIUM IRON SILICATE	1-C, 2-E, 10-B-P, 10-A-P
015	ANGLESITE [*Lead Vitriol^]	*	054	Pb.68%	LEAD SULFATE	1-J, 1-S-P, 4-R-P, 12-A-P
016	ANHYDRITE		149	Ca.29%	CALCIUM SULFATE	1-C, 10-B-P, 12-A-P

MINERAL INDEX & TEST CODES

Copyright 1993

CHART A

-A- I.D. #	-B- MINERAL NAME (& Field Name)	-C- CLUE CODE	-D- ORE OF %	-F- CHEMICAL NAME	-G- MINERAL ELEMENT TESTS QUALITATIVE ANALYSIS (P = Positive) (I = Indicated)
017	ANNABERGITE * [*Nickel Bloom ^]	:079	Coll.	HYDRATED NICKEL ARSENATE	2-H, 4-B-I,4-D-P, 3-A-P
018	ANTHROPHYLLITE [A]	:241	Mg.	HYDROUS MAGNESIUM IRON SILICATE	10-C-P, 10-A-P
019	ANTIMONY	:058	Sb.100%	NATIVE ELEMENT	1-H, 1-U-P, 3-C-P
020	APATITE (GROUP)	:180	Gem	PHOSPHATE OF CALCIUM with F. or Cl.	1-C, 1-L, 10-B-P, 14-A
021	APOPHYLLITE	:175	Coll.	HYDRATED POTASSIUM CALCIUM SILICATE	1-B, 1-C, 10-B-P
022	ARAGONITE	:156	Gem	CALCIUM CARBONATE	1-C 10-B-P, 9-B-P
023	ARFVEDSONITE [A]	:025	Coll.	HYDROUS SODIUM (FE.MG.AL.) SILICATE	1-A, 10-A-P, 10-C-P
024	ARGENTITE * [*Silver Glance ^]	:121	Ag.65%	SILVER SULFIDE	4-T,V,W-P, 8-A-P, 9-A-P
025	ARSENIC	:073	As.100%	NATIVE ELEMENT	1-G, 3-A-P
026	ARSENOPYRITE * [*Mispickel^]	:001	As.46%	IRON ARSENIC SULFIDE	10-A-P, 3-A-P, 8-A-P
027	ARTINITE	:146	Coll.	HYDRATED MAGNESIUM CARBONATE	10-C-P, 9-B-P
028	ASTROPHYLLITE	:247	Coll.	HYDROUS (K,Fe) TITANIUM CARBONATE	1-B, 2-E, 2-J, 9-B-P
029	ATACAMITE	:083	Cu.59%	HYDROUS COPPER CHLORIDE	1-F, 2-C, 3-M, 14-A
030	AUGITE [P]	:078	Coll.	CALCIUM (Mg,Fe,Ti,Al) SILICATE	2-J, 10-A-P, 10-B-P, 10-C-P
031	AURICHALCITE	:027	Coll	HYDROUS ZINC COPPER CARBONATE	1-T-P, 2-C, 9-B-P

MINERAL INDEX & TEST CODES

Copyright 1993

CHART A

-A- I.D. #	-B- MINERAL NAME (& Field Name)		-C- CLUE CODE	-D- ORE OF %	-F- CHEMICAL NAME	-G- MINERAL ELEMENT TESTS QUALITATIVE ANALYSIS (P = Positive) (I = Indicated)
032	AUTINITE	*	246	U	HYDRATED COPPER URANIUM PHOSPHATE	1-F, 1-L, 2-C, 2-L
	[*Radioactive ^]					
033	AXINITE		219	Gem	HYDROUS CALCIUM(Fe, Mn, Mg)SILICATE	2-F, 3-B-P, 10-A-P, 10-C-P
034	AZURITE	*	023	Cu.55%	HYDROUS COPPER CARBONATE	1-F, 9-B-P
	[*Blue Carbonate ^]					
035	BABINGTONITE		260	---	HYDROUS CALCIUM (Fe,Mn) SILICATE	10-A-P, 10-B-P, 2-F
036	BARITE	*	136	Ba.58%	BARIUM SULFATE	1-E, 12-A-P
	[*Heavy Spar^]					
037	BAUXITE		173	Al.	HYDROUS ALUMINUM OXIDE	*[-?-]
038	BENITOITE		276	Gem	BARIUM TITANIUM SILICATE	1-E, 2-J
039	BERTRANDITE		274	Be.	HYDROUS BERYLLIUM SILICATE	*[-?-]
040	BERYL		188	Be.5%	BERYLLIUM ALUMINUM SILICATE	*[-?-]
041	BETAFITE	*	277	Th.& U.	HYDROUS URANIUM (Ti,Ni) OXIDE	2-H, 2-J, 2-L
	[*Radioactive ^]					
042	BIOTITE	*	133	---	HYDROUS POTASSIUM ALUMINUM SILICATE	1-B
	[*Black Mica ^]					
043	BISMUTH		122	Bi.100%	NATIVE ELEMENT	1-R, 3-H-I
044	BISMUTHINITE		067	Bi.81%	BISMUTH SULFIDE	1-R, 8-A-P
045	BIXBYITE		006	Mn.	MANGANESE IRON OXIDE	2-E, 2-F, 10-A-P
046	BOLEITE		020	COL	HYDRATED LEAD COPPER CHLORIDE	4-R-P, 1-F, 2-C, 14-A
047	BORAX		232	IND.	HYDRATED SODIUM BORATE	1-A, 1-N

MINERAL INDEX & TEST CODES

Copyright 1993

CHART A

-A- I.D. #	-B- MINERAL NAME (& Field Name)		-C- CLUE CODE	-D- ORE OF %	-F- CHEMICAL NAME	-G- MINERAL ELEMENT TESTS QUALITATIVE ANALYSIS (P = Positive) (I = Indicated)
048	BORNITE [*Peacock Ore ˆ]	*	:075	Cu.63%	COPPER IRON SULFIDE	1-F, 2-C, 10-A-P
049	BOULANGERITE [*Feather Ore ˆ]	*	:007	Pb.55%	LEAD ANTIMONY SULFIDE	3-C-P, 4-R-P, 8-A-P
050	BOURNONITE [*Cog Wheel Ore ˆ]	*	:063	Cu.13%	LEAD COPPER ANTIMONY SULFIDE	4-R-P, 1-F, 3-C-P, 8-A-P
051	BRAZILIANITE		:040	Gem	HYDROUS SODIUM ALUMINUM PHOSPHATE	1-A, 1-L
052	BROCHANTITE		:084	Cu.	HYDROUS COPPER SULFATE	1-F, 12-A-P
053	BRONZITE	[P]	:056	Coll	MAGNESIUM IRON SILICATE	4-D-P, 10-A-P
054	BROOKITE		:250	Ti.	TITANIUM OXIDE	2-J
055	BRUCITE		:209	Ind.	MAGNESIUM HYDROXIDE	[10-C-P
056	CALCITE [*Calc Spar ˆ]	*	:145	Ca.40%	CALCIUM CARBONATE	[9-B-P, 10-B-P
057	CANCRINITE		:194	---	COMPLEX SILICATE	*[-?-]
058	CARNALLITE		:125	K.14.%	HYDRATED POTASSIUM MG. CHLORIDE	1-B, 10-C-P, 14-A
059	CARNOTITE		:243	U.& V.	HYDRATED POTASSIUM URANIUM VANADATE	1-B, 2-L, 2-M
060	CASSITERITE [*Tinstone ˆ]	*	:217	Sn.78%	TIN OXIDE	*[-?-]
061	CELESTITE		:148	Sr.59%	STRONTIUM SULFATE	1-D, 12-A-P
062	CERARGYRITE [*Horn Silver ˆ]	*	:233	Ag.75%	SILVER CHLORIDE	5-A-I, 14-A
063	CERRUSITE [*White Lead Ore ˆ]	*	:147	Pb.77%	LEAD CARBONATE	4-R-P, 9-B-P

MINERAL INDEX & TEST CODES

Copyright 1993

CHART A

-A- I.D. #	-B- MINERAL NAME (& Field Name)		-C- CLUE CODE	-D- ORE OF %	-F- CHEMICAL NAME	-G- MINERAL ELEMENT TESTS QUALITATIVE ANALYSIS [(P= Positive) (I= Indicated)]
064	CHABASITE	*	126	Coll.	HYDRATED CALCIUM ALUMINUM SILICATE	1-C
	[*Cubic Zeolite ^]					
065	CHALCANTHITE	*	203	Cu.25%	HYDRATED COPPER SULFATE	1-F, 12-A-P
	[Copper Vitriol ^]					
066	CHALCEDONY	*	144	Gem	SILICON DIOXIDE	*[-?-]
	[*Micro Quartz ^					
067	CHALCOCITE	*	062	Cu.79%	COPPER SULFIDE	1-F, 12-A-P
	[*Copper Glance ^]					
068	CHALCOPYRITE	*	094	*Cu.34%	COPPER IRON SULFIDE	1-F, 10-A-P, 12-A-P
	[*Copper Pyrites ^]			[^*Au,Ag.]		
069	CHLORITOID		089	Coll.	HYDROUS IRON (MG.,AL.) SILICATE	20E, 10-C-P
070	CHONDRODITE		215	Coll.	HYDROUS MAGNESIUM IRON SILICATE	2-E, 10-C-P
071	CHROMITE	*	031	Cr.46%	IRON CHROMIUM OXIDE	2-A, 2-E
	[*Chrome Iron Ore ^]					
072	CHRYSOBERYL		273	Be. 7.1%	BERYLLIUM ALUMINUM OXIDE	*[-?-]
073	CHRYSOCOLLA		200	Cu.	HYDROUS COPPER SILICATE	1-F
074	CHRYSOTILE	*	139	IND	HYDROUS NICKEL MAGNESIUM SILICATE	2-H, 10-C-P
	[*Asbestos^]					
075	CINNABAR		120	Hg.86%	MERCURY SULFIDE	5-B-I, 9-A-P
076	CLINOCHLORE	*	091	Coll.	HYDROUS MAGNESIUM (Fe,Al) SILICATE	2-E, 5-K-I
	[*Chlorite ^]					
077	CLINOZOISITE		071	Coll.	HYDROUS CALCIUM ALUMINUM SILICATE	1-C
078	COBALTITE		074	Co.35%	COBALT ARSENIC SULFIDE	4-A-I, 4-C-I, 4-L-I
079	COLEMANITE		171	B.	HYDRATED CALCIUM BORATE	1-C, 1-N

MINERAL INDEX & TEST CODES

Copyright 1993

CHART A

-A- I.D. #	-B- MINERAL NAME (& Field Name)		-C- CLUE CODE	-D- ORE OF %	-F- CHEMICAL NAME	-G- MINERAL ELEMENT TESTS QUALITATIVE ANALYSIS (P = Positive) (I = Indicated)
080	COLUMBITE [*Niobite ^]	*	014	Nb.& Ta.	IRON MANGANESE (Nb, Ta.) OXIDE	2-D. 2-E, 2-F
081	CONICHALCITE		082	Coll.	HYDROUS CALCIUM COPPER ARSENATE	1-C, 1-F, 1-G
082	COPIAPITE		261	Coll.	HYDRATED IRON SULFATE	2-E, 12-A-P
083	COPPER		108	Cu. 100%	NATIVE ELEMENT	1-F
084	CORDIERITE [*Iolite ^]	*	184	Gem	MAGNESIUM ALUMINUM SILICATE	5-K-I
085	CORUNDUM		225	Ind.	ALUMINUM OXIDE	*[-?-]
086	COVELLITE [*Indigo Copper ^]		068	Cu. 66%	COPPER SULFIDE	1-F, 8-A-P
087	CRISTOBALITE [*Obsidian ^]	*	165	Gem	SILICON DIOXIDE	*[-?-]
088	CROCIDOLITE [*Riebeckite ^]	* [A]	128	Gem	HYDROUS SODIUM (Fe,Mg) SILICATE	1-A, 2-E, 5-K-I
089	CROCOITE [*Red Lead Ore ^]	*	101	Pb.	LEAD CHROMATE	1-J, 2-A
090	CRYOLITE		202	Al. 12%	SODIUM ALUMINUM FLUORIDE	1-A
091	CUMMINGTONITE	[A]	271	Coll.	HYDROUS MAGNESIUM SILICATE	10-C-P
092	CUPRITE [*Ruby Copper ^]	*	110	Cu. 88%	COPPER OXIDE	1-F
093	CYANOTRICHITE [*Lettsomite ^]	*	019	Coll.	HYDRATED COPPER ALUMINUM SULFATE	1-F, 12-A-P
094	DANBURITE		182	Coll.	CALCIUM BOROSILICATE	1-C, 1-N
095	DATOLITE		226	B.	HYDROUS CALCIUM BOROSILICATE	1-C, 1-N

MINERAL INDEX & TEST CODES

Copyright 1993

CHART A

-A- I.D. #	-B- MINERAL NAME (& Field Name)		-C- CLUE CODE	-D- ORE OF %	-F- CHEMICAL NAME	-G- MINERAL ELEMENT TESTS QUALITATIVE ANALYSIS (P = Positive) (I = Indicated)
096-	DESCLOIZITE		:099	V. 11%	HYDROUS LEAD ZINC VANADATE	1-J, 1-M, 2-M
097-	DIAMOND		:064	Gem	CARBON	12-A-P versus quartz
098-	DIASPORE		:221	Al.45%	ALUMINUM HYDROXIDE	*[-?-]
099-	DIOPSIDE	[P]	:195	Gem	CALCIUM MAGNESIUM SILICATE	1-C, 5-K-I
100-	DIOPTASE [*Copper Emerald^]	*	:080	Cu.44%	HYDROUS COPPER SILICATE	1-F
101-	DOLOMITE		:161	Mg.13%	CALCIUM MAGNESIUM CARBONATE	1-C, 5-K-I, 9-B-P
102-	DUMORTIERITE		:127	IND.	ALUMINUM BOROSILICATE	1-N
103-	ENARGITE		:008	Cu.48%	COPPER ARSENIC SULFIDE	1-F, 1-G, 8-A-P
104-	ENSTATITE	[P]	:057	Coll.	MAGNESIUM SILICATE	10-C-P
105-	EOSPHORITE		:179	Coll.	HYDRATED MANGANESE ALUM. PHOSPHATE	1-L, 2-F
106-	EPIDOTE		:072	Gem	HYDROUS CALCIUM (AL.,FE.) SILICATE	1-C, 2-E
107-	EPSOMITE [*Bitter Salt^]	*	:237	IND.	HYDRATED MAGNESIUM SULFATE	5-K-I, 12-A-P
108-	ERYTHRITE [*Cobalt Bloom ^]	*	:115	Co.*	HYDRATED COBALT ARSENATE	2-B, 1-G
109-	EUCLASE		:266	Gem	HYDROUS BERYLLIUM ALUM. SILICATE	*[-?-]
110-	FERBERITE [*Wolframite ^]	*	:016	V.60%	IRON MANGANESE TUNGSTATE	2-E, 2-F, 2-K
111-	FERGUSONITE	(REO)	:037	Y.	YTTRIUM NIOBIUM OXIDE	2-D

MINERAL INDEX & TEST CODES

Copyright 1993

CHART A

-A- I.D. #	-B- MINERAL NAME (& Field Name)		-C- CLUE CODE	-D- ORE OF %	-F- CHEMICAL NAME	-G- MINERAL ELEMENT TESTS QUALITATIVE ANALYSIS (P= Positive) (I= Indicated)
112-	FLUORITE	*	169	IND.	CALCIUM FLUORIDE	1-C
	[*Fluor Spar ^]					
113-	FRANKLINITE		116	Zn.20%	ZINC MANGANESE IRON OXIDE	1-M, 1-E, 2-F
114-	FUCHITE	(Mica)*	043	Coll.	HYDROUS POTASSIUM ALUMINUM SILICATE	1-B
	[*Muscovite ^]					
115-	GADOLINITE	(REO)	077	Y.38%	BERYLLIUM IRON YTTRIUM SILICATE	2-E
116-	GALENA	*	059	Pb.86%	LEAD SULFIDE	1-J, 8-A-P
	[*Lead Glance ^]					
117-	GARNIERITE	*	087	Ni.36%	HYDROUS NICKEL MAGNESIUM SILICATE	2-H, 10-C-P
	[*A Generic Term ^]					
118-	GIBBSITE		137	Al.	ALUMINUM HYDROXIDE	*[-?-]
119-	GLAUBERITE		134	IND.	SODIUM CALCIUM SULFATE	1-A, 1-C, 12-A-P
120-	GLAUCODOT		003	Co.25%	COBALT IRON ARSENIC	1-G, 2-B, 2-E
121-	GLAUCONITE	(Mica)	270	IND.	HYDROUS POTASSIUM (Al,Fe,Mg.) SILICATE	1-B, 2-E, 5-K-I
122-	GLAUCOPHANE	[A]	076	Coll.	HYDROUS SODIUM (Mg,Al.) SILICATE	1-A, 5-K-I
123-	GOETHITE	*	036	Fe.	IRON HYDROXIDE	2-E
	[*Needle Ironstone ^]					
124-	GOLD		244	Au.100%	GOLD - NATIVE ELEMENT	3-P-I, 6-A-P
125-	GRAPHITE		013	C.100%	CARBON - NATIVE ELEMENT	*[-?-]
126-	GREENOCKITE		100	Cd.	CADMIUM SULFIDE	3-B-P, 8-A-P
127-	GROSSULAR	*	222	Gem	CALCIUM ALUMINUM SILICATE	1-C
	[*Garnet ^]					

MINERAL INDEX & TEST CODES

Copyright 1993

CHART A

I.D. # -A-	MINERAL NAME (& Field Name) -B-	CLUE CODE -C-	ORE OF % -D-	CHEMICAL NAME -F-	MINERAL ELEMENT TESTS QUALITATIVE ANALYSIS (P= Positive) (I= Indicated) -G-
128-	GYPSUM	:236	Ca.23%	HYDRATED CALCIUM SULFATE	1-C, 12-A-P
129-	HALITE * [*Rock Salt ^]	:124	Na.39%	SODIUM CHLORIDE	1-A, 14-A
130-	HARMOTOME * [*Cross Stone ^]	:174	Coll.	HYDRATED BARIUM (K,Al.) SILICATE	1-B, 1-E
131-	HAUERITE	:105	Coll.	MANGANESE SULFIDE	2-F, 9-A-P
132-	HAUSMANNITE	:117	Mn.72%	MANGANESE OXIDE	2-F
133-	HAUYNITE (HAUYNE)	:038	Coll.	SODIUM (CA,AL.) SILICATE SULFATE	1-A, 1-C, 12-A-P
134-	HEDENBERGITE [P]	:034	Coll.	CALCIUM IRON SILICATE	1-C, 2-E
135-	HEMATITE * [*Bloodstone^]	:112	Fe.70%	IRON OXIDE	2-E
136-	HEMIMORPHITE * [*Calamine ^]	:177	Zn.54%	HYDRATED ZINC SILICATE	1-M
137-	HERDERITE	:272	Coll.	CALCIUM BERYLLIUM PHOSPHATE	1-C, 1-L
138-	HEULANDITE * [*Lamellar Zeolite ^]	:162	Coll.	HYDRATED SODIUM (Ca,Al.) SILICATE	1-A, 1-C
139-	HORNBLENDE [A]	:191	Coll.	COMPLEX AMPHIBOLE GROUP SILICATE	[-?-]
140-	HYDROMAGNESITE	:235	Coll.	HYDRATED MAGNESIUM CARBONATE	10-C-P, 9-B-P
141-	HYDROZINCITE * [*Zinc Bloom ^]	:189	Zn.	HYDROUS ZINC CARBONATE	1-M, 9-B-P
142-	ILMENITE * [*Titanoferrite ^]	:015	Ti.35%	IRON TITANIUM OXIDE	2-E, 2-J
143-	ILVAITE	:004	Coll.	HYDROUS CALCIUM IRON SILICATE	1-C, 2-E

MINERAL INDEX & TEST CODES

Copyright 1993

CHART A

-A- I.D. #	-B- MINERAL NAME (& Field Name)		-C- CLUE CODE	-D- ORE OF %	-F- CHEMICAL NAME	-G- MINERAL ELEMENT TESTS QUALITATIVE ANALYSIS (P = Positive) (I = Indicated)
144-	JADITE	* [P]	046	Gem	SODIUM ALUMINUM IRON SILICATE	1-A, 2-E
	[*Precious Jade ^]					
145-	JAMESONITE	*	055	Pb.	LEAD IRON ANTIMONY SULFIDE	1-H, 1-J, 2-E, 9-A-P
	[*Brittle Feather Ore ^]					
146-	JOHANNSENITE	[P]	267	Coll.	CALCIUM MANGANESE SILICATE	1-C, 2-F
147-	KAOLINITE		240	Al.20%	HYDROUS ALUMINUM SILICATE	*[-?-]
148-	KERNITE		129	B.	HYDRATED SODIUM BORATE	1-A, 1-N
149-	KYANITE	*	216	Ind.	ALUMINUM SILICATE	*[-?-]
	[*Disthene ^]					
150-	LABRADORITE	*	193	Ind.	SODIUM CALCIUM ALUMINUM SILICATE	1-A, 1-C
	[*Spectrolite ^]					
151-	LARUNEITE		052	Gem	COMPLEX SILICATE	*[-?-]
152-	LAUMONTITE	*	039	Coll.	HYDRATED CALCIUM ALUMINUM SILICATE	1-C
	[*Fibrous Zeolite ^]					
153-	LAZULITE	*	190	Ind.	HYDROUS MAGNESIUM ALUMINUM PHOSPHATE	5-K-I, 1-L
	[*Blue Spar ^]					
154-	LAZURITE	*	022	Gem	SODIUM CALCIUM (Al)SILICATE SULFATE	1-A, 1-C, 12-A-P
	[*Lapis Lazuli ^]					
155-	LEPIDOCROCITE		098	Fe.	HYDROUS IRON OXIDE	2-E
156-	LEPIDOLITE	*	140	Li.3.6%	HYDROUS POTASSIUM (Li,Al.) SILICATE	1-B, 1-T
	[*Lithium Mica ^]					
157-	LEUCITE		204	K.18%	POTASSIUM ALUMINUM SILICATE	1-B
158-	LIMONITE	*	033	Fe.62%	HYDRATED IRON OXIDE	2-E
	[*Brown Iron Ore ^]					
159-	LINARITE		021	Coll.	HYDROUS LEAD COPPER SULFATE	1-F, 1-J, 12-A-P

MINERAL INDEX & TEST CODES

Copyright 1993

CHART A

-A- I.D. #	-B- MINERAL NAME (& Field Name)	-C- CLUE CODE	-D- ORE OF %	-F- CHEMICAL NAME	-G- MINERAL ELEMENT TESTS QUALITATIVE ANALYSIS (P = Positive) (I = Indicated)
160-	MAGNESITE *	167	Mg.28.8%	MAGNESIUM CARBONATE	5-K-I, 9-B-P
	[*Bitter Spar ^]				
161-	MAGNETITE	010	Fe.72%	IRON OXIDE	2-E
162-	MALACHITE	085	Cu.57%	HYDROUS COPPER CARBONATE	1-F, 9-B-P
163-	MANGANITE *	030	Mn.62%	HYDRATED MANGANESE OXIDE	2-F
	[*Brown Man Ore ^]				
164-	MARCASITE *	095	Ind.	IRON SULFIDE	2-E, 9-A-P
	[*White Pyrites ^]				
165-	MARGERITE *	163	Coll.	HYDROUS CALCIUM ALUMINUM SILICATE	1-C
	[*Brittle Mica ^]				
166-	MELILITE*(Field Term)	264	Coll.	CALCIUM MAGNESIUM ALUMINUM SILICATE	1-C, 5-K-I
	[*For Akermanite ^]				
167-	MERCURY *	263	Hg.100%	NATIVE ELEMENT	8-A-P
	[*Quicksilver ^]				
168-	MICROCLINE *	213	Gem	POTASSIUM ALUMINUM SILICATE	1-B
	[*Amazonite ^]				
169-	MICROLITE	254	Coll.	HYDROUS SODIUM CALCIUM TANTALUM OXIDE	1-A, 1-C
170-	MILARITE	187	Gem	HYDRATED POTASSIUM (Be,Al.)SILICATE	1-B
171-	MILLERITE *	093	Ni.	NICKEL SULFIDE	2-H, 9-A-P
	[*Hair Pyrites ^]				
172-	MIMETITE *	152	Pb.	LEAD ARSENATE CHLORIDE	1-G, 1-J, 14-A
	[*Campylite ^]				
173-	MOLYBDENITE	026	Mo.60%	MOLYBDENUM SULFIDE	1-S, 8-A-P
174-	MONAZITE (REO)	230	Ce\Th12%	RARE EARTH METALS PHOSPHATE	1-L
175-	MONTMORILLONITE	242	Ind	HYDRATED SODIUM (Ca,Al) SILICATE	1-A, 1-C

MINERAL INDEX & TEST CODES

Copyright 1993

CHART A

-A- I.D. #	-B- MINERAL NAME (& Field Name)		-C- CLUE CODE	-D- ORE OF %	-F- CHEMICAL NAME	-G- MINERAL ELEMENT TESTS QUALITATIVE ANALYSIS (P = Positive)(I = Indicated)
176-	MUSCOVITE [*Isinglass ^]	*	:041	Ind.	HYDROUS POTASSIUM ALUMINUM SILICATE	1-B
177-	NATUROLITE		:181	Coll.	HYDRATED SODIUM ALUMINUM SILICATE	1-A
178-	NEPHELINE		:205	Ind.	SODIUM POTASSIUM ALUMINUM SILICATE	1-A, 1-B
179-	NEPTUNITE		:113	Coll.	SODIUM POTASSIUM (Fe,Ta) SILICATE	1-A, 1-B, 2-E
180-	NICCOLITE [*Arsenical Nickel ^]		:035	Ni.43%	NICKEL ARSENIDE	2-H, 1-G
181-	NICKEL-IRON [*Taenite^]	*	:259	Coll.	NATIVE ELEMENTS - Ni-Fe INTERGROWTH	2-E, 2-H
182-	OLIVENITE		:245	Coll.	HYDROUS COPPER ARSENATE	1-F, 1-G
183-	OLIVINE [*Fosterite,Peridot ^]	*	:045	Gem	MAGNESIUM IRON SILICATE	5-K-I, 2-E
184-	OMPHACITE	[P]	:231	Coll.	CALCIUM MAGNESIUM (Fe,Al) SILICATE	1-C, 2-E, 5-K-I
185-	OPAL		:207	Gem	HYDROUS SILICON OXIDE	*[-?-]
186-	ORPIMENT [*Tinsel ^]	*	:248	As.	ARSENIC SULFIDE	1-G, 4-A-I
187-	ORTHOCLASE [*K-Feldspar ^]	*	:239	Ind.	POTASSIUM ALUMINUM SILICATE	1-B
188-	PARAGASITE [*Hornblende ^]	* [A]	:257	Coll.	COMPLEX SILICATE	*[-?-]
189-	PECTOLITE		:178	Coll.	HYDROUS SODIUM CALCIUM SILICATE	1-A, 1-C
190-	PENNINITE [*Green Chlorite ^]	*	:081	Coll.	HYDROUS MAGNESIUM (Fe,Al) SILICATE	2-E, 5-K-I
191-	PENTLANDITE		:028	Ni.40%	IRON NICKEL SULFIDE	2-E, 2-H, 4-A-I

MINERAL INDEX & TEST CODES

Copyright 1993

CHART A

-A- I.D. #	-B- MINERAL NAME (& Field Name)	-C- CLUE CODE	-D- ORE OF %	-F- CHEMICAL NAME	-G- MINERAL ELEMENT TESTS / QUALITATIVE ANALYSIS (P = Positive) (I = Indicated)
192-	PEROVSKITE (REO)	252	Nb.\Ce.	CALCIUM TITANIUM OXIDE	1-C, 2-J
193-	PETALITE	044	Li.2.2%	LITHIUM ALUMINUM SILICATE	1-T
194-	PHENAKITE *	192	Gem	BERYLLIUM SILICATE	*[?-]
	[*Phenacite^]				
195-	PHILLIPSITE *	042	Coll.	HYDRATED POTASSIUM(Na,Al)SILICATE	1-B, 1-A
	[*Cross Stone ^]				
196-	PHLOGOPITE *	132	Ind.	HYDROUS POTASSIUM ALUMINUM SILICATE	1-B
	[*Amber Mica ^]				
197-	PHOSGENITE *	234	Coll.	LEAD CARBONATE CHLORIDE	1-J, 9-B-P, 14-A
	[*Horn Lead ^]				
198-	PHOSPHORPHYLLITE	269	Gem	HYDRATED ZINC (Fe,Mn.) SILICATE	1-M, 2-E, 2-F
199-	PIEDMONTITE *	111	Coll.	HYDROUS CALCIUM (Al,Mn) SILICATE	1-C, 2-F
	[*Mangan Epidote ^]				
200-	PLATINUM	060	Pt.	NATIVE ELEMENT	6-B-P
201-	POLYBASITE	005	Ag.75%	SILVER COPPER ANTIMONY SULFIDE	1-F, 1-H, 3-N-I, 4A-I
202-	PREHNITE	212	Coll.	HYDROUS CALCIUM ALUMINUM SILICATE	1-C
203-	PROUSTITE *	119	Ag.65%	SILVER ARSENIC SULFIDE	1-G, 4-A-I, 4-N-I
	[*Ruby Silver ^]				
204-	PSILOMELANE *	012	Mn.63%	HYDROUS BARIUM MANGANESE OXIDE	1-E, 2-F
	[Romanechite ^]				
205-	PURPURITE	109	Coll.	MANGANESE IRON PHOSPHATE	2-E, 2-F
206-	PYRARGYRITE *	107	Ag.59%	SILVER ANTIMONY SULFIDE	1-H, 3-N-I, 4-A-I
	[*Dark Ruby Silver ^]				
207-	PYRITE *	092	S. 53%	IRON SULFIDE	2-E, 4-A-I
	[*Iron Pyrite,Fools Gold^]				

MINERAL INDEX & TEST CODES

Copyright 1993

CHART A

-A- I.D. #	-B- MINERAL NAME (& Field Name)		-C- CLUE CODE	-D- ORE OF %	-F- CHEMICAL NAME	-G- MINERAL ELEMENT TESTS QUALITATIVE ANALYSIS (P = Positive) (I = Indicated)
208-	PYROLUSITE		:024	Mn.63%	MANGANESE OXIDE	[5-K-I]
209-	PYROMORPHITE		:154	Pb.76%	LEAD PHOSPHATE CHLORIDE	1-J, 1-L, 14-A]
210-	PYROPE [*Bohemian Garnet ^]	*	:114	Gem	MAGNESIUM ALUMINUM SILICATE	5-K-I]
211-	PYROPHYLLITE		:262	Ind.	HYDROUS ALUMINUM SILICATE	*[-?-]]
212-	PYRRHOTITE [*Magnetic Pyrite ^]	*	:002	S.36%	IRON SULFIDE [Check for Ni,Co,Pt]	2-B, 2-E, 2-H, 6-B-P, 4-A-I]
213-	QUARTZ [*Rock Crystal ^]	**	:143	Si.46%	SILICON DIOXIDE [Coarse Crystals]	*[-?-]]
214-	REALGAR		:102	Ind.	ARSENIC SULFIDE	1-G, 4-A-I]
215-	RHODOCHROSITE [*Manganese Spar ^]	*	:104	Mn.47%	MANGANESE CARBONATE	2-F, 9-B-P]
216-	RHODONITE [*Fowlerite ^]	*	:208	Mn.41%	MANGANESE IRON MAGNESIUM SILICATE	2-E, 2-F, 5-K-I]
217-	RICHTERTITE	[A]	:251	Coll.	HYDROUS SODIUM MAGNESIUM SILICATE	1-A, 5-K-I]
218-	ROSASITE		:086	Cu.	HYDROUS COPPER ZINC CARBONATE	1-F, 1-M, 9-B-P]
219-	RUTILE		:029	Ti.60%	TITANIUM OXIDE	2-J]
220-	SAMARSKITE	(REO)	:118	Y. 28%	RARE EARTH NIOBIUM TANTALUM OXIDE	2-D]
221-	SANADINE [*K Feldspar ^]	*	:210	Coll.	POTASSIUM ALUMINUM SILICATE	1-B]
222-	SCAPOLITE		:198	Gem	COMPLEX TECTOSILICATE	*[-?-]]
223-	SCHEELITE		:229	W. 63%	CALCIUM TUNGSTATE	1-C, 2-K]

MINERAL INDEX & TEST CODES

Copyright 1993

CHART A

-A- I.D. #	-B- MINERAL NAME (& Field Name)	-C- CLUE CODE	-D- ORE OF %	-F- CHEMICAL NAME	-G- MINERAL ELEMENT TESTS QUALITATIVE ANALYSIS (P = Positive) (I = Indicated)
224-	SCOLECITE * [*Fibrous Zeolite ^]	:268	Coll.	HYDRATED CALCIUM ALUMINUM SILICATE	1-C
225-	SCORODITE	:164	Coll.	HYDRATED IRON ARSENATE	2-E, 1-G
226-	SEPIOLITE * [*Meerschaum ^]	:172	Ind.	HYDROUS MAGNESIUM SILICATE	5-K-I
227-	SERPENTINE	:138	Ind.	HYDROUS MAGNESIUM IRON SILICATE	2-E, 5-K-I
228-	SIDERITE * [*Chalybite ^]	:160	Fe.48%	IRON CARBONATE	2-E, 9-B-P
229-	SILLIMANITE * [*Fibrolite ^]	:051	Ind.	ALUMINUM SILICATE	*[-?-]
230-	SILVER	:123	Ag.100%	NATIVE ELEMENT	4-V-P, 4-W-P
231-	SKUTTERUDITE * [*Smaltite ^]	:009	Co.28%	COBALT ARSENIDE	2-B, 1-G
232-	SMITHSONITE * [*Zinc Spar ^]	:228	Zn.52%	ZINC CARBONATE	1-M, 9-B-P
233-	SODALITE	:050	Gem	SODIUM (AL) SILICATE CHLORIDE	1-A, 14-A
234-	SPHALERITE * [*Zinc Blende ^]	:253	Zn.67%	ZINC IRON SULFIDE	1-M, 2-E, 4-A-I
235-	SPINEL	:201	Gem	MAGNESIUM ALUMINUM OXIDE	5-K-I
236-	SPODUMENE [P]	:227	Li.3.4%	LITHIUM ALUMINUM SILICATE	1-T
237-	STAUROLITE * [*Rood Stone ^]	:070	Gem	HYDROUS IRON (MG,AL) SILICATE	2-E, 5-K-I
238-	STEPHANITE * [*Brittle Silver ^]	:011	Ag.68%	SILVER ANTIMONY SULFIDE	3-N-I, 1-H, 4-A-I
239-	STIBICONITE	:256	Coll.	HYDROUS ANTIMONY OXIDE	1-H

MINERAL INDEX & TEST CODES

Copyright 1993

CHART A

-A-	-B-		-C-	-D-	-F-	-G-
I.D. #	MINERAL NAME (& Field Name)		CLUE CODE	ORE OF %	CHEMICAL NAME	MINERAL ELEMENT TESTS QUALITATIVE ANALYSIS (P = Positive) (I = Indicated)
240-	STIBNITE	*	:066	Sb.71%	ANTIMONY SULFIDE	1-H, 4-A-I
	[*Antimonite ^]					
241-	STILBITE	*	:155	Coll.	HYDRATED SODIUM (Ca,Al) SILICATE	1-A, 1-C
	[Bundle Zeolite ^]					
242-	STRONTIANITE		:158	Sr.59%	STRONTIUM CARBONATE	1-D, 9-B-P
243-	SULFUR		:142	S.100%	NATIVE ELEMENT	3-E-P
244-	SYLVANITE	*	:069	Au.24%	SILVER GOLD TELLURIDE	3-J-I, 3-N-I, 3-P-I
	[*Graphic Tellurium ^]					
245-	SILVITE		:166	K.52%	POTASSIUM CHLORIDE	1-B, 14-A
246-	TALC*		:131	Ind.	HYDROUS MAGNESIUM SILICATE	5-K-I
	[*^ Soapstone,Steatite,Freestone]					
247-	TARBUTTITE		:168	Coll.	HYDROUS ZINC PHOSPHATE	1-L, 1-M
248-	TETRAHEDRITE-	*	:065	Cu.53%	COPPER IRON SULFIDE	1-F, 2-E, 4-A-I
	-TENNANTITE-[*^Gray Copper]					
249-	THOMSONITE	*	:048	Gem	HYDRATED SODIUM (Ca,Al) SILICATE	1-A, 1-C
	[*Comptonite ^]					
250-	THORIANITE		:096	U & Th.	THORIUM URANIUM OXIDE	2-L
251-	TITANITE	*	:183	Ti.& Gem	CALCIUM TITANIUM SILICATE	1-C, 2-J
	[*Sphene ^]					
252-	TOPAZ		:220	Gem	HYDROUS ALUMINUM SILICATE	*[-?-]
253-	TORBERNITE	*	:088	U.	HYDRATED COPPER URANIUM PHOSPHATE	1-L, 2-C, 2-L
	[*Copper Uranite ^]					
254-	TOURMALINE		:238	Gem	COMPLEX BORON CYCLOSILICATE	1-N
255-	TREMOLITE	[A]	:197	Ind.	HYDROUS CALCIUM MAGNESIUM SILICATE	1-C, 5-K-I

MINERAL INDEX & TEST CODES

Copyright 1993

CHART A

-A- I.D. #	-B- MINERAL NAME (& Field Name)	-C- CLUE CODE	-D- ORE OF %	-F- CHEMICAL NAME	-G- MINERAL ELEMENT TESTS QUALITATIVE ANALYSIS (P = Positive) (I = Indicated)
256-	TRIDYMITE *	:153	Ind.	SILICON DIOXIDE	*[-?-]
	[*Quartz ^]				
257-	TRONA	:135	Ind.	HYDRATED SODIUM CARBONATE	1-A, 9-B-P
258-	TURQUOISE *	:090	Gem	HYDRATED COPPER ALUMINUM PHOSPHATE	1-F, 1-L
	[*Callaite ^]				
259-	ULEXITE *	:265	Coll.	HYDRATED SODIUM CALCIUM BORATE	1-A, 1-C, 1-N
	[*Cotton Balls ^][*T.V. stone]				
260-	URANINITE *	:017	U.83%	URANIUM OXIDE	2-L
	[*Pitchblende ^]				
261-	UVAROVITE *	:097	Gem	CALCIUM CHROMIUM SILICATE	1-C, 2-A
	[*Garnet ^]				
262-	VANADINITE	:255	V. 10%	LEAD VANADATE CHLORIDE	1-J, 2-M, 14-A
263-	VARISCITE	:170	Gem	HYDRATED ALUMINUM PHOSPHATE	1-L
264-	VERMICULITE	:258	Ind.	HYDRATED MAGNESIUM (FE,AL) SILICATE[	5-K-I, 2-E
265-	VESUVIANITE *	:218	Gem	HYDROUS CALCIUM (MG,AL) SILICATE	1-C, 5-K-I
	[*IDOCRASE ^]				
266-	VIVIANITE *	:199	Ind.	HYDRATED IRON PHOSPHATE	2-E, 1-L
	[*Blue Iron ^]				
267-	VOLBORTHITE	:275	Coll.	HYDRATED COPPER VANADATE	2-C, 2-M
268-	WAVELLITE	:150	HYDRATED	ALUMINUM PHOSPHATE	1-L
269-	WILLEMITE	:049	Zn.58%	ZINC SILICATE	1-M
270-	WITHERITE (toxic)	:157	Ba.69%	BARIUM CARBONATE	1-E, 9-B-P
271-	WOLFRAMITE *	:018	W.	MANGANESE IRON TUNGSTATE	2-E, 2-F, 2-K
	[*See Also Ferberite ^]				

MINERAL INDEX & TEST CODES

Copyright 1993

CHART A

I.D. #	MINERAL NAME (& Field Name)	CLUE CODE	ORE OF %	CHEMICAL NAME	MINERAL ELEMENT TESTS QUALITATIVE ANALYSIS (P = Positive) (I = Indicated)
-A-	-B-	-C-	-D-	-F-	-G-
272-	WOLLASTONITE * [*Table Spar ^]	176	Ind.	CALCIUM SILICATE	1-C
273-	WULFENITE * [*Yellow Lead Ore ^]	141	Mo.26%	LEAD MOLYBDATE	1-J, 1-S
274-	ZINCITE * [*Red Oxide Of Zinc ^]	103	Zn.80%	ZINC MANGANESE OXIDE	1-M, 2-F
275-	ZIRCON	047	Zr.49%	ZIRCONIUM SILICATE	*[-?-]
276-	ZINNWALDITE * [*Lithionite ^]	130	Li.1.8%	HYDROUS POTASSIUM Li Al SILICATE	1-B, 1-T
277-	ZOISITE	053	Gem	HYDROUS CALCIUM ALUMINUM SILICATE	1-C

[TESTS MARKED *[-?-] Unknown or not worth effort.]

NOTE: Where a CHEMICAL NAME shows as "POTASSIUM Li Al SILICATE", or as "POTASSIUM (Li, Al) SILICATE" it means that potassium is the usual element found, but lithium or aluminum can substitute for potassium in the same mineral. YOU SHOULD TEST FOR ALL THREE ELEMENTS TO GAIN CERTAINTY.

SEE COOKBOOK CHEMISTRY NOTES, PART FOUR, LAST PAGE FOR ADDITIONAL ELEMENT TESTS.

FIRST SAFETY RULE - WHEN TESTING, ALWAYS WEAR PROTECTIVE CLOTHING, GLOVES AND GLASSES.

SECOND SAFETY RULE - NEVER, NEVER, LICK THE SPOON !

CHAPTER 65

STREAK COLOR – SPECIFIC GRAVITY

HARDNESS – LUSTER

CHART B HEADINGS

COLUMN | **EXPLANATION**

A. – MINERAL I.D. NUMBER = This number identifies the same specific mineral throughout ALL charts.

B. – MINERAL NAME & FIELD NAME = Identifies the mineral name.
Note: Some minerals are better known by their field name. Example: Silver Glance for Argentite. (The charts use the U.S.A. mineral name.)

C. – CLUE CODE = This code number gets you in COLOR STREAK, Chart C for quick pick-up of streak, hardness, specific gravity, luster & mineral color as it relates to the mineral with that clue code.

D. – ORE % = Once you identify your mineral you want to know if it has a market value. Ore % lets you know if your market is for an ore, gem, or collector's item.

E. – OTHER CLUES = This tells you the minerals that react to black light of long or short wavelengths. Radioactivity is indicated.

F. – BREAK COLOR = Mineral color must be determined on a FRESH BREAK. This column lets you know what colors are usual for that mineral I.D. number.

G. – STREAK COLOR = ONCE YOUR HAVE DETERMINED THE POWDER COLOR OF YOUR MINERAL, THE SOLE PURPOSE OF CHART B IS TO GET YOU INTO CHART C TO LET YOU KNOW THE PROBABLE MINERALS WITHIN THAT SPECIFIC COLOR STREAK. OTHER PHYSICAL TESTS CAN ELIMINATE MOST PROBABLES.

YOU THEN CONFIRM YOUR CONCLUSION WITH TESTS FOR CHEMICAL ELEMENTS PRESENT.

H. – MOHS HARDNESS = This column tells you the HARDNESS RANGE of your mineral. It is another confirming physical test for elimination purposes.

J. – SPECIFIC GRAVITY = Column tells you the SPECIFIC GRAVITY RANGE of your mineral. It is another physical test for elimination purposes.

K. – SURFACE LUSTER = Column tells you the LUSTER OF A MINERAL SURFACE as it appears or reflects light.

STREAK—HARDNESS—SPECIFIC GRAVITY

LaRune's ROCKPECKER Notes - CHART B. Page 1 of 17 Copyright 1993

I.D. # MINERAL NAME		CLUE CODE	ORE OF %	FLUOR RADIO	BREAK COLOR	STREAK COLOR	MOHS HARDNESS	SPECIFIC GRAVITY	SURFACE LUSTER	(& Field Name)	Other Colors
001-ACMITE	* [P]	032	Coll.	---	Br.*	BROWN	6.0-6.5	3.5-3.6	VIT t/l	[*Aegirine ^]	[*Green^^]
02-ACTINOLITE	*	196	Gem	---	Gn.	WHITE	5.0-6.0	3.1-3.3	VIT tp/tl	[Nephrite ^]	
003-ADAMITE		151	Col.	F-uv.	Gn.*	WHITE	3.5	4.3-4.5	VIT tp/tl		[*Pink, Yellow, Violet ^^]
004-ADULARIA	*	211	Gem	----	Wh.*	WHITE	6.0	2.5-2.6	VIT tp/tl	[*Feldspar ^]	[*Colorless ^^]
005-ALBITE	*	214	IND.	----	Wh.*	WHITE	6.0-6.5	2.6	VIT tl	[*Feldspar ^]	[*Colorless ^^]
006-ALLANITE	* (REO)	061	Ce.5%	Rad.	Br.*	GRAY	5.5-6.0	3.5-4.2	S-MET opq	[*Orthite ^]	[*Black ^^]
007-ALMANDINE	*	224	Gem	----	Rd-Br.	WHITE	6.5-7.5	4.32	SRK opq/tl	[*Garnet ^]	
008-ALUNITE	*	159	Coll.	F-uv	Wh.*	WHITE	3.5-4.0	2.75	VIT tp/tl	[*Alum Stone ^]	[*Red,Yellow,Gray ^^]
009-ALURGITE	*	106	Coll.	----	Rd.	PINK-WH	2.0-2.5	2.9-3.0	VIT tp/tl	[*Mica ^]	
010-AMBLYGONITE		206	Li.4.7%	F-uv	Wh.*	WHITE	5.5-6.0	3.1	VIT tp/tl		[*Yellow,Pink,Green,Blue ^^]
011-ANALCITE	*	185	Coll.	----	Wh.*	WHITE	5.0-5.5	2.2-2.3	VIT tp/tl	[*Analcime ^]	[*Pink, Yellow,Colorless ^^]
012-ANATASE		249	Ti.	----	Bk.*	YELLOW	5.5-6.0	3.8-3.9	SRK tp/tl		[*Yellow,Blue ^^]
013-ANDALUSITE		186	Gem	---	Rd.*	WHITE	7.5	3.1-3.2	VIT tl/opq		[*Brown,Olive ^^]
014-ANDRADITE	*	223	Gem	---	Br.*	WHITE	6.5-7.5	3.86	SRK tp	[*Garnet ^]	[*Green,Yellow,Black,Red ^^]
015-ANGLESITE	*	054	Pb.68%	F-uv	Wh.*	C-LESS	2.7-3.0	6.3	SRK tp/tl	[*Lead Vitriol^]	[*Gray,Green,Brown,Black,Purple ^^]
016-ANHYDRITE		149	Ca.29%	---	Wh.*	WHITE	3.0-3.5	3.0	VIT tp/tl		[*Red,Gray ^^]
017-ANNABERGITE	*	079	Coll.	---	Gn.	GREEN	2.5-3.0	3.1	VIT tl	[*Nickel Bloom ^]	

STREAK—HARDNESS—SPECIFIC GRAVITY

LaRune's ROCKPECKER Notes - CHART B. Page 2 of 17 Copyright 1993

I.D. # – MINERAL NAME (& Field Name)		CLUE CODE	ORE OF %	FLUOR RADIO	BREAK COLOR	STREAK COLOR	MOHS HARDNESS	SPECIFIC GRAVITY	SURFACE LUSTER
018-ANTHROPHYLLITE	[A]	241	Mg.	---	Gy.*	WHITE	5.5-6.0	2.8-3.2	VIT tl
[*Green,Brown ^^]									
019-ANTIMONY		058	Sb.100%	---	Si-Wh.	GRAY	3.0-3.5	6.7	MET opq
020-APATITE (GROUP)		180	Gem	F-uv	Wh.*	WHITE	5.0	3.1-3.2	VIT tp/opq
[*Colorless,Yellow,Green,Brown,Red,Blue ^^]									
021-APOPHYLLITE		175	Coll.	F-uv	Wh.*	WHITE	4.5-5.0	2.3	VIT tp/tl
[*Yellow,Colorless,Pink,Green ^^]									
022-ARAGONITE		156	Gem	F-uv	Wh.*	WHITE	3.5-4.0	2.9-3.0	VIT tp/tl
[*Red,Colorless ^^]									
023-ARFVEDSONITE	[A]	025	Coll.	---	B-Gy.	GN.-BK.	5.5-6.0	3.4-3.5	VIT opq
024-ARGENTITE	*	121	Ag.65%	---	Si-Gy	LEAD-GY	2.0-2.5	7.3	MET opq
[*Silver Glance ^]									
025-ARSENIC		073	As.100%	---	BK.	GRAY	3.5	5.4-5.9	DULL opq
026-ARSENOPYRITE	*	001	As.46%	---	Si.*	BLACK	5.0-6.0	5.9-6.2	MET opq
[*Mispickel^] [*White-Gray ^^]									
027-ARTINITE		145	Coll.	---	Wh.*	WHITE	2.0	2.03	SKY tp
[*Gray ^^]									
028-ASTROPHYLLITE		247	Coll.	---	Gold	YELLOW	3.5-4.5	3.3-3.4	VIT tp/tl
029-ATACAMITE		083	Cu.59%	---	Gr.	GREEN	3.0-3.5	3.76	VIT tp
030-AUGITE	[P]	078	Coll.	---	Bk.*	GRAY-GN	5.0-6.0	3.2-3.5	VIT tl
[*Green-Black,Brown ^^]									
031-AURICHALCITE		027	Coll	---	Be.	BLUE-GN	2.0-2.5	4.2	PLY tl
032-AUTINITE	*	246	U	F-uv.	Gn-Yw	YELLOW	2.0-2.5	3.2	VIT tl
[*Radioactive ^]									
033-AXINITE		219	Gem	F-uv	Vi-Br.*	WHITE	6.5-7.0	3.2-3.3	VIT tl
[*Brown,Green,Gray,Yellow,Blue ^^]									
034-AZURITE	*	023	Cu.55%	---	Be.	BLUE	3.5-4.0	3.83	VIT tl/tp
[*Blue Carbonate ^]									

STREAK—HARDNESS—SPECIFIC GRAVITY

LaRune's ROCKPECKER Notes - CHART B. Page 3 of 17 Copyright 1993

I.D. #	MINERAL NAME (& Field Name)		CLUE CODE	ORE OF %	FLUOR RADIO	BREAK COLOR	STREAK COLOR	MOHS HARDNESS	SPECIFIC GRAVITY	SURFACE LUSTER
035	BABINGTONITE		260	Coll.	---	Bk.*	?	5.5-6.0	3.4	VIT tl/opq
						[*Brown,Blue ^^]				
036	BARITE	*	136	Ba.58%	F-uv	Yw.*	WHITE	2.5-3.5	4.48	VIT tl/tp
	[*Heavy Spar^]					[*Red,Green,Black,Colorless ^^]				
037	BAUXITE		173	Al.	---	Wh.*	WHITE	1.0-3.0	2.3-2.7	ERY
						[*Yellow,Red,Brown ^^]				
038	BENITOITE		276	Gem	F-uv	Blue	?	6.2-6.5	3.65	SRK tp/tl
039	BERTRANDITE		274	Be.	---	Yw.	?	6.0-7.0	2.6	VIT tp/tl
040	BERYL		188	Be.5%	F-uv	Gn-Wh*	WHITE	7.5-8.0	2.7-2.9	VIT tp/tl
						[*Yw-Wh,Be-Gn,Pk,Be,Gn ^^]				
041	BETAFITE	*	277	Th.& U.	---	Bk.*	YW-WH.	5.0	3.7-5.0	ERY opq
	[*Radioactive ^]					[*Yellow,Brown-Green ^^]				
042	BIOTITE	*	133	Coll.	---	Bk.*	WHITE	2.5-3.0	2.8-3.2	VIT tp tl
	[*Black Mica ^]					[*Brown,Green ^^]				
043	BISMUTH		122	Bi.100%	---	Pk.	SI-WH.	2.0-2.5	9.7-9.8	MET opq
044	BISMUTHINITE		067	Bi.81%	---	Gy.	GRAY	2.0	6.8-7.2	MET opq
045	BIXBYITE		006	Mn.	---	Bk.	BLACK	6.0-6.5	4.9-5.0	MET opq
046	BOLEITE		020	COL	---	Dark Be	BLUE	3.5	4.8-5.1	PRL tl
047	BORAX		232	IND.	---	Wh.*	WHITE	2.0-2.5	1.7	GSY tp/tl
						[*Colorless,Yellow,Blue ^^]				
048	BORNITE	*	075	Cu.63%	---	Rd-Br*	GRAY-BK	3.0	4.9-5.3	MET opq
	[*Peacock Ore ^]					[*Iridescent Purple-Blue ^^]				
049	BOULANGERITE	*	007	Pb.55%	----	Gy.	BLACK	2.5-3.0	5.8-6.2	MET opq
	[*Feather Ore ^]									
050	BOURNONITE	*	063	Cu.13%	----	Gy.,Bk	GRAY	2.5-3.0	5.7-5.9	MET opq
	[*Cog Wheel Ore ^]									
051	BRAZILIANITE		040	Gem		Gn-Yw	C-LESS	5.5	2.98	VIT tp/tl

STREAK–HARDNESS–SPECIFIC GRAVITY

LaRune's ROCKPECKER Notes - CHART B. Page 4 of 17 Copyright 1993

I.D. #	MINERAL NAME (& Field Name)	CLUE CODE	ORE OF %	FLUOR RADIO	BREAK COLOR	STREAK COLOR	MOHS HARDNESS	SPECIFIC GRAVITY	SURFACE LUSTER
052	BROCHANTITE	084	Cu.		Gn.	GREEN	3.5-4.0	3.97	VIT tp/tl
053	BRONZITE [P]	056	Coll.		Gn-Br	GRAY	5.0	3.2-3.5	S-MET tl
054	BROOKITE	250	Ti.		Bk.,Br	YELLOW	5.5-6.0	4.1	SRK tp/tl
055	BRUCITE	209	Ind.	F/uv	Gn.*	WHITE	2.5	2.4	PRL tp/tl
					[*Blue,Pink-Yellow,Brown ^^]				
056	CALCITE *	145	Ca.40%	F/uv	Wh.*	WHITE	3.0	2.7	VIT tl
	[*Calc Spar ^]				[*Yellow, Pink, Green ^^]				
057	CANCRINITE	194	Coll.	---	Yw.*	WHITE	5.0-6.0	2.4-2.5	VIT tp/tl
					[*White,Red,Violet,Colorless ^^]				
058	CARNALLITE	125	K.14.%	---	Rd,C/L	WHITE	2.5	1.6	VIT tp
059	CARNOTITE	243	U.& V.	Rad.	Yw.	YELLOW	1.0-2.0	4.7-5.0	DULL tl/opq
060	CASSITERITE *	217	Sn.78%	---	Br,Bk	WHITE	6.0-7.0	6.0-7,1	MET tl/opq
	[*Tinstone ^]								
061	CELESTITE	148	Sr.59%	F/uv	Wh.*	WHITE	3.0-3.5	3.96	VIT tp/tl
					[*Blue,Yellow,Colorless ^^]				
062	CERARGYRITE *	233	Ag.75%	---	Gy-Gn	WHITE	2.5	5.5	RSN tl
	[*Horn Silver ^]								
063	CERRUSITE *	147	Pb.77%	F/uv	Wh-Gy	WHITE	3.0-3.5	6.5	SRK tp/tl
	[*White Lead Ore ^]								
064	CHABASITE *	126	Coll.	----	Wh.*	WHITE	4.0-5.0	2.0-2.1	VIT tp/tl
	[*Cubic Zeolite ^]				[*Green,Red,Brown ^^]				
065	CHALCANTHITE *	203	Cu.25%	---	Sky Be.	WHITE	2.5	2.3	VIT tp/tl
	[Copper Vitriol ^]								
066	CHALCEDONY *	144	Gem	F/uv	VAR**	WHITE	7.0	2.5-2.6	WAXY tp/tl
	[*Micro Quartz ^][**Agate,Onyx,Sardonyx,Sard,Carnelian,Chrysoprase,Jasper,Prase,Plasma,Bloodstone,*]								
067	CHALCOCITE *	062	Cu.79%	---	Gy.	GRAY	2.5-3.0	5.5-5.8	MET opq
	[*Copper Glance ^]								

STREAK—HARDNESS—SPECIFIC GRAVITY

LaRune's ROCKPECKER Notes - CHART B. Page 5 of 17 Copyright 1993

I.D. #	MINERAL NAME (& Field Name)	CLUE CODE	ORE OF %	FLUOR RADIO	BREAK COLOR	STREAK COLOR	MOHS HARDNESS	SPECIFIC GRAVITY	SURFACE LUSTER
068	CHALCOPYRITE *	094	*Cu.34%	---	Yw.	GN-BK	3.5-4.0	4.2-4.3	MET opq
	[*Copper Pyrites ^]		[^*Au,Ag.]						
069	CHLORITOID	089	Coll.	---	Yw-Gn.	GREEN	6.5	3.5-3.8	VIT tl
070	CHONDRODITE	215	Coll.	---	Yw.*	WHITE	6.0-6.5	3.14	VIT tl
					[*Red-Brown ^^]				
071	CHROMITE *	031	Cr.46%	---	Bk.	BROWM	5.5	4.5-4.8	S-MET opq
	[*Chrome Iron Ore ^]								
072	CHRYSOBERYL	273	Be. 7.1%	---	Gn.*	WHITE	8.5	3.7	SRK tp/tl
					[*Colorless,Green, Gray,Brown ^^]				
073	CHRYSOCOLLA	200	Cu.	----	Gn,Be	WHITE	2.0-4.0	2.0-2.3	VIT tl
074	CHRYSOTILE *	139	IND	----	Gy-Wh*	WHITE	2.5-4.0	2.55	SRK opq
	[*Asbestos^]				[*Yellow,Brown,Green,Opaque ^^]				
075	CINNABAR	120	Hg.86%	---	Rd.	SCARLET	2.0-2.5	8.10	SRK tl
076	CLINOCHLORE *	091	Coll.	---	Be-Gn	GREEN	2.0-2.5	2.6-2.8	PRL tl
	[*Chlorite ^]								
077	CLINOZOISITE	071	Coll.	---	Gy.*	GRAY	6.5	3.3-3.5	VIT tl tp
					[*Pink,Green ^^]				
078	COBALTITE	074	Co.35%	---	Gy-Pur	GRAY-BK.	5.5	6.3	MET opq
079	COLEMANITE	171	B.	---	Wh.*	WHITE	4.0-4.5	2.4	VIT tp/tl
					[*Red, Pink ^^]				
080	COLUMBITE *	014	Nb.& Ta.	---	Bk.	BK-BR.	6.0	5.3-8.1	S-MET opq
	[*Niobite ^]								
081	CONICHALCITE	082	Coll.	---	Gn.	GREEN	4.5	4.3	VIT tl
082	COPIAPITE	261	Coll.	---	Olive*	----	2.5-3.0	2.0-2.8	PRL tl
					[*Orange, Yellow ^^]				
083	COPPER	108	Cu. 100%	---	Rd-Or	RED	2.5-3.0	8.5-9.0	MET opq

STREAK—HARDNESS—SPECIFIC GRAVITY

LaRune's ROCKPECKER Notes - CHART B. Page 6 of 17 Copyright 1993

I.D. #	MINERAL NAME (& Field Name)	CLUE CODE	ORE OF %	FLUOR RADIO	BREAK COLOR	STREAK COLOR	MOHS HARDNESS	SPECIFIC GRAVITY	SURFACE LUSTER
084-	CORDIERITE *	184	Gem	---	Gy.*	WHITE	7.0-7.5	2.6-2.7	VIT tl
	[*Iolite ^]				[*Blue,Green,Black^^]				
085-	CORUNDUM	225	Ind.	F-uv	Gy.*	WHITE	9.0	3.9-4.1	SRK tl tp
					[*Colorless,Purple,Green,Yellow,Red,Brown^^]				
086-	COVELLITE	068	Cu. 66%		Gy.	dk-Be.	1.5-2.0	4.6-4.8	S-MET opq
	[*Indigo Copper ^]								
087-	CRISTOBALITE *	165	Gem	---	Wh.*	WHITE	7.0	2.3	VIT opq
	[*Obsidian ^]				[*Gray,Colorless^^]				
088-	CROCIDOLITE * [A]	128	Gem	---	Be-Bk.	WHITE	4.0	3.0-3.4	SRK tl
	[*Riebeckite ^]								
089-	CROCOITE *	101	Pb.	----	Or-Rd.	OR-YL.	2.5-3.0	5.9-6.0	SRK tl
	[*Red Lead Ore ^]								
090-	CRYOLITE	202	Al. 12%	---	Wh.*	WHITE	2.5	2.9-3.0	RSN tp
					[*Colorless,Red-Brown^^]				
091-	CUMMINGTONITE [A]	271	Coll.	---	Br.	----	5.0-6.0	3.1-3.7	SKY tl
092-	CUPRITE *	110	Cu. 88%	---	Rd.	RED	3.5-4.0	5.8-6.2	SRK tp
	[*Ruby Copper ^]								
093-	CYANOTRICHITE *	019	Coll.	---	Sky-Be	BLUE	1.0	2.75	SKY tp tl
	[*Lettsomite ^]								
094-	DANBURITE	182	Coll.	---	Yw.	WHITE	7.0-7.3	2.9-3.0	VIT tp tl
095-	DATOLITE	226	B.	---	Wh.*	WHITE	5.0-5.5	2.9-3.0	VIT tp tl
					[*Green,Colorless^^]				
096-	DESCLOIZITE	099	V. 11%	---	Rd.*	ORANGE	3.5	5.5-6.2	VIT tp tl
					[*Black-Brown,Yellow-Brown^^]				
097-	DIAMOND	064	Gem	F/uv	Gy.*	GRAY	10.0	3.52	SRK tp
					[*Colorless,Yellow,Brown,Green,Black^^]				
098-	DIASPORE	221	Al.45%	F/uv	Pk.*	WHITE	6.5-7.0	3.3-3.5	VIT tp tl
					[*Green,Gray^^]				
099-	DIOPSIDE [P]	195	Gem	F/uv	Wh.*	WHITE	5.0-6.0	3.1-3.3	VIT tp tl
					[*Purple,Brown,Blue,Yellow, Green^^]				
100-	DIOPTASE *	080	Cu.44%	---	Gn.	GREEN	5.0	3.3	VIT tp tl
	[*Copper Emerald^]								

STREAK—HARDNESS—SPECIFIC GRAVITY

LaRune's ROCKPECKER Notes - CHART B. Page 7 of 17 Copyright 1993

I.D. #	MINERAL NAME (& Field Name)	CLUE CODE	ORE OF %	FLUOR RADIO	BREAK COLOR	STREAK COLOR	MOHS HARDNESS	SPECIFIC GRAVITY	SURFACE LUSTER
101-	DOLOMITE	161	Mg.13%	F/uv	Wh.*	WHITE	3.5-4.0	2.8-2.9	VIT tp tl
	[*Pink,Yellow,Colorless^^]								
102-	DUMORTIERITE	127	IND.	F/uv	Vi.*	WHITE	7.0	3.2-3.4	VIT tl
	[*Blue,Red-Brown^^]								
103-	ENARGITE	008	Cu.48%	---	Bk.	BLACK	3.0	4.5	MET opq
104-	ENSTATITE [P]	057	Coll.	---	Gn.*	GRAY	5.5	3.1-3.2	VIT tl
	[*Yellow^^]								
105-	EOSPHORITE	197	Coll.	---	Wh.*	WHITE	5.0-6.0	2.9	SRK tl tp
	[*Gray-Green,Colorless^^]								
106-	EPIDOTE	072	Gem	---	Gn-Yl*	GRAY	6.0-7.0	3.3-3.5	VIT tl
	[*Green,Yellow,Brown,Red,Gray,Black^^]								
107-	EPSOMITE *	237	IND.	F/uv	Wh.	WHITE	2.0-2.5	1.67	VIT tp tl
	[*Bitter Salt^]								
108-	ERYTHRITE *	115	Co.*	---	Rd.	RED	1.5-2.5	3.07	VIT tp tl
	[*Cobalt Bloom ^]								
109-	EUCLASE	266	Gem	---	Wh.	WHITE	7.5	3.1	VIT tp tl
110-	FERBERITE *	016	V.60%	---	Br.-Rd	BLACK	5.0-5.5	7.1-7.5	S-MET opq
	[*Wolframite ^]								
111-	FERGUSONITE (REO)	037	Y.	Rad	Br.*	BROWN	5.5-6.5	5.6-5.8	VIT opq
	[*Yellow,Gray^^]								
112-	FLUORITE *	169	IND.	F/uv	Cl.*	WHITE	4.0	3.1-3.3	VIT tp tl
	[*Fluor Spar ^]==[*Yellow,Green,Blue,Black^^]								
113-	FRANKLINITE	116	Zn.20%	---	Bk-Rd	RED-BR.	5.5-6.5	5.0-5.2	MET opq
114-	FUCHITE (Mica)*	043	Coll.	---	Gn.	Cless	2.0-2.5	2.88	VIT tp tl
	[*Muscovite ^]								
115-	GADOLINITE (REO)	077	Y.38%	Rad.	Gn.*	GRAY-GN	6.5-7.0	3.1-3.3	VIT tp
	[*Brown^^]								
116-	GALENA *	059	Pb.86%	---	Gy.	GRAY	2.5-2.8	7.2-7.6	MET opq
	[*Lead Glance ^]								
117-	GARNIERITE *	087	Ni.36%	---	Gn.	GREEN	2.0-4.0	2.3-2.5	GSY tl
	[*A Generic Term ^]								

STREAK—HARDNESS—SPECIFIC GRAVITY

LaRune's ROCKPECKER Notes - CHART B. Page 8 of 17

Copyright 1993

I.D. #	MINERAL NAME (& Field Name)	CLUE CODE	ORE OF %	FLUOR RADIO	BREAK COLOR	STREAK COLOR	MOHS HARDNESS	SPECIFIC GRAVITY	SURFACE LUSTER
118-	GIBBSITE	137	Al.	---	Wh.*	WHITE	2.5-3.5	2.4	VIT opq
	[*Brown^^]								
119-	GLAUBERITE	134	IND.	F/uv	Wh.*	WHITE	2.5-3.0	2.81	GSY tp
	[*Yellow,Red^^]								
120-	GLAUCODOT	003	Co.25%	---	Wh.*	BLACK	5.0	6.1	MET opq
	[*Gray-White^^]								
121-	GLAUCONITE (Mica)	270	IND.	----	Be-Gn	----	2.0-2.3	2.5-2.8	GSY tl
122-	GLAUCOPHANE [A]	076	Coll.	----	Be-Gy*	GRAY-Be	6.0-6.5	3.1-3.3	VIT tl
	[*Black,Blue^^]								
123-	GOETHITE *	036	Fe.	---	Bn.	BROWN-Yw	5.0-5.5	4.3	SKY tl
	[*Needle Ironstone ^]								
124-	GOLD	244	Au.100%	---	Yw.	YELLOW	2.5-3.0	15.3-19.3	MET opq
125-	GRAPHITE	013	C.100%	---	Bk.	BLACK	1.0-2.0	2.2	S-MET opq
126-	GREENOCKITE	100	Cd.	---	Yw-Or	ORANGE	3.0-3.5	4.9	SRK tl
127-	GROSSULAR *	222	Gem	F/uv	Cl.*	WHITE	6.5-7.0	3.59	VIT tp
	[*Garnet ^][*Green,Brown,Orange,Pink,Red,Black^^]								
128-	GYPSUM	236	Ca.23%	F/uv	Wh.*	WHITE	2.0	2.35	VIT tp
	[*Colorless,Brown,Gray,Yellow^^]								
129-	HALITE *	124	Na.39%	F/uv	Wh.*	WHITE	2.5	2.1-2.2	VIT tp tl
	[*Rock Salt ^]==[*Yellow,Red,Brown,Blue,Black ^]								
130-	HARMOTOME *	174	Coll.	---	Wh.*	WHITE	4.5	2.3-2.5	VIT tp tl
	[*Cross Stone ^]=====[*Gray,Yellow,Red,Brown^^]								
131-	HAUERITE	105	Coll.	---	Bn.-Bk	PINK-Bn	3.5-4.5	3.5	S-MET opq
132-	HAUSMANNITE	117	Mn.72%	---	Bk.	RED-BN.	5.0-5.5	4.7-4.8	S-MET tl
133-	HAUYNITE (HAUYNE)	038	Coll.	---	Be.*	Cl.<Be.	5.5-6.0	2.4-2.5	VIT tp tl
	[*Green,White^^]								
134-	HEDENBERGITE [P]	034	Coll.	---	Gn.*	Lt-BROWN	5.0-6.0	3.5-3.6	VIT opq
	[*Brown^^]								

STREAK—HARDNESS—SPECIFIC GRAVITY

LaRune's ROCKPECKER Notes - CHART B. Page 9 of 17 Copyright 1993

```
=============================================================================================:
I.D.  MINERAL NAME        CLUE  ORE OF   FLUOR  BREAK  STREAK    MOHS      SPECIFIC   SURFACE   :
 #    (& Field Name)      CODE    %      RADIO  COLOR  COLOR   HARDNESS    GRAVITY    LUSTER    :
=============================================================================================:
135- HEMATITE       *     :112 : Fe.70% : ---  : : Bk.   : RED     : 5.5-6.5  : 5.2-5.3   : MET opq    :
=======[*Bloodstone^]=========================================================================:
136- HEMIMORPHITE *       :177 : Zn.54% : F/uv : : Wh.*  : WHITE : 4.5-5.0  : 3.1-3.2   : VIT tp tl  :
==[*Calamine ^]===[*Colorless,Brown,Yellow,Blue-Green^^]======================================:
137- HERDERITE            :272 : Coll.  : F/uv : :Gn-Wh*:  ?      : 5.0-5.5  : 2.9-3.0   : VIT tl tp  :
=====================================[*Purple,Yellow^^]=======================================:
138- HEULANDITE      *    :162 : Coll.  : ---- : : Yw.*  : WHITE : 3.5-4.0  : 2.1-2.2   : VIT tp tl  :
===[*Lamellar Zeolite ^]===========[*Green,Red-Orange^^]======================================:
139- HORNBLENDE      [A]  :191 : Coll.  : ---  : : Gn-Bk: WHITE : 5.0-6.0  : 3.0-3.4   : VIT tl     :
=============================================================================================:
140- HYDROMAGNESITE       :235 : Coll.  : ---  : :  Wh.  : WHITE : 3.5      : 2.24      : SRK tp     :
=============================================================================================:
141- HYDROZINCITE    *    :189 :  Zn.   : F/uv : :Wh\Gy : WHITE : 2.0-2.5  : 3.2-3.8   : VIT tl     :
=========[*Zinc Bloom ^]======================================================================:
142- ILMENITE        *    :015 : Ti.35% : ---  : :Bk\Bn :BN.-RED: 5.0-6.0  : 4.5-5.0   : MET opq    :
=====[*Titanoferrite ^]=======================================================================:
143- ILVAITE              :004 : Coll.  : ---  : :  Bn.  : BLACK : 5.5-6.0  : 3.8-4.1   : VIT tl     :
=============================================================================================:
144- JADITE        *  [P] :046 : Gem    : ---  : : Wh.*  : Cl\Wh : 6.5-7.0  : 3.3-3.5   : VIT tl     :
===[*Precious Jade ^]==[*Colorless,Yellow-White,Green^^]======================================:
145- JAMESONITE      *    :055 :  Pb.-  : ---- : : Gy.   :GRAY-BK.: 2.5      : 5.63      : MET opq    :
==[*Brittle Feather Ore ^]====================================================================:
146- JOHANNSENITE    [P]  :267 : Coll>  : ---- : :Gn-Br.:  ?      : 5.0-6.0  : 3.44-3.46: VIT tl     :
=============================================================================================:
147- KAOLINITE            :240 : Al.20% : ---- : :Yw-Gy*: WHITE  : 2.0-2.5  : 2.5-2.6   : PRL tl     :
============================================[*Brown ^]========================================:
148- KERNITE              :129 :  B.    : ---- : :Wh\Cl.: WHITE  : 2.5      : 1.9       : VIT  tl    :
=============================================================================================:
149- KYANITE        *     :216 :  Ind.  : ---- : :Wh.*  : WHITE  : 6.0-7.0  : 3.67      : VIT tp tl  :
=========[*Disthene ^]============[*Green,Gray,Blue ^]========================================:
150- LABRADORITE     *    :193 :  Ind.  : ---- : :Be\Gn.: WHITE  : 5.0-6.0  : 2.62-2.75: VIT tp tl  :
=======[*Spectrolite ^]=======================================================================:
151- LARUNEITE            :052 :  Gem   : ---- : : Wh.* :GREEN-WH: 5.6-6.1  : 3.4-3.5   : VIT tp tl  :
==================================[*Green,Colorless ^]========================================:
```

STREAK—HARDNESS—SPECIFIC GRAVITY

LaRune's ROCKPECKER Notes - CHART B. Page 10 of 17 Copyright 1993

I.D. #	MINERAL NAME (& Field Name)	CLUE CODE	ORE OF %	FLUOR RADIO	BREAK COLOR	STREAK COLOR	MOHS HARDNESS	SPECIFIC GRAVITY	SURFACE LUSTER
152-	LAUMONTITE *	039	Coll.	----	Wh.*	CLESS	3.5-4.0	2.2-2.3	VIT tp
	[*Fibrous Zeolite ^]				[*Yellow,Pink ^]				
153-	LAZULITE *	190	Ind.	----	Be.	WHITE	5.0-6.0	3.0-3.1	VIT tp
	[*Blue Spar ^]								
154-	LAZURITE *	022	Gem	----	Be.	BLUE	5.0-5.3	2.38-2.42	VIT opq
	[*Lapis Lazuli ^]								
155-	LEPIDOCROCITE	098	Fe.	----	Rd	ORANGE	5.0-5.5	4.0	SUB-SRK tl
156-	LEPIDOLITE *	140	Li.3.6%	F/uv	Pk \Vi	WHITE	2.5-4.0	2.8-2.9	PRL tl
	[*Lithium Mica ^]								
157-	LEUCITE	204	K.18%	----	Wh\Gy	WHITE	5.5-6.0	2.5	VIT tp
158-	LIMONITE *	033	Fe.62%	---	Yw-Bn	BROWN	5.0-5.5	3.5-4.0	ERY opq
	[*Brown Iron Ore ^]								
159-	LINARITE	021	Coll.	---	Be.	BLUE	2.5	5.35	SRK tl
160-	MAGNESITE *	167	Mg.28.8%	---	Wh.*	WHITE	3.5-4.5	3.0	VIT tp tl
	[*Bitter Spar ^]				[*Yellow,Gray^^]				
161-	MAGNETITE	010	Fe.72%	---	Bk-Be	BLACK	5.5-6.5	5.2	MET opq
162-	MALACHITE	085	Cu.57%	---	Gn.	GREEN	3.5-4.0	4.0	VIT opq
163-	MANGANITE *	030	Mn.62%	---	Bk.	BROWN	4.0	4.3-4.4	S-MET opq
	[*Brown Man Ore ^]								
164-	MARCASITE *	095	Ind.	---	Yw.-Gn	GR.-BK	6.0-6.5	4.8-4.9	MET opq
	[*White Pyrites ^]								
165-	MARGERITE *	163	Coll.	---	Wh. *	WHITE	3.5-4.0	2.9-3.1	PRL tp tl
	[*Brittle Mica ^]				[*Gray,Pink ^]				
166-	MELILITE*(Field Term)	264	Coll.	---	Wh.*	?	5.0	3.4-3.5	VIT tp tl
	[*For Akermanite ^]				[*Yellow,Brown-Red,Gray-Green ^]				
167-	MERCURY *	263	Hg.100%	---	Si.-Wh.	?	-0-	13.6	MET opq
	[*Quicksilver ^]								
168-	MICROCLINE *	213	Gem	---	Wh.*	WHITE	6.0-6.5	2.56-2.59	VIT tp tl
	[*Amazonite ^]				[*Pink,Red-Brown,Blue-Green,Green ^]				

STREAK—HARDNESS—SPECIFIC GRAVITY

LaRune's ROCKPECKER Notes - CHART B. Page 11 of 17 Copyright 1993

I.D. #	MINERAL NAME (& Field Name)	CLUE CODE	ORE OF %	FLUOR RADIO	BREAK COLOR	STREAK COLOR	MOHS HARDNESS	SPECIFIC GRAVITY	SURFACE LUSTER
169-	MICROLITE	254	Coll.	----	Yw.*	YELLOW	5.0-5.5	4.2-6.4	VIT tp tl op
					[*Brown, Colorless ^]				
170-	MILARITE	187	Gem	----	Wh.*	WHITE	5.5-6.0	2.6	VIT tp tl
					[*Green,Colorless ^]				
171-	MILLERITE *	093	Ni.	---	Yw.	GN-BK	3.0-3.5	5.2-5.6	MET opq
	[*Hair Pyrites ^]								
172-	MIMETITE *	152	Pb.	---	Yw-Br*	WHITE	3.5-4.0	7.1	VIT tp tl
	[*Campylite ^]				[*Orange,Colorless ^]				
173-	MOLYBDENITE	026	Mo.60%	---	Be-Gy	BE-GY	1.0-1.5	4.7	MET opq
174-	MONAZITE (REO)	230	Ce\Th12%	---	Yw-Br	WHITE	5.0-5.5	4.8-5.5	RSN tp tl
175-	MONTMORILLONITE	242	Ind	---	Wh-Gy	WHITE	1.0	1.7-2.7	DUL opq
176-	MUSCOVITE *	041	Ind.	---	Si-Wh*	CL	2.0-2.25	2.7-2.8	PRL tp tl
	[*Isinglass ^]				[*Yellow,Brown^^]				
177-	NATUROLITE	181	Coll.	F/uv	Wh.*	WHITE	5.0-5.5	2.9-3.0	VIT tp tl
					[*Pink, Yellow, Colorless^^]				
178-	NEPHELINE	205	Ind.	F/uv	Wh.*	WHITE	5.5-6.0	2.5-2.6	GSY tl
					[*Gray, Yellow,Red,Green^^]				
179-	NEPTUNITE	113	Coll.	---	Bn.*	RD-BN	5.0-6.0	3.2	VIT opq
					[*Black^^]				
180-	NICCOLITE	035	Ni.43%	---	Cu-Rd	BN-BK	5.0-5.5	7.8	MET opq
	[*Arsenical Nickel ^]								
181-	NICKEL-IRON *	259	Coll.	---	Gr\Bk.	?	4.0-5.0	7.3-8.8	MET opq
	[*Taenite^]								
182-	OLIVENITE	245	Coll.	---	Yw.*	YELLOW	3.0	4.3	SRK tp tl
					[*Green,Brown-Yellow,Gray,White^^]				
183-	OLIVINE *	045	Gem	---	Olive*	CL	6.5-7.0	3.3-4.2	VIT tp tl
	[*Fosterite,Peridot ^]				[*Yellow,Brown^^]				
184-	OMPHACITE [P]	231	Coll.	---	Gr.	WHITE	5.0-6.0	3.3-3.4	VIT tl
185-	OPAL	207	Gem	F/uv	Wh.*	WHITE	5.5-6.5	1.9-2.5	GSY tp tl
					[*Blue,Black^^]				

STREAK—HARDNESS—SPECIFIC GRAVITY

LaRune's ROCKPECKER Notes - CHART B. Page 12 of 17 Copyright 1993

I.D. #	MINERAL NAME (& Field Name)		CLUE CODE	ORE OF %	FLUOR RADIO	BREAK COLOR	STREAK COLOR	MOHS HARDNESS	SPECIFIC GRAVITY	SURFACE LUSTER
186-	ORPIMENT	*	248	As.	---	Or-Yw*	YELLOW	1.5-2.0	3.48	GSY tp tl
	[*Tinsel ^]					[*Gold^^]				
187-	ORTHOCLASE	*	239	Ind.	---	Wh.*	WHITE	6.0	3.0-3.3	VIT tp tl
	[*K-Feldspar ^]					[*Colorless,Yellow,Pink,Blue,Gray^^]				
188-	PARAGASITE	* [A]	257	Coll.	---	Bn.*	YELLOW	5.0-6.0	3.1-3.2	VIT tp tl
	[*Hornblende ^]					[*Green,Gray^^]				
189-	PECTOLITE		178	Coll.	F/uv	Wh.\Gy.	WHITE	5.0	2.8-2.9	VIT tl
190-	PENNINITE	*	081	Coll.	---	Gn.	GREEN	2.0-2.5	2.5-2.8	VIT tl
	[*Green Chlorite ^]									
191-	PENTLANDITE		028	Ni.40%	---	Bn-Yw	BROWN	3.5-4.0	4.6-5.2	MET opq
192-	PEROVSKITE	(REO)	252	Nb.\Ce.	Rad.	Yw.*	YELLOW	5.5	4.0	SRK tl
						[*Red-Brown,Black-Gray^^]				
193-	PETALITE		044	Li.2.2%	----	Wh.*	C-LESS	6.0-6.5	2.3-2.5	VIT tp tl
						[*Colorless,Gray^^]				
194-	PHENAKITE	*	192	Gem	----	Wh.*	WHITE	7.5-8.0	2.9-3.0	VIT tp
	[*Phenacite^]					[*Pink, Colorless,Yellow^^]				
195-	PHILLIPSITE	*	042	Coll.	---	Wh.*	C-LESS	4.0-4.5	2.1-2.2	VIT tp tl
	[*Cross Stone ^]					[*Colorless,Gray,Yellow^^]				
196-	PHLOGOPITE	*	132	Ind.	F/lw	Yw.*	WHITE	2.5-3.0	2.7-2.0	PRL tp
	[*Amber Mica ^]					[*Light Brown ^^]				
197-	PHOSGENITE	*	234	Coll.	F/uv	Cl.*	WHITE	2.7-3.0	6.0-6.3	GSY tp tl
	[*Horn Lead ^]					[*Amber-Yellow ^^]				
198-	PHOSPHORPHYLLITE		269	Gem	F/uv	Be-Gn*	?	3.0-3.5	3.0-3.1	VIT tp tl
						[*Colorless ^^]				
199-	PIEDMONTITE	*	111	Coll.	---	Pu-Bn*	RED	6.5	3.4	VIT tl
	[*Mangan Epidote ^]					[*Red ^^]				
200-	PLATINUM		060	Pt.	---	Si-Gy	GRAY	4.0-4.5	14.0-19.0	MET opq
201-	POLYBASITE		005	Ag.75%	---	Bk.	BLACK	2.0-3.0	6.0-6.2	MET opq
202-	PREHNITE		212	Coll.	---	Gn-Wh*	WHITE	6.0-6.5	2.9-2.95	VIT tl
						[*Colorless,Gray,Yellow ^^]				

STREAK—HARDNESS—SPECIFIC GRAVITY

LaRune's ROCKPECKER Notes - CHART B. Page 13 of 17 Copyright 1993

I.D. #	MINERAL NAME (& Field Name)		CLUE CODE	ORE OF %	FLUOR RADIO	BREAK COLOR	STREAK COLOR	MOHS HARDNESS	SPECIFIC GRAVITY	SURFACE LUSTER
203-	PROUSTITE	*	119	Ag.65%	---	Dk.-Rd	SCARLET	2.0-2.5	5.6	SRK tl
	[*Ruby Silver ^]									
204-	PSILOMELANE	*	012	Mn.63%	---	Bk.	BLACK	5.0-7.0	3.7-4.7	MET opq
	[Romanechite ^]									
205-	PURPURITE		109	Coll.	---	Pur-Rd	RED	4.0-4.5	3.4	GSY tl
206-	PYRARGYRITE	*	107	Ag.59%	---	Bk-Rd.	RED-PURP	2.5	5.85	S-MET tl
	[*Dark Ruby Silver ^]									
207-	PYRITE *		092	S. 53%	---	Gold	GREEN-BK	6.0-6.5	5.0-5.2	MET opq
	[*Iron Pyrite,Fools Gold^]									
208-	PYROLUSITE		024	Mn.63%	---	Bk.	BLUE-BK	6.0-6.5	4.7-5.0	S-MET opq
209-	PYROMORPHITE		154	Pb.76%	---	Gn.*	WHITE	3.5-4.0	6.7	SRK tl
	[*Yellow,Brown,Colorless,Yellow-Red ^^]									
210-	PYROPE	*	114	Gem	---	Dk.Rd.	RED	6.5-7.5	3.58	VIT tp
	[*Bohemian Garnet ^]									
211-	PYROPHYLLITE		262	Ind.	F/lw	Yw-Wh*	WHITE	1.0-2.0	2.8	GSY tl
	[*Green, Green-Brown^^]									
212-	PYRRHOTITE	*	002	S.36%	---	Gold-Bn	BLACK	3.5-4.5	4.6	MET opq
	[*Magnetic Pyrite ^]									
213-	QUARTZ	*	143	Si.46%	F/uv	Wh.*	WHITE	7.0	2.65	VIT tp tl
	[*Rock Crystal ^]==[*Colorless,Brown,Black,^^^ Yellow,Blue,Violet,Green,Pink]									
214-	REALGAR		102	Ind.	---	Or-Rd	OR-YW.	1.5-2.0	3.5-3.6	RSN tp tl
215-	RHODOCHROSITE	*	104	Mn.47%	---	Pk.*	PINK	3.5-4.5	3.3-3.6	VIT tl
	[*Manganese Spar ^]==[*Rose-Red,Raspberry-Red,^^^Gray,Brown]									
216-	RHODONITE	*	208	Mn.41%	---	Pk.*	WHITE	5.5-6.5	3.4-3.7	VIT tp tl
	[*Fowlerite ^]=====[*Red-Brown,Pink-Red ^^]									
217-	RICHTERTITE	[A]	251	Coll.	---	Yw.*	YELLOW	5.0-6.0	2.9-3.4	VIT tp tl
	[*Purple-Red ^^]									
218-	ROSASITE		086	Cu.	---	Gn.*	GREEN	4.5	4.0-4.2	VIT tl
	[*Blue-Green ^^]									
219-	RUTILE		029	Ti.60%	---	Bn.*	BROWN	6.0-6.5	4.3	MET opq tl
	[*Yellow,Red,Black ^^]									

STREAK—HARDNESS—SPECIFIC GRAVITY

LaRune's ROCKPECKER Notes - CHART B. Page 14 of 17 Copyright 1993

I.D. #	MINERAL NAME (& Field Name)	CLUE CODE	ORE OF %	FLUOR RADIO	BREAK COLOR	STREAK COLOR	MOHS HARDNESS	SPECIFIC GRAVITY	SURFACE LUSTER
220-	SAMARSKITE (REO)	118	Y. 28%	Rad.	Bk.	RED-BN.	5.0-6.0	5.7	RSN opq
221-	SANADINE *	210	Coll.	---	Wh.*	WHITE	6.0	2.5-2.6	VIT tl opq
	[*K Feldspar ^] [*Colorless,Gray,Yellow ^^]								
222-	SCAPOLITE	198	Gem	F/uv	Wh.*	WHITE	5.0-6.5	2.5-2.7	VIT tp tl
	[*Colorless,Purple,Blue,Gray,Pink ^^]								
223-	SCHEELITE	229	W. 63%	F/uv	Wh.*	WHITE	4.5-5.0	5.9-6.1	VIT tp tl
	[*Yellow,Green,Red-Gray^^]								
224-	SCOLECITE *	268	Coll.	---	Wh.*	WHITE	5.0-5.5	2.2-2.4	VIT tp
	[*Fibrous Zeolite ^] [*Colorless ^^]								
225-	SCORODITE	164	Coll.	---	Gn.*	WHITE	3.5-4.0	3.3	VIT tp tl
	[*Blue,Violet,Colorless,Yellow^^]								
226-	SEPIOLITE *	172	Ind.	---	Wh.*	WHITE	2.0-2.25	2.0	DUL opq
	[*Meerschaum ^] [*Yellow ^^]								
227-	SERPENTINE	138	Ind.	F/uv	Gn.*	WHITE	2.5-4.0	2.5-2.6	PRL opq
	[*Yellow-White ^^]								
228-	SIDERITE *	160	Fe.48%	---	Yw.*	WHITE	3.5-4.0	3.7-3.9	VIT tp tl
	[*Chalybite ^] [*Green,Brown,Black ^^]								
229-	SILLIMANITE *	051	Ind.	---	Wh.*	CLESS	6.0-7.0	3.2	VIT tp tl
	[*Fibrolite ^] [*Blue-Gray,Brown,Green^^]								
230-	SILVER	123	Ag.100%	---	Si.	SI.-WH.	2.5-3.0	10-0-12.0	MET opq
231-	SKUTTERUDITE *	009	Co.28%	---	Si-W	BLACK	6.0	6.5-6.8	MET opq
	[*Smaltite ^]								
232-	SMITHSONITE *	228	Zn.52%	---	Wh.*	WHITE	5.5	4.3-4.5	VIT tl
	[*Zinc Spar ^] [*Blue,Green,Yellow,Pink,Violet ^^]								
233-	SODALITE	050	Gem	F/uv	Be.*	CLESS	5.5-6.0	2.27-2.33	VIT tp tl
	[*White,Gray,Red,Green ^^]								
234-	SPHALERITE*	253	Zn.67%	F/uv	Yw.*	YELLOW	3.5-4.0	3.9-4.2	SRK tp tl
	[*Zinc Blende^] [*Colorless,Red-Brown,Pink,Black,Green^^]								
235-	SPINEL	201	Gem	F/uv	Wh.*	WHITE	8.0	3.5-4.1	VIT tp
	[*Colorless,Blue,Black,Pink^^]								
236-	SPODUMENE [P]	227	Li.3.4%	F/uv	Wh.*	WHITE	6.5-7.0	3.1-3.2	VIT tp
	[*Violet,Yellow,Pink,Green,Gray^^]								

STREAK—HARDNESS—SPECIFIC GRAVITY

LaRune's ROCKPECKER Notes - CHART B. Page 15 of 17 Copyright 1993

I.D. #	MINERAL NAME (& Field Name)	CLUE CODE	ORE OF %	FLUOR RADIO	BREAK COLOR	STREAK COLOR	MOHS HARDNESS	SPECIFIC GRAVITY	SURFACE LUSTER
237-	STAUROLITE *	070	Gem	---	Rd-Bn*	GRAY	7.0-7.5	3.73	VIT tp tl
	[*Rood Stone ^] [*Yellow-Brown, Black^^]								
238-	STEPHANITE *	011	Ag.68%	---	Bk.	BLACK	2.0-2.5	6.3	MET opq
	[*Brittle Silver ^]								
239-	STIBICONITE	256	Coll.	---	Yw.*	YELLOW	4.0-5.5	5.6	PRL opq
	[*Yellow-White, Orange, Brown, Black ^^]								
240-	STIBNITE *	066	Sb.71%	---	Gy.	GRAY	2.0	4.6-4.7	MET opq
	[*Antimonite ^]								
241-	STILBITE *	155	Coll.	----	Wh.*	WHITE	3.5-4.0	2.1-2.2	VIT tp tl
	[Bundle Zeolite ^] [*Red-Brown, Gray ^^]								
242-	STRONTIANITE	158	Sr.59%	----	Wh.*	WHITE	3.5-4.0	3.7	VIT tp tl
	[*Colorless, Gray, Green, Pink ^^]								
243-	SULFUR	142	S.100%	----	Yw.*	WHITE	1.5-2.5	2.0-2.1	GSY tp tl
	[*Brown, Black ^^]								
244-	SYLVANITE *	069	Au.24%	----	Si.-Wh.	GRAY	1.5-2.0	8.0-8.3	MET opq
	[*Graphic Tellurium ^]								
245-	SILVITE	166	K.52%	----	Wh.*	WHITE	2.0	1.99	VIT tp
	[*Red, Purple, Blue, Yellow-Red ^^]								
246-	TALC*	131	Ind.	----	Wh:.*	WHITE	1.0	2.58-2.83	GSY tl
	[*^ Soapstone, Steatite, Freestone] [*Green-White, Gray, ^^Brown]								
247-	TARBUTTITE	168	Coll.	----	Wh.*	WHITE	3.7	4.1	VIT tp tl
	[*Green, Yellow, Pink ^^]								
248-	TETRAHEDRITE- *	065	Cu.53%	----	Gy.*	GRAY	3.0-4.5	4.5-5.2	MET opq
	TENNANTITE-[*^Gray Copper] [*Purple, Blue, Brown ^^]								
249-	THOMSONITE *	048	Gem	----	Wh.*	CLESS	5.0-5.5	2.3-2.4	VIT tp tl
	[*Comptonite ^] [*Yellow, Pink, Brown, Green ^^]								
250-	THORIANITE	096	U & Th.	Rad.	Bk\Gy	GR-GY.	6.5	9.7-9.8	MET s-opq
251-	TITANITE*	183	Ti.& Gem	----	Wh.*	WHITE	5.0-5.5	3.45	VIT tp tl
	[*Sphene ^] [*Yellow-Green, Brown, Black, Pink, Colorless^^]								
252-	TOPAZ	220	Gem	----	Bn.*	WHITE	8.0	3.26-3.36	VIT tp tl
	[*Colorless, Red-Yellow, Violet, Green, Blue, Yellow ^^]								
253-	TORBERNITE *	088	U.	Rad.	Gn.	GREEN	2.0-2.5	3.3	PRL tp tl
	[*Copper Uranite ^]								

STREAK HARDNESS—SPECIFIC GRAVITY

LaRune's ROCKPECKER Notes - CHART B. Page 16 of 17 Copyright 1993

I.D. #	MINERAL NAME (& Field Name)	CLUE CODE	ORE OF %	FLUOR RADIO	BREAK COLOR	STREAK COLOR	MOHS HARDNESS	SPECIFIC GRAVITY	SURFACE LUSTER
254-	TOURMALINE	:238	Gem	F/uv	Pk.*	CLESS	7.0	3.0-3.25	VIT tp tl
	[*Colorless,Black,Green,Brown,Blue,Yellow ^^]								
255-	TREMOLITE [A]	:197	Ind.	----	Wh *	WHITE	5.0-6.0	2.9	SRK tp tl
	[*Colorless,Gray-Green ^^]								
256-	TRIDYMITE *	:153	Ind.	----	Wh.	WHITE	7.0	2.27	VIT tp tl
	[*Quartz ^]								
257-	TRONA	:135	Ind.	----	Wh.*	WHITE	2.5-3.0	2.17	ERY tp tl
	[*Yellow-White ^^]								
258-	TURQUOISE *	:090	Gem	----	Gn.*	GREEN	5.0-6.0	2.6-2.8	WAXY tl
	[*Callaite ^] [*Blue-Green ^^]								
259-	ULEXITE *	:265	Coll.	----	Wh.	?	1.0	2.0	SKY tp
	[*Cotton Balls ^][*T.V. stone]								
260-	URANINITE *	:017	U.83%	Rad.	Bk.	BLACK	5.5	7.5-9.7	S-MET opq
	[*Pitchblende ^]								
261-	UVAROVITE *	:097	Gem	----	Gn.	GN.-WH.	6.5-8.0	3.4-3.8	VIT tp
	[*Garnet ^]								
262-	VANADINITE	:255	V. 10%	----	Yw.*	YELLOW	2.75-3.0	6.8-7.1	RSN tp
	[*Red-Orange,Brown ^^]								
263-	VARISCITE	:170	Gem	----	Gn.*	WHITE	4.0-5.0	2.2-2.5	VIT tl
	[*Blue-Green,Colorless ^^]								
264-	VERMICULITE	:258	Ind.	----	Yw.*	YELLOW	2.0-3.0	2.1-2.7	PRL tl
	[*Gray,Green,Brown,Black^^]								
265-	VESUVIANITE *	:218	Gem	----	Bn.*	WHITE	6.0-7.0	6.0-7.1	VIT tp tl
	[*Idocrase ^]==[*Yellow,Black,Red,Green-Brown^^]								
266-	VIVIANITE *	:199	Ind.	----	Gn.*	WHITE	1.5-2.0	2.6-2.7	VIT tp
	[*Blue Iron ^]===[*Green,Blue,Violet,Black^^]								
267-	VOLBORTHITE	:275	Coll.	----	Olive	WHITE	3.0	3.5-3.8	VIT tp tl
268-	WAVELLITE	:150	Ind.	F/uv	Wh.*	WHITE	3.3-4.0	2.3-2.4	VIT tl
	[*Yellow,Green,Brown,Blue ^^]								
269-	WILLEMITE	:049	Zn.58%	F/uv	Yw-Gn*	CLESS	5.5	4.0	VIT tl
	[*Blue-Red,Colorless,Brown,Red,Black^^]								
270-	WITHERITE (toxic)	:157	Ba.69%	F/uv	Wh.*	WHITE	3.5-4.0	4.2-4.3	VIT tp tl
	[*Colorless,Gray,Yellow,Brown,Green ^^]								

STREAK—HARDNESS—SPECIFIC GRAVITY

LaRune's ROCKPECKER Notes - CHART B. Page 17 of 17 Copyright 1993

I.D. #	MINERAL NAME (& Field Name)		CLUE CODE	ORE OF %	FLUOR RADIO	BREAK COLOR	STREAK COLOR	MOHS HARDNESS	SPECIFIC GRAVITY	SURFACE LUSTER
271-	WOLFRAMITE	*	018	W.	----	Bn-Bk.	BK.-BN.	4.0-4.5	7.1-7.5	S-MET opq
	[*See Also Ferberite ^]									
272-	WOLLASTONITE	*	176	Ind.	F/uv	Wh.*	WHITE	4.5-5.0	2.8-3.1	VIT tl
	[*Table Spar ^] [*Green,Colorless,Gray ^^]									
273-	WULFENITE	*	141	Mo.26%	----	Yw.*	WHITE	2.7-3.0	6.5-7.0	RSN tp tl
	[*Yellow Lead Ore ^] [*Red-Orange,Red,Gray ^^]									
274-	ZINCITE	*	103	Zn.80%	----	Or\Rd.	OR-YW.	4.0-4.5	5.4-5.7	SRK tl tp
	[*Red Oxide Of Zinc ^]									
275-	ZIRCON		047	Zr.49%	F/uv	Yw.*	CLESS	7.5	4.2-4.7	SRK tp tl
	[*Colorless,Red,Brown,Gray,Green ^^]									
276-	ZINNWALDITE	*	130	Li.1.8%	----	Yw.*	WHITE	2.5-3.0	2.9-3.1	PRL tp tl
	[*Lithionite ^] [*Green,Silver-Gray,Brown ^^]									
277-	ZOISITE		053	Gem	F/uv	Wh.*	CLESS	6.0-6.5	3.2-3.3	VIT tp
	[*Violet-Blue,Blue,Green,Pink ^^]									

CHAPTER 66

COLOR OF STREAK SELECTION

CHART C

CHART C HEADINGS

THIS CHART IS ARRANGED BY COLOR OF STREAK. THE CLUE CODE COLOR NUMBER [Column C] IS ARRANGED SEQUENTIALLY. WITH THE MINERAL I.D NUMBER AND CLUE CODE NUMBER YOU CAN HOP FROM CHART TO CHART.

ENTER THE CHART BY STREAK COLOR. ELIMINATE MINERALS THAT DO NOT MATCH YOUR HARDNESS, LUSTER OR SPECIFIC GRAVITY VALUES. THIS SHOULD LEAVE YOU WITH A MATCH FOR YOUR UNKNOWN MINERAL.

COLUMN EXPLANATION

A. - MINERAL I.D. NUMBER = This number identifies the same specific mineral throughout ALL CHARTS.

B. - MINERAL NAME & FIELD NAME = Identifies the mineral name.
Note: Some minerals are better known by their field name. Example: Silver Glance for Argentite. (The charts use the U.S.A. mineral name.)

C. - CLUE CODE = This code number gets you in COLOR STREAK, Chart C for quick pick-up of streak, hardness, specific gravity, luster & mineral color as it relates to the mineral with that clue code.

D. - ORE % = Once you identify your mineral you want to know if it has a market value. Ore % lets you know if your market is for an ore, gem, or collector's item.

E. - FLUOR/RADIO = This tells you the minerals that react to black light of long or short wavelengths. Radioactivity is indicated.

F. - BREAK COLOR = Mineral color must be determined on a FRESH BREAK. This column lets you know what colors are usual for that mineral I.D. number.

G. - STREAK COLOR = ONCE YOUR HAVE DETERMINED THE POWDER COLOR OF YOUR MINERAL, THE SOLE PURPOSE OF CHART C IS TO LET YOU KNOW THE PROBABLE MINERALS WITHIN THAT SPECIFIC COLOR STREAK. OTHER PHYSICAL TESTS CAN ELIMINATE MOST PROBABLES. YOU THEN CONFIRM YOUR CONCLUSION WITH TESTS FOR CHEMICAL ELEMENTS PRESENT.

H. - MOHS HARDNESS = This column tells you the HARDNESS RANGE of your mineral. It is another confirming physical test for elimination purposes.

J. - SPECIFIC GRAVITY = Column tells you the SPECIFIC GRAVITY RANGE of your mineral. It is another physical test for elimination purposes.

K. - SURFACE LUSTER = Column tells you the LUSTER OF A MINERAL SURFACE as it appears or reflects light.

Copyright 1993 Page 1

COLOR STREAK CHART C

A	B	C	D	E	F	G	H	J	K
I.D. #	MINERAL NAME (& Field Name)	CLUE CODE	ORE OF %	TEST #	BREAK COLOR	STREAK COLOR	MOHS HARDNESS	SPECIFIC GRAVITY	SURFACE LUSTER

SHADES OF BLACK STREAK

I.D. #	MINERAL NAME (& Field Name)	CLUE CODE	ORE OF %	TEST #	BREAK COLOR	STREAK COLOR	MOHS HARDNESS	SPECIFIC GRAVITY	SURFACE LUSTER
026	ARSENOPYRITE *	001	As.46%	---	Si.*	BLACK	5.0-6.0	5.9-6.2	MET opq
	[*Mispickel^]				[*White-Gray ^^]				
212	PYRRHOTITE *	002	S.36%	---	Gold-Bn	BLACK	3.5-4.5	4.6	MET opq
	[*Magnetic Pyrite ^]								
120	GLAUCODOT	003	Co.25%	---	Wh.*	BLACK	5.0	6.1	MET opq
					[*Gray-White^^]				
143	ILVAITE	004	Coll.	---	Bn.	BLACK	5.5-6.0	3.8-4.1	VIT tl
201	POLYBASITE	005	Ag.75%	---	Bk.	BLACK	2.0-3.0	6.0-6.2	MET opq
045	BIXBYITE	006	Mn.	---	Bk.	BLACK	6.0-6.5	4.9-5.0	MET opq
049	BOULANGERITE *	007	Pb.55%	---	Gy.	BLACK	2.5-3.0	5.8-6.2	MET opq
	[*Feather Ore ^]								
103	ENARGITE	008	Cu.48%	---	Bk.	BLACK	3.0	4.5	MET opq
231	SKUTTERUDITE *	009	Co.28%	---	Si-W	BLACK	6.0	6.5-6.8	MET opq
	[*Smaltite ^]								
161	MAGNETITE	010	Fe.72%	---	Bk-Be	BLACK	5.5-6.5	5.2	MET opq
238	STEPHANITE *	011	Ag.68%	---	Bk.	BLACK	2.0-2.5	6.3	MET opq
	[*Brittle Silver ^]								
204	PSILOMELANE *	012	Mn.63%	---	Bk.	BLACK	5.0-7.0	3.7-4.7	MET opq
	[Romanechite ^]								
125	GRAPHITE	013	C.100%	---	Bk.	BLACK	1.0-2.0	2.2	S-MET opq
080	COLUMBITE *	014	Nb.& Ta.	---	Bk.	BK-BR.	6.0	5.3-8.1	S-MET opq
	[*Niobite ^]								
142	ILMENITE *	015	Ti.35%	---	Bk\Bn	BK-RED	5.0-6.0	4.5-5.0	MET opq
	[*Titanoferrite ^]								
110	FERBERITE *	016	V.60%	---	Br-Rd	BLACK	5.0-5.5	7.1-7.5	S-MET opq
	[*Wolframite ^]								

Copyright_1993 Page 2

CHART C

```
= A ====== B ============ C  == D ====== E ====== F ===== G ======= H =========== J ======= K =======:
I.D.   MINERAL NAME       CLUE  ORE OF    TEST    BREAK  STREAK     MOHS          SPECIFIC  SURFACE    :
#    (& Field Name)       CODE    %        #      COLOR   COLOR     HARDNESS       GRAVITY   LUSTER    :
=====================================================================================================:
260-URANINITE        *   :017 : U.83% : Rad. : : Bk.   : BLACK  :  5.5       : 7.5-9.7  : S-MET opq   :
========[*Pitchblende ^]=============================================================================:
271-WOLFRAMITE       *   :018 : W.    : ---  : : Bn-Bk: BK-BN. :  4.0-4.5   : 7.1-7.5  :  S-MET opq  :
==[*See Also Ferberite ^]============================================================================:
```

SHADES OF BLUE

```
                             ---------------------STREAK
=======================================================>^^^^^^^^<====================================:
093-CYANOTRICHITE    *   :019 : Coll. : ---  : :Sky-Be: BLUE   :  1.0       : 2.75     :  SKY tp tl  :
========[*Lettsomite ^]==============================================================================:
046-BOLEITE              :020 : Coll. : ---  : :Dk-Be : BLUE   :  3.5       : 4.8-5.1  :  PRL tl     :
=====================================================================================================:
159-LINARITE             :021 : Coll. : ---  : : Be.  : BLUE   :  2.5       : 5.35     :  SRK tl     :
=====================================================================================================:
154-LAZURITE       *     :022 : Gem   : ---- : : Be.  : BLUE   :  5.0-5.3   : 2.38-2.42:  VIT opq    :
======[*Lapis Lazuli ^]==============================================================================:
034-AZURITE          *   :023 : Cu.55% : --- : : Be.  : BLUE   :  3.5-4.0   : 3.83     :  VIT tl/tp  :
====[*Blue Carbonate  ^]=============================================================================:
208-PYROLUSITE           :024 : Mn.63% : --- : : Bk.  : BLUE-BK:  6.0-6.5   : 4.7-5.0  :  S-MET opq  :
=====================================================================================================:
023-ARFVEDSONITE     [A]:025 : Coll.  : --- : : Gn-Bk: BLUE-GY:  5.5-6.0   : 3.4-3.5  :  VIT opq    :
=====================================================================================================:
173-MOLYBDENITE          :026 : Mo.60% : --- : : Be-Gy : BE-GY :  1.0-1.5   : 4.7      :  MET opq    :
=====================================================================================================:
031-AURICHALCITE         :027 : Coll   : --- : : Be.  : BLUE-GN:  2.0-2.5   : 4.2      :  PRL tl     :
=====================================================================================================:
```

SHADES OF BROWN

```
                   ----------------------------------STREAK
=======================================================>^^^^^^^^<====================================:
191-PENTLANDITE          :028 : Ni.40% : --- : : Bn-Yw : BROWN :  3.5-4.0   : 4.6-5.2  :  MET opq    :
=====================================================================================================:
219-RUTILE               :029 : Ti.60% : --- : : Bn.*  : BROWN :  6.0-6.5   : 4.3      :  MET opq tl :
===============================[*Yellow,Red,Black ^^]================================================:
163-MANGANITE        *   :030 : Mn.62% : --- : : Bk.   : BROWN :  4.0       : 4.3-4.4  :  S-MET opq  :
=======[*Brown Man Ore ^]=============================================================================:
```

Copyright 1993 Page 3

CHART C

A I.D. #	B MINERAL NAME (& Field Name)		C CLUE CODE	D ORE OF %	E TEST #	F BREAK COLOR	G STREAK COLOR	H MOHS HARDNESS	J SPECIFIC GRAVITY	K SURFACE LUSTER
071	CHROMITE	*	031	Cr.46%	---	Bk.	BROWN	5.5	4.5-4.8	S-MET opq
	[*Chrome Iron Ore ^]									
001	ACMITE	* [P]	032	Coll.	---	Br.*	BROWN	6.0-6.5	3.5-3.6	VIT t/l
	[*Aegirine ^]					[*Green^^]				
158	LIMONITE	*	033	Fe.62%	---	Yw-Bn	BROWN	5.0-5.5	3.5-4.0	ERY opq
	[*Brown Iron Ore ^]									
134	HEDENBERGITE	[P]	034	Coll.	---	Gn.*	Lt-BROWN	5.0-6.0	3.5-3.6	VIT opq
						[*Brown^^]				
180	NICCOLITE		035	Ni.43%	---	Cu-Rd	BN-BK	5.0-5.5	7.8	MET opq
	[*Arsenical Nickel ^]									
123	GOETHITE	*	036	Fe.	---	Bn.	BN-YW	5.0-5.5	4.3	SKY tl
	[*Needle Ironstone ^]									
111	FERGUSONITE (REO)		037	Y.	Rad	Br.*	BROWN	5.5-6.5	5.6-5.8	VIT opq
						[*Yellow,Gray^^]				

SHADES OF COLORLESS

STREAK ^^^^^^^^

A I.D. #	B MINERAL NAME (& Field Name)		C CLUE CODE	D ORE OF %	E TEST #	F BREAK COLOR	G STREAK COLOR	H MOHS HARDNESS	J SPECIFIC GRAVITY	K SURFACE LUSTER
133	HAUYNITE (HAUYNE)		038	Coll.	---	Be.*	Cl.\Be.	5.5-6.0	2.4-2.5	VIT tp tl
						[*Green,White^^]				
152	LAUMONTITE	*	039	Coll.	----	Wh.*	CLESS	3.5-4.0	2.2-2.3	VIT tp
	[*Fibrous Zeolite ^]					[*Yellow,Pink ^]				
051	BRAZILIANITE		040	Gem		Gn-Yw	CLESS	5.5	2.98	VIT tp/tl
176	MUSCOVITE	*	041	Ind.	---	Si-Wh*	CLESS	2.0-2.25	2.7-2.8	PRL tp tl
	[*Isinglass ^]					[*Yellow,Brown^^]				
195	PHILLIPSITE	*	042	Coll.	---	Wh.*	CLESS	4.0-4.5	2.1-2.2	VIT tp tl
	[*Cross Stone ^]					[*Colorless,Gray,Yellow^^]				
114	FUCHITE (Mica)	*	043	Coll.	---	Gn.	CLESS	2.0-2.5	2.88	VIT tp tl
	[*Muscovite ^]									
193	PETALITE		044	Li.2.2%	----	Wh.*	CLESS	6.0-6.5	2.3-2.5	VIT tp tl
						[*Colorless,Gray^^]				
183	OLIVINE	*	045	Gem	---	Olive*	CLESS	6.5-7.0	3.3-4.2	VIT tp tl
	[*Fosterite,Peridot ^]					[*Yellow,Brown^^]				

```
Copyright 1993          Page 4                                  CHART  C
= A ====== B ============ C  == D ====== E ====== F ===== G ======= H =========== J ======= K =======:
I.D.   MINERAL NAME      CLUE  ORE OF   TEST    BREAK   STREAK   MOHS         SPECIFIC  SURFACE    :
#    (& Field Name)      CODE    %        #     COLOR   COLOR    HARDNESS      GRAVITY   LUSTER     :
====================================================================================================:
144-JADEITE        *  [P] :046 : Gem    : ---  : : Wh.*  : CL\WH :  6.5-7.0  : 3.3-3.5  :  VIT tl    :
===[*Precious Jade ^]==[*Colorless,Yellow-White,Green^^]============================================:
275-ZIRCON                :047 : Zr.49% : F/uv : : Yw.*  : CLESS :  7.5      : 4.2-4.7  :  SRK tp tl :
===================[*Colorless,Red,Brown,Gray,Green ^^]=============================================:
249-THOMSONITE      *     :048 : Gem    : ---- : : Wh.*  : CLESS :  5.0-5.5  : 2.3-2.4  :  VIT tp tl :
=========[*Comptonite ^]==[*Yellow,Pink,Brown,Green ^^]=============================================:
269-WILLEMITE             :049 : Zn.58% : F/uv : :Yw-Gn* : CLESS :  5.5      : 4.0      :  VIT tl    :
================[*Blue-Red,Colorless,Brown,Red,Black^^]=============================================:
233-SODALITE              :050 : Gem    : F/uv : : Be.*  : CLESS :  5.5-6.0  : 2.27-2.33:  VIT tp tl :
============================[*White,Gray,Red,Green ^^]=============================================:
229-SILLIMANITE     *     :051 : IND.   : ---  : : Wh.*  : CLESS :  6.0-7.0  : 3.2      :  VIT tp tl :
==========[*Fibrolite ^]=====[*Blue-Gray,Brown,Green^^]=============================================:
151-LARUNEITE             :052 : Gem    : ---- : : Wh.*  : CLESS :  5.6-6.1  : 3.4-3.5  :  VIT tp tl :
===================================[*Green,Colorless ^]=============================================:
277-ZOISITE               :053 : Gem    : F/uv : : Wh.*  : CLESS :  6.0-6.5  : 3.2-3.3  :  VIT tp    :
=====================[*Violet-Blue,Blue,Green,Pink ^^]=============================================:
015-ANGLESITE      *      :054 : Pb.68% :F-uv  : : Wh.*  : CLESS :  2.7-3.0  : 6.3      :  SRK tp/tl :
=====[*Lead Vitriol^]=[*Gray,Green,Brown,Black,Purple ^^]===========================================:
```

SHADES OF GRAY

```
                   ------------------------------------STREAK
==========================================================>^^^^^^^^<=====================================:
145-JAMESONITE     *     :055 : Pb.    : ---- : : Gy.   : GY-BK.:  2.5      : 5.63     :  MET opq   :
==[*Brittle Feather Ore ^]==========================================================================:
053-BRONZITE         [P]:056 : Coll.  :      : : Gn-Br : GRAY  :  5.0      : 3.2-3.5  :  S-MET tl  :
====================================================================================================:
104-ENSTATITE      [P] :057 : Coll.  : ---  : : Gn.*  : GRAY  :  5.5      : 3.1-3.2  :  VIT tl    :
==========================================[*Yellow^^]===============================================:
019-ANTIMONY             :058 : Sb.100%: ---  : : Si-Wh : GRAY  :  3.0-3.5  : 6.7      :  MET opq   :
====================================================================================================:
116-GALENA         *     :059 : Pb.86% : ---  : : Gy.   : GRAY  :  2.5-2.8  : 7.2-7.6  :  MET opq   :
======[*Lead Glance ^]==============================================================================:
200-PLATINUM             :060 : Pt.    : ---  : : Si-Gy : GRAY  :  4.0-4.5  :14.0-19.0 :  MET opq   :
====================================================================================================:
```

Copyright 1993 Page 5

CHART C

A I.D. #	B MINERAL NAME (& Field Name)	C CLUE CODE	D ORE OF %	E TEST #	F BREAK COLOR	G STREAK COLOR	H MOHS HARDNESS	J SPECIFIC GRAVITY	K SURFACE LUSTER
006	ALLANITE * (REO)	061	Ce.5%	Rad.	Br.*	GRAY	5.5-6.0	3.5-4.2	S-MET opq
	[*Orthite ^]				[*Black ^^]				
067	CHALCOCITE *	062	Cu.79%	---	Gy.	GRAY	2.5-3.0	5.5-5.8	MET opq
	[*Copper Glance ^]								
050	BOURNONITE *	063	Cu.13%	----	Gy.\Bk	GRAY	2.5-3.0	5.7-5.9	MET opq
	[*Cog Wheel Ore ^]								
097	DIAMOND	064	Gem	F/uv	Gy.*	GRAY	10.0	3.52	SRK tp
					[*Colorless, Yellow, Brown, Green, Black^^]				
248	TETRAHEDRITE- *	065	Cu.53%	----	Gy.*	GRAY	3.0-4.5	4.5-5.2	MET opq
	-TENNANTITE-[*^Gray Copper]				[*Purple, Blue, Brown ^^]				
240	STIBNITE *	066	Sb.71%	---	Gy.	GRAY	2.0	4.6-4.7	MET opq
	[*Antimonite ^]								
044	BISMUTHINITE	067	Bi.81%	---	Gy.	GRAY	2.0	6.8-7.2	MET opq
086	COVELLITE	068	Cu.66%		Dk.Be	GRAY	1.5-2.0	4.6-4.8	S-MET opq
	[*Indigo Copper ^]								
244	SYLVANITE *	069	Au.24%	----	Si-Wh.	GRAY	1.5-2.0	8.0-8.3	MET opq
	[*Graphic Tellurium ^]								
237	STAUROLITE *	070	Gem	---	Rd-Bn*	GRAY	7.0-7.5	3.73	VIT tp tl
	[*Rood Stone ^]				[*Yellow-Brown, Black^^]				
077	CLINOZOISITE	071	Coll.	---	Gy.*	GRAY	6.5	3.3-3.5	VIT tl tp
					[*Pink, Green ^^]				
106	EPIDOTE	072	Gem	---	Gn-Yl*	GRAY	6.0-7.0	3.3-3.5	VIT tl
					[*Green, Yellow, Brown, Red, Gray, Black^^]				
025	ARSENIC	073	As.100%	---	Bk.	GRAY	3.5	5.4-5.9	DULL opq
078	COBALTITE	074	Co.35%	---	Gy-Pur	GY-BK.	5.5	6.3	MET opq
048	BORNITE *	075	Cu.63%	---	Rd-Br*	GY-BK	3.0	4.9-5.3	MET opq
	[*Peacock Ore ^]				[*Iridescent Purple-Blue ^^]				
122	GLAUCOPHANE [A]	076	Coll.	----	Be-Gy*	GY-BE	6.0-6.5	3.1-3.3	VIT tl
					[*Black, Blue^^]				
115	GADOLINITE (REO)	077	Y.38%	Rad.	Gn.*	GY-GN	6.5-7.0	3.1-3.3	VIT tp
					[*Brown^^]				

Copyright 1993 Page 6

CHART C

I.D. #	MINERAL NAME (& Field Name)	CLUE CODE	ORE OF %	TEST #	BREAK COLOR	STREAK COLOR	MOHS HARDNESS	SPECIFIC GRAVITY	SURFACE LUSTER
= A	B	C	D	E	F	G	H	J	K
030	AUGITE	[P]:078	Coll.	---	Bk.*	GY-GN	5.0-6.0	3.2-3.5	VIT tl
	[*Green-Black,Brown ^^]								

SHADES OF GREEN

----------------------------------STREAK

I.D. #	MINERAL NAME (& Field Name)	CLUE CODE	ORE OF %	TEST #	BREAK COLOR	STREAK COLOR	MOHS HARDNESS	SPECIFIC GRAVITY	SURFACE LUSTER
017	ANNABERGITE *	:079	Coll.	---	Gn.	GREEN	2.5-3.0	3.1	VIT tl
	[*Nickel Bloom ^]								
100	DIOPTASE *	:080	Cu.44%	---	Gn.	GREEN	5.0	3.3	VIT tp tl
	[*Copper Emerald^]								
190	PENNINITE *	:081	Coll.	---	Gn.	GREEN	2.0-2.5	2.5-2.8	VIT tl
	[*Green Chlorite ^]								
081	CONICHALCITE	:082	Coll.	---	Gn.	GREEN	4.5	4.3	VIT tl
029	ATACAMITE	:083	Cu.59%	---	Gr.	GREEN	3.0-3.5	3.76	VIT tp
052	BROCHANTITE	:084	Cu.		Gn.	GREEN	3.5-4.0	3.97	VIT tp/tl
162	MALACHITE	:085	Cu.57%	---	Gn.	GREEN	3.5-4.0	4.0	VIT opq
218	ROSASITE	:086	Cu.	---	Gn.*	GREEN	4.5	4.0-4.2	VIT tl
					[*Blue-Green ^^]				
117	GARNIERITE *	:087	Ni.36%	---	Gn.	GREEN	2.0-4.0	2.3-2.5	GSY tl
	[*A Generic Term ^]								
253	TORBERNITE *	:088	U.	Rad.	Gn.	GREEN	2.0-2.5	3.3	PRL tp tl
	[*Copper Uranite ^]								
069	CHLORITOID	:089	Coll.	---	Yw-Gn.	GREEN	6.5	3.5-3.8	VIT tl
258	TURQUOISE *	:090	Gem	----	Gn.*	GREEN	5.0-6.0	2.6-2.8	WAXY tl
	[*Callaite ^]				[*Blue-Green ^^]				
076	CLINOCHLORE *	:091	Coll.	---	Be-Gn	GREEN	2.0-2.5	2.6-2.8	PRL tl
	[*Chlorite ^]								
207	PYRITE *	:092	S.53%	---	Gold	GN-BK	6.0-6.5	5.0-5.2	MET opq
	[*Iron Pyrite,Fool's Gold^]								

Copyright 1993 Page 7

CHART C

I.D. #	MINERAL NAME (& Field Name)		CLUE CODE	ORE OF %	TEST #	BREAK COLOR	STREAK COLOR	MOHS HARDNESS	SPECIFIC GRAVITY	SURFACE LUSTER
A	B		C	D	E	F	G	H	J	K
171	MILLERITE	*	093	Ni.	---	Yw.	GN-BK	3.0-3.5	5.2-5.6	MET opq
	[*Hair Pyrites ^]									
068	CHALCOPYRITE	*	094	*Cu.34%	---	Yw.	GN-BK	3.5-4.0	4.2-4.3	MET opq
	[*Copper Pyrites ^]			[^*Au,Ag.]						
164	MARCASITE	*	095	IND.	---	Yw.-Gn	GN-BK	6.0-6.5	4.8-4.9	MET opq
	[*White Pyrites ^]									
250	THORIANITE		096	U & Th.	Rad.	Bk\Gy	GN-GY	6.5	9.7-9.8	MET s-opq
261	UVAROVITE	*	097	Gem	----	Gn.	GN-WH	6.5-8.0	3.4-3.8	VIT tp
	[*Garnet ^]									

SHADES OF ORANGE

I.D. #	MINERAL NAME (& Field Name)		CLUE CODE	ORE OF %	TEST #	BREAK COLOR	STREAK COLOR	MOHS HARDNESS	SPECIFIC GRAVITY	SURFACE LUSTER
155	LEPIDOCROCITE		098	Fe.	----	Rd	ORANGE	5.0-5.5	4.0	SUB-SRK tl
096	DESCLOIZITE		099	V.11%	---	Rd.*	ORANGE	3.5	5.5-6.2	VIT tp tl
					[*Black-Brown, Yellow-Brown^^]					
126	GREENOCKITE		100	Cd.	---	Yw-Or	ORANGE	3.0-3.5	4.9	SRK tl
089	CROCOITE	*	101	Pb.	----	Or-Rd	OR-YW	2.5-3.0	5.9-6.0	SRK tl
	[*Red Lead Ore ^]									
214	REALGAR		102	IND.	---	Or-Rd	OR-YW	1.5-2.0	3.5-3.6	RSN tp tl
274	ZINCITE	*	103	Zn.80%	----	Or\Rd	OR-YW	4.0-4.5	5.4-5.7	SRK tl tp
	[*Red Oxide Of Zinc ^]									

SHADES OF PINK

I.D. #	MINERAL NAME (& Field Name)		CLUE CODE	ORE OF %	TEST #	BREAK COLOR	STREAK COLOR	MOHS HARDNESS	SPECIFIC GRAVITY	SURFACE LUSTER
215	RHODOCHROSITE	*	104	Mn.47%	---	Pk.*	PINK	3.5-4.5	3.3-3.6	VIT tl
	[*Manganese Spar ^]			[*Rose-Red, Red, Gray, Brown^^]						
131	HAUERITE		105	Coll.	---	Bn-Bk	PK-BN	3.5-4.5	3.5	S-MET opq
009	ALURGITE	*	106	Coll.	----	Rd.	PK-WH	2.0-2.5	2.9-3.0	VIT tp/tl
	[*Mica ^]									

Copyright 1993 Page 8

CHART C

A I.D. #	B MINERAL NAME (& Field Name)	C CLUE CODE	D ORE OF %	E TEST #	F BREAK COLOR	G STREAK COLOR	H MOHS HARDNESS	J SPECIFIC GRAVITY	K SURFACE LUSTER
	SHADES OF RED					STREAK			
206	PYRARGYRITE *	107	Ag.59%	---	Bk-Rd	RED-PUR	2.5	5.85	S-MET tl
	[*Dark Ruby Silver ^]								
083	COPPER	108	Cu.100%	---	Rd-Or	RED	2.5-3.0	8.5-9.0	MET opq
205	PURPURITE	109	Coll.	---	Pur-Rd	RED	4.0-4.5	3.4	GSY tl
092	CUPRITE *	110	Cu.88%	---	Rd.	RED	3.5-4.0	5.8-6.2	SRK tp
	[*Ruby Copper ^]								
199	PIEDMONTITE *	111	Coll.	---	Pu-Bn*	RED	6.5	3.4	VIT tl
	[*Mangan Epidote ^]				[*Red ^^]				
135	HEMATITE *	112	Fe.70%	---	Bk.	RED	5.5-6.5	5.2-5.3	MET opq
	[*Bloodstone^]								
179	NEPTUNITE	113	Coll.	---	Bn.*	RD-BN	5.0-6.0	3.2	VIT opq
					[*Black^^]				
210	PYROPE *	114	Gem	---	Dk.Rd.	RED	6.5-7.5	3.58	VIT tp
	[*Bohemian Garnet ^]								
108	ERYTHRITE *	115	Co.	---	Rd.	RED	1.5-2.5	3.07	VIT tp tl
	[*Cobalt Bloom ^]								
113	FRANKLINITE	116	Zn.20%	---	Bk-Rd	RD-BN	5.5-6.5	5.0-5.2	MET opq
132	HAUSMANNITE	117	Mn.72%	---	Bk.	RD-BN	5.0-5.5	4.7-4.8	S-MET tl
220	SAMARSKITE (REO)	118	Y.28%	Rad.	Bk.	RD-BN.	5.0-6.0	5.7	RSN opq
	SHADES OF SCARLET					STREAK			
203	PROUSTITE *	119	Ag.65%	---	Dk.Rd	SCARLET	2.0-2.5	5.6	SRK tl
	[*Ruby Silver ^]								
075	CINNABAR	120	Hg.86%	---	Rd.	SCARLET	2.0-2.5	8.10	SRK tl

Copyright 1993 Page 9

CHART C

SHADES OF SILVER

A I.D. # / B MINERAL NAME (& Field Name)	C CLUE CODE	D ORE OF %	E TEST #	F BREAK COLOR	G STREAK COLOR	H MOHS HARDNESS	J SPECIFIC GRAVITY	K SURFACE LUSTER
024-ARGENTITE *	121	Ag.65%	---	Si-Gy	LEAD-GY	2.0-2.5	7.3	MET opq
[*Silver Glance ^]								
043-BISMUTH	122	Bi.100%	---	Pk.	SI-WH	2.0-2.5	9.7-9.8	MET opq
230-SILVER	123	Ag.100%	---	Si.	SI-WH.	2.5-3.0	10-0-12.0	MET opq

SHADES OF WHITE

A I.D. # / B MINERAL NAME (& Field Name)	C CLUE CODE	D ORE OF %	E TEST #	F BREAK COLOR	G STREAK COLOR	H MOHS HARDNESS	J SPECIFIC GRAVITY	K SURFACE LUSTER
129-HALITE *	124	Na.39%	F/uv	Wh.*	WHITE	2.5	2.1-2.2	VIT tp tl
[*Rock Salt ^]				[*Yellow,Red,Brown,Blue,Black ^]				
058-CARNALLITE	125	K.14.%	---	Rd\CL	WHITE	2.5	1.6	VIT tp
064-CHABASITE *	126	Coll.	----	Wh.*	WHITE	4.0-5.0	2.0-2.1	VIT tp/tl
[*Cubic Zeolite ^]				[*Green,Red,Brown ^^]				
102-DUMORTIERITE	127	IND.	F/uv	Vi.*	WHITE	7.0	3.2-3.4	VIT tl
				[*Blue,Red-Brown^^]				
088-CROCIDOLITE * [A]	128	Gem	---	Be-Bk.	WHITE	4.0	3.0-3.4	SRK tl
[*Riebeckite ^]								
148-KERNITE	129	B.	----	Wh\Cl	WHITE	2.5	1.9	VIT tl
276-ZINNWALDITE *	130	Li.1.8%	----	Yw.*	WHITE	2.5-3.0	2.9-3.1	PRL tp tl
[*Lithionite ^]				[*Green,Silver-Gray,Brown ^^]				
246-TALC*	131	IND.	----	Wh:.*	WHITE	1.0	2.58-2.83	GSY tl
[*^ Soapstone,Steatite,Freestone]				[*Green-White,Gray,^^Brown]				
196-PHLOGOPITE *	132	IND.	F/lw	Yw.*	WHITE	2.5-3.0	2.7-2.0	PRL tp
[*Amber Mica ^]				[*Light Brown ^^]				
042-BIOTITE *	133	Coll.	---	Bk.*	WHITE	2.5-3.0	2.8-3.2	VIT tp tl
[*Black Mica ^]				[*Brown,Green ^^]				
119-GLAUBERITE	134	IND.	F/uv	Wh.*	WHITE	2.5-3.0	2.81	GSY tp
				[*Yellow,Red^^]				

Copyright 1993 Page 10

CHART C

```
= A ====== B ============ C  == D ====== E ====== F ===== G ======= H =========== J ======= K =======:
I.D.   MINERAL NAME      CLUE  ORE OF   TEST    BREAK   STREAK   MOHS         SPECIFIC   SURFACE    :
#    (& Field Name)      CODE    %        #     COLOR    COLOR   HARDNESS      GRAVITY    LUSTER     :
=====================================================================================================:
257-TRONA                :135 : IND.    : ---- : : Wh.*  : WHITE :  2.5-3.0  : 2.17     :  ERY tp tl  :
=====================================[*Yellow-White ^^]===============================================:
036-BARITE            *  :136 : Ba.58% :F-uv   : : Yw.*  : WHITE :  2.5-3.5  : 4.48     :  VIT tl/tp  :
==========[*Heavy Spar^]=[*Red,Green,Black,Colorless ^^]=============================================:
118-GIBBSITE             :137 : Al.     : ---   : : Wh.*  : WHITE :  2.5-3.5  : 2.4      :  VIT opq    :
============================================[*Brown^^]===============================================:
227-SERPENTINE           :138 : IND.    : F/uv : : Gn.*  : WHITE :  2.5-4.0  : 2.5-2.6  :  PRL opq    :
===================================[*Yellow-White ^^]================================================:
074-CHRYSOTILE        *  :139 : IND     : ---- : :Gy-Wh* : WHITE :  2.5-4.0  : 2.55     :  SRK opq    :
===========[*Asbestos^]===[*Yellow,Brown,Green,Opaque ^^]============================================:
156-LEPIDOLITE     *     :140 : Li.3.6%: F/uv : : Pk\Vi : WHITE :  2.5-4.0  : 2.8-2.9  :  PRL tl     :
======[*Lithium Mica ^]==============================================================================:
273-WULFENITE         *  :141 : Mo.26% : ---- : : Yw.*  : WHITE :  2.7-3.0  : 6.5-7.0  :  RSN tp tl  :
=====[*Yellow Lead Ore ^]=====[*Red-Orange,Red,Gray ^^]==============================================:
243-SULFUR               :142 : S.100% : ---- : : Yw.*  : WHITE :  1.5-2.5  : 2.0-2.1  :  GSY tp tl  :
=====================================[*Brown,Black ^^]===============================================:
213-QUARTZ           *   :143 : Si.46% : F/uv : : Wh.*  : WHITE :  7.0      : 2.65     :  VIT tp tl  :
============[*See #278^]==[*Colorless,Brown,Black,^^^ Yellow,Blue,Violet,Green,Pink ]================:
066-CHALCEDONY  *        :144 : Gem     : F/uv : :See 279: WHITE :  7.0      : 2.5-2.6  :  WAXY tp/tl :
=[*Micro Quartz ^]==[See # 279 for varieties]========================================================:
056-CALCITE           *  :145 : Ca.40% : F/uv : : Wh.*  : WHITE :  3.0      : 2.7      :  VIT tl     :
==========[*Calc Spar ^]======[*Yellow, Pink, Green  ^^]=============================================:
027-ARTUNITE             :146 : Coll.  : ---   : : Wh.*  : WHITE :  2.0      : 2.03     :  SKY tp     :
=============================================[*Gray ^^]==============================================:
063-CERUSSITE         *  :147 : Pb.77% : F/uv : : Wh-Gy : WHITE :  3.0-3.5  : 6.5      :  SRK tp/tl  :
=====[*White Lead Ore ^]=============================================================================:
061-CELESTITE            :148 : Sr.59% : F/uv : : Wh.*  : WHITE :  3.0-3.5  : 3.96     :  VIT tp/tl  :
===========================[*Blue,Yellow,Colorless ^^]===============================================:
016-ANHYDRITE            :149 : Ca.29% : ---   : : Wh.*  : WHITE :  3.0-3.5  : 3.0      :  VIT tp/tl  :
==========================================[*Red,Gray ^^]=============================================:
268-WAVELLITE            :150 : IND.   : F/uv : : Wh.*  : WHITE :  3.3-4.0  : 2.3-2.4  :  VIT tl     :
=========================[*Yellow,Green,Brown,Blue ^^]===============================================:
003-ADAMITE              :151 : Coll.  : F-uv.: : Gn.*  : WHITE :  3.5      : 4.3-4.5  :  VIT tp/tl  :
==============================[*Pink, Yellow, Violet ^^]=============================================:
```

Copyright 1993 Page 11

CHART C

A	B	C	D	E	F	G	H	J	K
I.D. #	MINERAL NAME (& Field Name)	CLUE CODE	ORE OF %	TEST #	BREAK COLOR	STREAK COLOR	MOHS HARDNESS	SPECIFIC GRAVITY	SURFACE LUSTER
172	MIMETITE *	152	Pb.	---	Yw-Br*	WHITE	3.5-4.0	7.1	VIT tp tl
	[*Campylite ^]				[*Orange,Colorless ^]				
256	TRIDYMITE *	153	IND.	----	Wh.	WHITE	7.0	2.27	VIT tp tl
	[*Quartz ^]								
209	PYROMORPHITE	154	Pb.76%	---	Gn.*	WHITE	3.5-4.0	6.7	SRK tl
					[*Yellow,Brown,Colorless,Yellow-Red ^^]				
241	STILBITE *	155	Coll.	----	Wh.*	WHITE	3.5-4.0	2.1-2.2	VIT tp tl
	[Bundle Zeolite ^]				[*Red-Brown,Gray ^^]				
022	ARAGONITE	156	Gem	F-uv	Wh.*	WHITE	3.5-4.0	2.9-3.0	VIT tp/tl
					[*Red,Colorless ^^]				
270	WITHERITE (toxic)	157	Ba.69%	F/uv	Wh.*	WHITE	3.5-4.0	4.2-4.3	VIT tp tl
					[*Colorless,Gray,Yellow,Brown,Green ^^]				
242	STRONTIANITE	158	Sr.59%	----	Wh.*	WHITE	3.5-4.0	3.7	VIT tp tl
					[*Colorless,Gray,Green,Pink ^^]				
008	ALUNITE *	159	Coll.	F-uv	Wh.*	WHITE	3.5-4.0	2.75	VIT tp/tl
	[*Alum Stone ^]				[*Red,Yellow,Gray ^^]				
228	SIDERITE *	160	Fe.48%	---	Yw.*	WHITE	3.5-4.0	3.7-3.9	VIT tp tl
	[*Chalybite ^]				[*Green,Brown,Black ^^]				
101	DOLOMITE	161	Mg.13%	F/uv	Wh.*	WHITE	3.5-4.0	2.8-2.9	VIT tp tl
					[*Pink,Yellow,Colorless^^]				
138	HEULANDITE *	162	Coll.	----	Yw.*	WHITE	3.5-4.0	2.1-2.2	VIT tp tl
	[*Lamellar Zeolite ^]				[*Green,Red-Orange^^]				
165	MARGERITE *	163	Coll.	---	Wh. *	WHITE	3.5-4.0	2.9-3.1	PRL tp tl
	[*Brittle Mica ^]				[*Gray,Pink ^]				
225	SCORODITE	164	Coll.	---	Gn.*	WHITE	3.5-4.0	3.3	VIT tp tl
					[*Blue,Violet,Colorless,Yellow^^]				
087	CRISTOBALITE *	165	Gem	---	Wh.*	WHITE	7.0	2.3	VIT opq
	[*Obsidian ^]				[*Gray,Colorless^^]				
245	SILVITE	166	K.52%	----	Wh.*	WHITE	2.0	1.99	VIT tp
					[*Red,Purple,Blue,Yellow-Red ^^]				
160	MAGNESITE	167	Mg.28%	---	Wh.*	WHITE	3.5-4.5	3.0	VIT tp tl
	[*Bitter Spar ^]				[*Yellow,Gray^^]				
247	TARBUTTITE	168	Coll.	----	Wh.*	WHITE	3.7	4.1	VIT tp tl
					[*Green,Yellow,Pink ^^]				
112	FLUORITE *	169	IND.	F/uv	Cl.*	WHITE	4.0	3.1-3.3	VIT tp tl
	[*Fluor Spar ^]				[*Yellow,Green,Blue,Black^^]				

Copyright 1993 Page 12

CHART C

A	B	C	D	E	F	G	H	J	K
I.D. #	MINERAL NAME (& Field Name)	CLUE CODE	ORE OF %	TEST #	BREAK COLOR	STREAK COLOR	MOHS HARDNESS	SPECIFIC GRAVITY	SURFACE LUSTER
263	VARISCITE	170	Gem	----	Gn.*	WHITE	4.0-5.0	2.2-2.5	VIT tl
					[*Blue-Green, Colorless ^^]				
079	COLEMANITE	171	B.	---	Wh.*	WHITE	4.0-4.5	2.4	VIT tp/tl
					[*Red, Pink ^^]				
226	SEPIOLITE *	172	IND.	---	Wh.*	WHITE	2.0-2.25	2.0	DUL opq
	[*Meerschaum ^]				[*Yellow ^^]				
037	BAUXITE	173	Al.	---	Wh.*	WHITE	1.0-3.0	2.3-2.7	ERY
					[*Yellow, Red, Brown ^^]				
130	HARMOTOME *	174	Coll.	---	Wh.*	WHITE	4.5	2.3-2.5	VIT tp tl
	[*Cross Stone ^]				[*Gray, Yellow, Red, Brown^^]				
021	APOPHYLLITE	175	Coll.	F-uv	Wh.*	WHITE	4.5-5.0	2.3	VIT tp/tl
					[*Yellow, Colorless, Pink, Green ^^]				
272	WOLLASTONITE *	176	IND.	F/uv	Wh.*	WHITE	4.5-5.0	2.8-3.1	VIT tl
	[*Table Spar ^]				[*Green, Colorless, Gray ^^]				
136	HEMIMORPHITE *	177	Zn.54%	F/uv	Wh.*	WHITE	4.5-5.0	3.1-3.2	VIT tp tl
	[*Calamine ^]				[*Colorless, Brown, Yellow, Blue-Green^^]				
189	PECTOLITE	178	Coll.	F/uv	Wh.\Gy:	WHITE	5.0	2.8-2.9	VIT tl
105	EOSPHORITE	179	Coll.	---	Pk.*	WHITE	5.0-6.0	3.0	SRK tl tp
					[*Gray-Green, Colorless^^]				
020	APATITE (GROUP)	180	Gem	F-uv	Wh.*	WHITE	5.0	3.1-3.2	VIT tp/opq
					[*Colorless, Yellow, Green, Brown, Red, Blue ^^]				
177	NATROLITE	181	Coll.	F/uv	Wh.*	WHITE	5.0-5.5	2.9-3.0	VIT tp tl
					[*Pink, Yellow, Colorless^^]				
094	DANBURITE	182	Coll.	---	Yw.	WHITE	7.0-7.3	2.9-3.0	VIT tp tl
251	TITANITE*	183	Ti.& Gem:	----	Wh.*	WHITE	5.0-5.5	3.45	VIT tp tl
	[*Sphene ^]				[*Yellow-Green, Brown, Black, Pink, Colorless^^]				
084	CORDIERITE *	184	Gem	---	Gy.*	WHITE	7.0-7.5	2.6-2.7	VIT tl
	[*Iolite ^]				[*Blue, Green, Black^^]				
011	ANALCITE *	185	Coll.	----	Wh.*	WHITE	5.0-5.5	2.2-2.3	VIT tp/tl
	[*Analcime ^]				[*Pink, Yellow, Colorless ^^]				
013	ANDALUSITE	186	Gem	---	Rd.*	WHITE	7.5	3.1-3.2	VIT tl/opq
					[*Brown, Olive ^^]				

Copyright 1993 Page 13

CHART C

= A ====== B ============ C == D ====== E ====== F ===== G ======= H =========== J ======= K =======:

I.D. #	MINERAL NAME (& Field Name)	CLUE CODE	ORE OF %	TEST #	BREAK COLOR	STREAK COLOR	MOHS HARDNESS	SPECIFIC GRAVITY	SURFACE LUSTER
170	MILARITE	187	Gem	----	Wh.*	WHITE	5.5-6.0	2.6	VIT tp tl
	[*Green,Colorless ^]								
040	BERYL	188	Be.5%	F-uv	Gn-Wh*	WHITE	7.5-8.0	2.7-2.9	VIT tp/tl
	[*Yw-Wh,Be-Gn,Pk,Be,Gn ^^]								
141	HYDROZINCITE *	189	Zn.	F/uv	Wh\Gy	WHITE	2.0-2.5	3.2-3.8	VIT tl
	[*Zinc Bloom ^]								
153	LAZULITE *	190	IND.	----	Be.	WHITE	5.0-6.0	3.0-3.1	VIT tp
	[*Blue Spar ^]								
139	HORNBLENDE [A]	191	Coll.	---	Gn-Bk	WHITE	5.0-6.0	3.0-3.4	VIT tl
194	PHENAKITE *	192	Gem	----	Wh.*	WHITE	7.5-8.0	2.9-3.0	VIT tp
	[*Phenacite^]======[*Pink, Colorless,Yellow^^]								
150	LABRADORITE *	193	IND.	----	Be\Gn	WHITE	5.0-6.0	2.62-2.75	VIT tp tl
	[*Spectrolite ^]								
057	CANCRINITE	194	Coll.	---	Yw.*	WHITE	5.0-6.0	2.4-2.5	VIT tp/tl
	[*White,Red,Violet,Colorless ^^]								
099	DIOPSIDE [P]	195	Gem	F/uv	Wh.*	WHITE	5.0-6.0	3.1-3.3	VIT tp tl
	[*Purple,Brown,Blue,Yellow, Green^^]								
002	ACTINOLITE *	196	Gem	---	Gn.	WHITE	5.0-6.0	3.1-3.3	VIT tp/tl
	[Nephrite ^]								
255	TREMOLITE [A]	197	IND.	----	Wh *	WHITE	5.0-6.0	2.9	SRK tp tl
	[*Colorless,Gray-Green ^^]								
222	SCAPOLITE	198	Gem	F/uv	Wh.*	WHITE	5.0-6.5	2.5-2.7	VIT tp tl
	[*Colorless,Purple,Blue,Gray,Pink ^^]								
266	VIVIANITE *	199	IND.	----	Gn.*	WHITE	1.5-2.0	2.6-2.7	VIT tp
	[*Blue Iron ^]===[*Green,Blue,Violet,Black^^]								
073	CHRYSOCOLLA	200	Cu.	----	Gn\Be	WHITE	2.0-4.0	2.0-2.3	VIT tl
235	SPINEL	201	Gem	F/uv	Wh.*	WHITE	8.0	3.5-4.1	VIT tp
	[*Colorless,Blue,Black,Pink^^]								
090	CRYOLITE	202	Al.12%	---	Wh.*	WHITE	2.5	2.9-3.0	RSN tp
	[*Colorless,Red-Brown^^]								
065	CHALCANTHITE *	203	Cu.25%	---	Sky Be	WHITE	2.5	2.3	VIT tp/tl
	[Copper Vitriol ^]								

Copyright 1993 Page 14

CHART C

A I.D. #	B MINERAL NAME (& Field Name)	C CLUE CODE	D ORE OF %	E TEST #	F BREAK COLOR	G STREAK COLOR	H MOHS HARDNESS	J SPECIFIC GRAVITY	K SURFACE LUSTER
157	LEUCITE	204	K.18%	----	Wh\Gy	WHITE	5.5-6.0	2.5	VIT tp
178	NEPHELINE	205	IND.	F/uv	Wh.*	WHITE	5.5-6.0	2.5-2.6	GSY tl
	[*Gray, Yellow,Red,Green^^]								
010	AMBLYGONITE	206	Li.4.7%	F-uv	Wh.*	WHITE	5.5-6.0	3.1	VIT tp/tl
	[*Yellow,Pink,Green,Blue ^^]								
185	OPAL	207	Gem	F/uv	Wh.*	WHITE	5.5-6.5	1.9-2.5	GSY tp tl
	[*Blue,Black^^]								
216	RHODONITE *	208	Mn.41%	---	Pk.*	WHITE	5.5-6.5	3.4-3.7	VIT tp tl
	[*Fowlerite ^] [*Red-Brown,Pink-Red ^^]								
055	BRUCITE	209	IND.	F/uv	Gn.*	WHITE	2.5	2.4	PRL tp/tl
	[*Blue,Pink-Yellow,Brown ^^]								
221	SANADINE *	210	Coll.	---	Wh.*	WHITE	6.0	2.5-2.6	VIT tl opq
	[*K Feldspar ^] [*Colorless,Gray,Yellow ^^]								
004	ADULARIA *	211	Gem	----	Wh.*	WHITE	6.0	2.5-2.6	VIT tp/tl
	[*Feldspar ^] [*Colorless ^^]								
202	PREHNITE	212	Coll.	---	Gn-Wh*	WHITE	6.0-6.5	2.9-2.95	VIT tl
	[*Colorless,Gray,Yellow ^^]								
168	MICROCLINE *	213	Gem	---	Wh.*	WHITE	6.0-6.5	2.56-2.59	VIT tp tl
	[*Amazonite ^] [*Pink,Red-Brown,Blue-Green,Green ^]								
005	ALBITE *	214	IND.	----	Wh.*	WHITE	6.0-6.5	2.6	VIT tl
	[*Feldspar ^] [*Colorless ^^]								
070	CHRONDRODITE	215	Gem	---	Wh.*	WHITE	6.5-7.0	3.1-3.2	VIT tp
	[*Green,Pink,Gray,Yellow ^^]								
149	KYANITE *	216	IND.	----	Wh.*	WHITE	6.0-7.0	3.67	VIT tp tl
	[*Disthene ^] [*Green,Gray,Blue ^]								
060	CASSITERITE *	217	Sn.78%	---	Br\Bk	WHITE	6.0-7.0	6.0-7,1	MET tl/opq
	[*Tinstone ^]								
265	VESUVIANITE *	218	Gem	----	Bn.*	WHITE	6.0-7.0	6.0-7.1	VIT tp tl
	[*Idocrase ^] [*Yellow,Black,Red,Green-Brown^^]								
033	AXINITE	219	Gem	F-uv	Vi-Br.*	WHITE	6.5-7.0	3.2-3.3	VIT tl
	[*Brown,Green,Gray,Yellow,Blue ^^]								
252	TOPAZ	220	Gem	----	Bn.*	WHITE	8.0	3.26-3.36	VIT tp tl
	[*Colorless,Red-Yellow,Violet,Green,Blue,Yellow ^^]								

Copyright 1993 Page 15

CHART C

I.D. # (& Field Name) MINERAL NAME	CLUE CODE	ORE OF %	TEST #	BREAK COLOR	STREAK COLOR	MOHS HARDNESS	SPECIFIC GRAVITY	SURFACE LUSTER
A / B	C	D	E	F	G	H	J	K
098-DIASPORE	221	Al.45%	F/uv	Pk.*	WHITE	6.5-7.0	3.3-3.5	VIT tp tl
[*Green,Gray^^]								
127-GROSSULAR *	222	Gem	F/uv	Cl.*	WHITE	6.5-7.0	3.59	VIT tp
[*Garnet ^][*Green,Brown,Orange,Pink,Red,Black^^]								
014-ANDRADITE *	223	Gem	---	Br.*	WHITE	6.5-7.5	3.86	SRK tp
[*Garnet ^][*Green,Yellow,Black,Red ^^]								
007-ALMANDINE *	224	Gem	----	Rd-Br	WHITE	6.5-7.5	4.32	SRK opq/tl
[*Garnet ^]								
085-CORUNDUM	225	IND.	F-uv	Gy.*	WHITE	9.0	3.9-4.1	SRK tl tp
[*Colorless,Purple,Green,Yellow,Red,Brown^^]								
095-DATOLITE	226	B.	---	Wh.*	WHITE	5.0-5.5	2.9-3.0	VIT tp tl
[*Green,Colorless^^]								
236-SPODUMENE [P]	227	Li.3.4%	F/uv	Wh.*	WHITE	6.5-7.0	3.1-3.2	VIT tp
[*Violet,Yellow,Pink,Green,Gray^^]								
232-SMITHSONITE *	228	Zn.52%	---	Wh.*	WHITE	5.5	4.3-4.5	VIT tl
[*Zinc Spar ^]=[*Blue,Green,Yellow,Pink,Violet ^^]								
223-SCHEELITE	229	W.63%	F/uv	Wh.*	WHITE	4.5-5.0	5.9-6.1	VIT tp tl
[*Yellow,Green,Red-Gray^^]								
174-MONAZITE (REO)	230	Ce\Th12%	---	Yw-Br	WHITE	5.0-5.5	4.8-5.5	RSN tp tl
184-OMPHACITE [P]	231	Coll.	---	Gr.	WHITE	5.0-6.0	3.3-3.4	VIT tl
047-BORAX	232	IND.	---	Wh.*	WHITE	2.0-2.5	1.7	GSY tp/tl
[*Colorless,Yellow,Blue ^^]								
062-CERARGYRITE *	233	Ag.75%	---	Gy-Gn	WHITE	2.5	5.5	RSN tl
[*Horn Silver ^]								
197-PHOSGENITE *	234	Coll.	F/uv	Cl.*	WHITE	2.7-3.0	6.0-6.3	GSY tp tl
[*Horn Lead ^]=[*Amber-Yellow ^^]								
140-HYDROMAGNESITE	235	Coll.	---	Wh.	WHITE	3.5	2.24	SRK tp
128-GYPSUM	236	Ca.23%	F/uv	Wh.*	WHITE	2.0	2.35	VIT tp
[*Colorless,Brown,Gray,Yellow^^]								
107-EPSOMITE *	237	IND.	F/uv	Wh.	WHITE	2.0-2.5	1.67	VIT tp tl
[*Bitter Salt^]								

Copyright 1993 Page 16

CHART C

```
= A ====== B ============ C  == D ====== E ====== F ===== G ======= H =========== J ======= K =======:
I.D.   MINERAL NAME      CLUE  ORE OF   TEST    BREAK   STREAK   MOHS         SPECIFIC   SURFACE    :
#    (& Field Name)      CODE    %        #     COLOR    COLOR   HARDNESS      GRAVITY    LUSTER    :
=====================================================================================================:
254-TOURMALINE           :238 : Gem    : F/uv : : Pk.*  : WHITE :  7.0      : 3.0-3.25 :  VIT tp tl  :
============[*Colorless,Black,Green,Brown,Blue,Yellow ^^]============================================:
187-ORTHOCLASE      *    :239 : IND.   : ---  : : Wh.*  : WHITE :  6.0      : 3.0-3.3  :  VIT tp tl  :
====[*K-Feldspar ^]=[*Colorless,Yellow,Pink,Blue,Gray^^]=============================================:
147-KAOLINITE            :240 : Al.20% : ---- : :Yw-Gy* : WHITE :  2.0-2.5  : 2.5-2.6  :  PRL tl     :
=============================================[*Brown ^]==============================================:
018-ANTHROPHYLLITE    [A]:241 : Mg.    : ---  : : Gy.*  : WHITE :  5.5-6.0  : 2.8-3.2  :  VIT tl     :
=========================================[*Green,Brown ^^]===========================================:
175-MONTMORILLONITE      :242 : IND.   : ---  : : Wh-Gy : WHITE :  1.0      : 1.7-2.7  :  DULL opq   :
=====================================================================================================:
```

SHADES OF YELLOW

```
                         ---------------------------------STREAK
=====================================================>^^^^^^^^<======================================:
059-CARNOTITE            :243 : U.& V. : Rad. : : Yw.   : YELLOW:  1.0-2.0  : 4.7-5.0  :  DULL tl/opq:
=====================================================================================================:
124-GOLD                 :244 : Au.100%: ---  : : Yw.   : YELLOW:  2.5-3.0  :15.3-19.3 :  MET opq    :
=====================================================================================================:
182-OLIVENITE            :245 : Coll.  : ---  : : Yw.*  : YELLOW:  3.0      : 4.3      :  SRK tp tl  :
=====================[*Green,Brown-Yellow,Gray,White^^]===============================================:
032-AUTUNITE        *    :246 : U.     :F-uv. : : Gn-Yw : YELLOW:  2.0-2.5  : 3.2      :  VIT tl     :
=======[*Radioactive ^]==============================================================================:
028-ASTROPHYLLITE        :247 : Coll.  : ---  : : Gold  : YELLOW:  3.5-4.5  : 3.3-3.4  :  VIT tp/tl  :
=====================================================================================================:
186-ORPIMENT        *    :248 : As.    : ---  : : Or-Yw*: YELLOW:  1.5-2.0  : 3.48     :  GSY tp tl  :
=============[*Tinsel ^]=======================[*Gold^^]=============================================:
012-ANATASE              :249 : Ti.    : ---- : : Bk.*  : YELLOW:  5.5-6.0  : 3.8-3.9  :  SRK tp/tl  :
=========================================[*Yellow,Blue ^^]===========================================:
054-BROOKITE             :250 : Ti.    :      : : Bk\Br : YELLOW:  5.5-6.0  : 4.1      :  SRK tp/tl  :
=====================================================================================================:
217-RICHTERTITE      [A] :251 : Coll.  : ---  : : Yw.*  : YELLOW:  5.0-6.0  : 2.9-3.4  :  VIT tp tl  :
=========================================[*Purple-Red ^^]============================================:
192-PEROVSKITE    (REO)  :252 : Nb.\Ce.: Rad. : : Yw.*  : YELLOW:  5.5      : 4.0      :  SRK tl     :
===============================[*Red-Brown,Black-Gray^^]=============================================:
234-SPHALERITE*          :253 : Zn.67% : F/uv : : Yw.*  : YELLOW:  3.5-4.0  : 3.9-4.2  :  SRK tp tl  :
[*Zinc Blende^][*Colorless,Red-Brown,Pink,Black,Green^^]=============================================:
```

Copyright 1993 Page 17

CHART C

A I.D. #	B MINERAL NAME (& Field Name)	C CLUE CODE	D ORE OF %	E TEST #	F BREAK COLOR	G STREAK COLOR	H MOHS HARDNESS	J SPECIFIC GRAVITY	K SURFACE LUSTER
169	MICROLITE	254	Coll.	----	Yw.*	YELLOW	5.0-5.5	4.2-6.4	VIT tp tl op
	[*Brown, Colorless ^]								
262	VANADINITE	255	V.10%	----	Yw.*	YELLOW	2.75-3.0	6.8-7.1	RSN tp
	[*Red-Orange,Brown ^^]								
239	STIBICONITE	256	Coll.	---	Yw.*	YELLOW	4.0-5.5	5.6	PRL opq
	[*Yellow-White,Orange,Brown,Black ^^]								
188	PARAGASITE * [A]	257	Coll.	---	Bn.*	YELLOW	5.0-6.0	3.1-3.2	VIT tp tl
	[*Hornblende ^] [*Green,Gray^^]								
264	VERMICULITE	258	IND.	----	Yw.*	YELLOW	2.0-3.0	2.1-2.7	PRL tl
	[*Gray,Green,Brown,Black^^]								

SHADES OF WHITE, YELLOW or ???

THESE STREAK COLORS ARE QUESTIONABLE ! ------------------STREAK

A I.D. #	B MINERAL NAME (& Field Name)	C CLUE CODE	D ORE OF %	E TEST #	F BREAK COLOR	G STREAK COLOR	H MOHS HARDNESS	J SPECIFIC GRAVITY	K SURFACE LUSTER
181	NICKEL-IRON *	259	Coll.	---	Gr\Bk.	?GRAY	4.0-5.0	7.3-8.8	MET opq
	[*Taenite^]								
035	BABINGTONITE	260	Coll.	---	Bk.*	?CLESS	5.5-6.0	3.4	VIT tl/opq
	[*Brown,Blue ^^]								
082	COPIAPITE	261	Coll.	---	Olive*	?CLESS	2.5-3.0	2.0-2.8	PRL tl
	[*Orange, Yellow ^^]								
211	PYROPHYLLITE	262	IND.	F/lw	Yw-Wh*	WHITE	1.0-2.0	2.8	GSY tl
	[*Green, Green-Brown^^]								
167	MERCURY *	263	Hg.100%	---	Si-Wh.	?SILVER	-0-	13.6	MET opq
	[*Quicksilver ^]								
166	MELILITE*(Field Term)	264	Coll.	---	Wh.*	?CLESS	5.0	3.4-3.5	VIT tp tl
	[*For Akermanite ^] [*Yellow,Brown-Red,Gray-Green ^]								
259	ULEXITE *	265	Coll.	----	Wh.	?CLESS	1.0	2.0	SKY tp
	[*Cotton Balls ^][*T.V. stone]								
109	EUCLASE	266	Gem	---	Wh.	WHITE	7.5	3.1	VIT tp tl
146	JOHANNSENITE [P]	267	Coll.	----	Gn-Br.	?WHITE	5.0-6.0	3.44-3.46	VIT tl
224	SCOLECITE *	268	Coll.	---	Wh.*	WHITE	5.0-5.5	2.2-2.4	VIT tp
	[*Fibrous Zeolite ^] [*Colorless ^^]								
198	PHOSPHOPHYLLITE	269	Gem	F/uv	Be-Gn*	?WHITE	3.0-3.5	3.0-3.1	VIT tp tl
	[*Colorless ^^]								

Copyright 1993 Page 18

CHART C

= A = I.D. #	B MINERAL NAME (& Field Name)	C CLUE CODE	D ORE OF %	E TEST #	F BREAK COLOR	G STREAK COLOR	H MOHS HARDNESS	J SPECIFIC GRAVITY	K SURFACE LUSTER
121	GLAUCONITE (Mica)	270	IND.	----	Be-Gn	?CLESS	2.0-2.3	2.5-2.8	GSY tl
091	CUMMINGTONITE [A]	271	Coll.	---	Br.	?CLESS	5.0-6.0	3.1-3.7	SKY tl
137	HERDERITE	272	Coll.	F/uv	Gn-Wh*	?CLESS	5.0-5.5	2.9-3.0	VIT tl tp
					[*Purple,Yellow^^]				
072	CHRYSOBERYL	273	Be.7.1%	---	Gn.*	WHITE	8.5	3.7	SRK tp/tl
					[*Colorless,Green, Gray,Brown ^^]				
39	BERTRANDITE	274	Be.	---	Yw.	?CLESS	6.0-7.0	2.6	VIT tp/tl
267	VOLBORTHITE	275	Coll.	----	Olive	WHITE	3.0	3.5-3.8	VIT tp tl
038	BENITOITE	276	Gem	F-uv	Blue	?CLESS	6.2-6.5	3.65	SRK tp/tl
041	BETAFITE *	277	Th.& U.	---	Bk.*	YW-WH	5.0	3.7-5.0	ERY opq
	[*Radioactive ^]				[*Yellow,Brown-Green ^^]				

^^^^^^^<<< ? QUESTIONABLE ?

213-QUARTZ VARIETIES -

278-QUARTZ -COARSE CRYSTAL VARIETIES: [BULL QUARTZ = White] [ROCK CRYSTAL = Colorless] [SMOKY = Brown]: [MORION =Black] [BLUE = Blue] [CITRINE = Yellow or Brown] [AMETHYST=Violet] [ROSE =Pink] [TIGER EYE = : Yellow or Yellow-Brown] [CAT'S-EYE = Green,Yellow,Brown] [RUTILATED = Yellow to Red-Brown] : [ADVENTURINE = Spangled Green or Yellow-Brown] :

066-CHALCEDONY VARIETIES -

279-CHALCEDONY -- MICRO-QUARTZ VARIETIES: [AGATE = Banded Variety,opq.] [ONYX = Black & White Layers] : [SARDONYX = Red & White Layers] [SARD = tl Brown to Yellow, Red-Brown] [CARNELIAN = tl Red to Orange- : Red] [CHRYSOPRASE = tl Apple-Green or Green-Yellow] [JASPER = opq Red,Red-Brown,Yellow-Brown] [PRASE =: Dull Green] [PLASMA -Dark Dull Green][HELIOTROPE = (Bloodstone) opq Green with Jasper Red Spots] : [MOSS AGATE =White with Red-Brown Branches] [CHERT = opq. Dull Black,Gray,Yellow] [FLINT = opq Gray] :

CHAPTER 67

277 MINERAL IDENTIFICATION DATA

CHART D

CHART D HEADINGS

COLUMN **EXPLANATION**

A. - MINERAL I.D. NUMBER = This number identifies the same specific mineral throughout ALL charts.

B. - MINERAL NAME & FIELD NAME = Identifies the mineral name. Note: Some minerals are better known by their field name. Example: Silver Glance for Argentite. (The charts use the U.S.A. mineral name.)

M. - LOOKS LIKE = Some minerals can fool you because they look alike. EXAMPLE: Quartz looks like calcite. This column alerts you to look-a-likes.

N. - ALTERS FROM = Some minerals are alteration forms of another mineral. Talc alters from olivine. This column alerts you to look for the mineral it altered from.

P. - ALTERS TO = Some minerals change to a different form or into a new mineral. Pyrite can alter to goethite and to limonite. This column alerts you to look for the alteration mineral.

R. - ASSOCIATED WITH THESE MINERALS = Experience and frequent commonality has led to a body of associations for many minerals. Where you find one, you can usually find the other...BUT NOT ALWAYS!

S. - REFRACTIVE INDEX = The degree to which light slows down in a material is the index of refraction. It is a method of identifying minerals. We use it to distinguish crystal quartz from a diamond.

T. - CRYSTAL SYSTEM = A crystal structure consists of a lattice arrangement combined with the shape of the atomic motifs. This boils downs to 7 basic crystal shapes. 2 of the 7 shapes (rhombohedral = hexagonal) are the same. This is how the count of 6 crystal systems is reached.

>>CHART D WILL SHOW 7 DIFFERENT LATTICES.<<

CRYSTAL SYSTEMS ARE BASED ON LATTICE SYMMETRY.

THIS IS BEYOND THE SCOPE OF NORMAL ROCKPECKER ACTIVITY.

U. - PROBABLE ENVIRONMENT OCCURRENCE -- Just as one mineral is associated with 2 or more other minerals in nature, the genesis or birth of the mineral can have more than one source-mother.

U COLUMN LISTS THE PLACE TO LOOK FIRST, THOUGH THERE ARE

OTHER EQUALLY IMPORTANT FIRST SOURCES.

MINERAL ASSOCIATIONS & NATURAL OCCURRENCE ENVIRONMENTS

 Copyright 1993 T.D. LaRune

-A- -B-	-M-	-N-	-P-	-R-	-S-	-T-	-U-
I.D. # MINERAL NAME (& Field Name)	LOOKS LIKE	ALTERS FROM	ALTERS TO	ASSOCIATED WITH THESE MINERALS	REFRA INDEX	CRYSTAL SYSTEM	PROBABLE ENVIRONMENT OCCURRENCE
001-ACMITE * [*Aegirine ^]	---	---	---	104-028-030	1.82	MO	BASIC IGNEOUS ROCKS. PHONOLITES
002-ACTINOLITE * [Nephrite ^]	272	---	227-106	056-099-106-272	1.64	MO	MAFIC METAMORPHIC GREEN SCHIST
003-ADAMITE	202	---	---	231	1.74	OR	(Rare) FOUND IN ZINC DEPOSITS
004-ADULARIA * [*Feldspar^]	---	---	147	005	1.52	MO	HYDROTHERMAL VEINS. IGNEOUS ROCKS
005-ALBITE * [*Feldspar^]	---	---	056-147	122-168-150	1.53	TR	IGNEOUS ROCKS, MICA SCHIST,GNEISS
006-ALLANITE * [*Orthite ^]	---	---	---	106-213-077-199-277	1.75	MO	PEGMATITES, PLACERS, LATERITES
007-ALMANDINE * [*Garnet^]	---	---	---	013-149-237-229	1.83	IS	CONTACT METAMORPHIC ROCK & SCHIST
008-ALUNITE * [*Alum Stone ^]	101	---	---	187	1.57	HE	NODULES IN SEDIMENTARY ROCKS
009-ALURGITE * [*Mica ^]	---	---	---	099-122-213	----	MO	MANGANESE-RICH METAMORPHIC ROCKS
010-AMBLYGONITE	---	---	258-268	020-156-236-254	1.60	TR	LITHIUM & PHOSPHATE-RICH PEGMATITE
011-ANALCITE * [*Analcime ^]	157	---		202-056	1.49	IS	INTRUSIVE IGNEOUS ROCK CAVITIES
012-ANATASE	---	142	219	054-219	2.60	TE	VEINS, FISSURES, SCHISTS. PLACERS
013-ANDALUSITE	---	---	229-149	176-149-229-084	1.62	OR	METAMORPHIC MICA SCHISTS, GNEISS
014-ANDRADITE * [*Garnet^]	---	---	---	255	1.85	IS	CONTACT METAMORPHIC LIMESTONES
015-ANGLESITE * [*Lead Vitriol^]	---	116	063	116-063-062-209	1.88	OR	OXIDATION ZONE LEAD DEPOSITS
016-ANHYDRITE	56-61	128	128	056-061-128-213	1.58	OR	METAMORPHOSED SEDIMENTARY EVAPORITE
017-ANNABERGITE * [*Nickel Bloom ^]	---	---	---	231	1.68	MO	CRUSTY ALTERATION NICKEL MINERALS
ANTHROPHYLLITE	---	---	---	091-272	1.66	OR	CRYSTALLINE SCHISTS, META-ROCKS
019-ANTIMONY	---	---	---	234-240-116-207	---	HE	HYDROTHERMAL LODES
020-APATITE	44-99	---	---	161-051	1.63	HE	CRYSTALS IN ALL ROCK CLASSES
021-APOPHYLLITE	---	---	---	056-095-202-241	1.54	TE	CAVITIES IN BASALT & TUFF
022-ARAGONITE	056	---	056	158-228-162-232	1.69	OR	IN SCHISTS & HOT SPRING DEPOSITS
023-ARFVEDSONITE	---	---	---	178	OPQ.	MO	PEGMATITES, METAMORPHIC SCHISTS
024-ARGENTITE * [*Silver Glance^]	067	---	201-238	230-203-206-116	OPQ	IS	LOW TEMP. HYDROTHERMAL VEINS

-A- -B- I.D. MINERAL NAME # (& Field Name)	-M- LOOKS LIKE	-N- ALTERS FROM	-P- ALTERS TO	-R- ASSOCIATED WITH THESE MINERALS	-S- REFRA INDEX	-T- CRYSTAL SYSTEM	-U- PROBABLE ENVIRONMENT OCCURRENCE
025-ARSENIC	---	---	---	180-230-078-017-003	OPQ	HE	HYDROTHERMAL VEINS, METADOLOMITE
026-ARSENOPYRITE *	078	---	225-108	271-223-173-060-124	OPQ	MO	HYDROTHERMAL VEINS, PEGMATITES
[*Mispickelˆ]				[*230-116-207-068ˆ]			
027-ARTINITE	---	---	---	055-140	----	MO	FRACTURES OF SERPENTINITES
028-ASTROPHYLLITE	---	---	---	001-275-251	----	TR	NEPTHELINE SYENITE CAVITIES
029-ATACAMITE	162	---	---	052-162	1.86	OR	DESERT COPPER SULFIDE ALTERATION
030-AUGITE	---	---	106-139	255-222-265-251	1.75	MO	INTRUSIVE & METAMORPHIC ROCKS
031-AURICHALCITE	---	---	---	034-092-162-232-150	1.74	MO	OXIDATION ZINC & COPPER SULFIDE
032-AUTINITE *	---	260	---	260	1.58	TE	HYDROTHERMAL VEINS & PEGMATITES
[*Radioactiveˆ]							
033-AXINITE	251	---	---	145-213-106-254-002	1.69	TR	HYDROTHERMAL VEINS & SCARNS
034-AZURITE *	159	---	162	162	1.76	MO	WEATHERING OF COPPER ORES
[*Blue Carbonateˆ]							
035-BABINGTONITE	---	---	---	056-202	----	TR	GRANITE, GNEISS, BASALT CAVITIES
036-BARITE *	55-61	---	270	056-101-112-230-116	1.64	OR	HYDROTHERMAL & REPLACEMENT VEINS
[*Heavy Sparˆ]							
037-BAUXITE	---	---	----	098-118	<VAR>	AM	CHEMICAL SEDIMENTARY ROCK
038-BENITOITE	---	---	----	227-179-177	1.76	HE	BRECCIATED BLUE SCHIST VEINS
039-BERTRANDITE	---	040	----	040	---	OR	PEGMATITES & HYDROTHERMAL BERYL
040-BERYL	020	---	147-176	005-213-168-010-080-*	1.58	HE	GRANITE PEGMATITES,MICA SCHISTS
				[*-236-176ˆ]			
041-BETAFITE *	---	---	----	056	----	MO	METAMORPHOSED LIMESTONE, PEGMATITE
[*Radioactiveˆ]							
042-BIOTITE *	196	---	264-106	213-187-176	---	MO	ALL ROCK GROUPS & PEGMATITES
[*Black Mica ˆ]							
043-BISMUTH	230	---	----	230--19-240-116-207	---	HE	HYDROTHERMAL VEINS
044-BISMUTHINITE	067	---	----	043-060-026-213-271	---	OR	HYDROTHERMAL VEINS
045-BIXBYITE	---	---	----	252-054	---	IS	CAVITIES ALTERED RHYOLITIC ROCK
046-BOLEITE	---	---	----	083	2.04	TE	WEATHERED COPPER SULFIDE PRODUCT
047-BORAX	---	---	----	129-259-079-128-056	1.47	MO	DESERT SALT PAN EVAPORITE
048-BORNITE *	---	---	067-068	067-068-103-207-026	---	IS	MAGMATIC SEGREGATION MAFIC ROCKS
[*Peacock Oreˆ]							

MINERAL ASSOCIATIONS & NATURAL OCCURRENCE ENVIRONMENTS

 Copyright 1993 T.D. LaRune

-A- I.D. #	-B- MINERAL NAME (& Field Name)	-M- LOOKS LIKE	-N- ALTERS FROM	-P- ALTERS TO	-R- ASSOCIATED WITH THESE MINERALS	-S- REFRA INDEX	-T- CRYSTAL SYSTEM	-U- PROBABLE ENVIRONMENT OCCURRENCE
049	BOULANGERITE * [*Feather Ore ^]	145	---	---	019	---	MO	DISSEMINATED VEINS OF Pb.,Zn.,Sb.
050	BOURNONITE * [*Cog Wheel Ore^]	---	---	162-034	116-234-248-068-207	---	OR	HYDROTHERMAL VEINS OF SILVER
051	BRAZILIANITE	---	---	---	213-168-020-40-254	1.61	MO	ALTERED PEGMATITE CAVITIES
052	BROCHANTITE	029	---	073	162-092-073-158-034	1.78	MO	OXIDATION ZONE OF COPPER DEPOSITS
053	BRONZITE	---	---	---	104	---	OR	IGNEOUS MAFIC & ULTRAMAFIC ROCKS
054	BROOKITE	012	---	219	012-219-135-251-004	2.60	OR	FISSURES IN GRANITE & GNEISSES
055	BRUCITE	128	227	---	227-056-246-160-022	1.57	HE	IN METAMORPHIC SERPENTINE & SCHIST
056	CALCITE * [*Calc Spar ^]	101	---	---	215-232-228	1.66	HE	SEDIMENTARY CHEMICAL PRECIPITATION
057	CANCRINITE	---	178	---	178-020-168-251	----	HE	INTRUSIVE ROCKS LOW IN SILICA
058	CARNALLITE	---	---	---	129	----	OR	EVAPORATIVE PRECIPITATE IN DOMES
059	CARNOTITE	---	---	---	032	1.90	MO	SANDSTONE & LIMESTONE IMPREGNATION
060	CASSITERITE * [*Tinstone ^]	---	---	---	212-026-044-112-173	2.00	TE	PEGMATITES & PLACERS
061	CELESTITE	036	---	242	243-128-016-242-056	1.62	OR	VOLCANIC ROCK HYDROTHERMAL MINERAL
062	CERARGYRITE * [*Horn Silver^]	---	024	---	230-024-206-116-036	---	IS	OXIDATION ZONE OF SILVER DEPOSITS
063	CERRUSITE * [*White Lead Ore^]	015	116	---	116-234-209-197	2.08	OR	OXIDATION ZONE OF LEAD DEPOSITS
064	CHABASITE * [*Cubic Zeolite^]	056	---	---	056-021-241	1.48	HE	VOLCANIC & IGNEOUS ROCK CAVITIES
065	CHALCANTHITE * [Copper Vitriol ^]	---	---	---	052	1.54	TR	OXIDIZED ZONE OF COPPER DEPOSITS
066	CHALCEDONY ** [**Agate, Onyx, Sardonyx, Sard, Carnelian, Chrysoprase, Jasper, Prase, Plasma, Bloodstone, Moss Agate, Chert, Flint **]	---	---	---	213-185	---	HE	WATER SOLUTION PRECIPITATE-BASALT
067	CHALCOCITE * [*Copper Glance ^]	024	---	086-034	034-083-092-068-248	---	OR	SECONDARY ENRICHMENT SULFIDE VEINS
068	CHALCOPYRITE* [*Copper Pyrites^]	124	---	067-086	207-116-234-067-056	---	TE	HYDROTHERMAL VEINS, META-VOLCANICS
069	CHLORITOID	---	---	----	085-213-122	1.72	MOTR	LOW GRADE REG-METAMORPHIC ROCKS
070	CHONDRODITE	014	---	227	265-227-235-212	1.61	MO	METAMORPHIC MARBLES & SKARN
071	CHROMITE * [*Chrome Iron Ore ^]	056	---	----	056-021-241	2.16	IS	ULTRABASIC ROCKS & SERPENTINE
072	CHRYSOBERYL	---	---	----	213 000	1.75	OR	PEGMATITE, GRANITE, SCHIST, GNEISS

 Copyright 1993 T.D. LaRune

-A- I.D. # (& Field Name) MINERAL NAME	-B- LOOKS LIKE	-M- ALTERS FROM	-N- -P- ALTERS TO	-R- ASSOCIATED WITH THESE MINERALS	-S- REFRA INDEX	-T- CRYSTAL SYSTEM	-U- PROBABLE ENVIRONMENT OCCURRENCE
073-CHRYSOCOLLA	258	48-68	---	034-083-092-162	1.40	MO	OXIDATION ZONE OF COPPER DEPOSITS
074-CHRYSOTILE * [*Asbestos^]	---	---	---	227	1.53	MO	SERPENTINE ROCK META-ALTERED
075-CINNABAR	092	---	167	207-164-240-213-185	---	HE	HYDROTHERMAL VEINS, SED-ROCKS
076-CLINOCHLORE * [*Chlorite ^]	---	---	---	190	1.62	MO	ESSENTIAL SCHIST MINERAL
077-CLINOZOISITE	---	30-139	---	005-002	1.70	MO	REGIONAL METAMORPHIC GREEN SCHIST
078-COBALTITE	026	---	---	231-180-230-228-056	---	OR	HI-TEMP HYDROTHERMAL VEINS
079-COLEMANITE	095	259	---	259-128-047-056-061	1.59	MO	LENSES IN CLAY, GYPSUM, BORAX
080-COLUMBITE * [*Niobite ^]	---	---	---	236-276-040	---	OR	LITHIUM-RICH PEGMATITES, PLACERS
081-CONICHALCITE	---	---	---	182	---	OR	OXIDATION ZONE OF COPPER DEPOSITS
082- COPIAPITE	---	207	---	207	---	TR	UPPER PORTION OF SULFIDE DEPOSITS
083- COPPER	---	---	092-162	124-230-162-092-068	---	IS	OXIDATION ZONE SULFIDE DEPOSITS
084-CORDIERITE * [*Iolite ^]	213	---	---	085	1.55	OR	CONTACT METAMORPHIC ROCKS
085- CORUNDUM	---		277-165	161-135-235-084	1.77	HE	PEGMATITES & METAMORPHIC ROCKS
086- COVELLITE * [*Indigo Copper ^]	---	68-103	---	067-048-103	---	HE	HYDROTHERMAL COPPER VEINS
087-CRISTOBALITE* [*Obsidian^]	---	---	---	213	1.48	TE	FELSIC IGNEOUS ROCK CAVITIES
088-CROCIDOLITE* [*Riebeckite ^]	---	---	---	213	---	MO	CRACKS IN FINE-GRAINED QUARTZ
089-CROCOITE * [*Red Lead Ore ^]	---	---		209-063-273-262	2.40	MO	HYDROTHERMAL LEAD ORE ALTERATION
090- CRYOLITE	---	---	---	228-168-234-207	1.34	MO	ALWAYS IN PEGMATITES
091-CUMMINGTONITE	---	---	---	018-272	1.66	MO	CONTACT METAMORPHIC SCHISTS
092-CUPRITE * [*Ruby Copper ^]	135	68-135	083-162	158-034-162-083-067	2.85	IS	OXIDATION ZONE COPPER DEPOSITS
93-CYANOTRICHITE* [*Lettsomite^]	---	083	---	052	1.62	OR	OXIDATION ZONE COPPER DEPOSITS-RARE
094- DANBURITE	---	---	---	005	1.63	OR	PNEUMATOLYTIC IN DOLOMITE & GNEISS
095- DATOLITE	066	---	---	202-056-021-241-011	1.65	MO	BASALT & SERPENTINITE CAVITIES
096- DESCLOIZITE	---	262	---	262-209-063-273	2.26	OR	OXIDIZED ZONE LEAD, ZINC, COPPER ORE

MINERAL ASSOCIATIONS & NATURAL OCCURRENCE ENVIRONMENTS

 Copyright 1993 T.D. LaRune

-A- I.D. #	-B- MINERAL NAME (& Field Name)	-M- LOOKS LIKE	-N- ALTERS FROM	-P- ALTERS TO	-R- ASSOCIATED WITH THESE MINERALS	-S- REFRA INDEX	-T- CRYSTAL SYSTEM	-U- PROBABLE ENVIRONMENT OCCURRENCE
097-	DIAMOND	213	---	---	099	2.42	IS	KIMBERLITE BRECCIAS, PLACERS
098-	DIASPORE	---	---	---	118-085-165-101	---	OR	IN CHLORITE SCHISTS, BAUXITE
099-	DIOPSIDE	---	---	---	097-255-222-265-251	1.65	MO	METAMORPHIC ROCK FISSURES
100-	DIOPTASE * [*Copper Emerald^]	---	---	---	162-052-073-056-213	1.69	HE	OXIDATION ZONE COPPER ORE, PLACERS
101-	DOLOMITE	056	56-022	099	116-234-056-213-036	1.68	HE	SEDIMENTARY CARBONATE ROCK
102-	DUMORTIERITE	233	---	---	213	1.72	OR	PEGMATITES, CONTACT META-ROCKS
103-	ENARGITE	---	---	248	207-234-116-067-213	---	OR	HYDROTHERMAL COPPER VEINS
104-	ENSTATITE	053	---	---	053	1.75	OR	MAFIC & ULTRAMAFIC ROCKS
105-	EOSPHORITE	---	---	---	020-213	1.64	MO	GRANITE PEGMATITES & ROSE QUARTZ
106-	EPIDOTE	002	---	---	083-265-127	1.75	MO	REGIONAL & CONTACT META-ROCKS
107-	EPSOMITE * [*Bitter Salt^]	237	---	---	128-016-129	----	OR	THERMAL SOLUTION PRECIPITATE
108-	ERYTHRITE * [*Cobalt Bloom ^]	---	---	---	078-231-017-260	1.66	MO	PINK COBALT MINERAL GUIDEPOST
109-	EUCLASE	---	---	---	252	1.65	MO	TOPAZ IN PEGMATITES, PLACERS
110-	FERBERITE * [*Wolframite ^]	---	---	---	271-060	2.25	MO	PEGMATITES, HYDRO-VEINS, PLACERS
111-	FERGUSONITE	---	---	---	275-174-115	---	TE	PEGMATITES, PLACERS, LATERITES
112-	FLUORITE * [*Fluor Spar ^]	056	---	---	056-116-234-271-173	1.43	IS	HYDROTHERMAL VEINS LEAD, AG., ZN.
113-	FRANKLINITE	161	---	---	274-269-161-216	---	IS	CONTACT METASOMATIC DOLOMITES
114-	FUCHITE * [*Muscovite ^]	---	---	---	042	1.60	MO	GUIDE TO PERIDOT & SERPENTINE ROCK
115-	GADOLINITE	---	---	---	111	1.79	MO	PEGMATITES, PLACERS, LATERITES
116-	GALENA * [*Lead Glance ^]	---	---	015-209	230-234-207-068-213	---	IS	HYDROTHERMAL ORE VEINS
117-	GARNIERITE * [* Generic Name ^]	017	---	---	246-227-071	---	MO	METAMORPHOSED NICKEL SERPENTINE
118-	GIBBSITE	---	85-168	---	039-098	---	MO	RESIDUAL SEDIMENTARY PRODUCT
119-	GLAUBERITE	---	---	128	128-129-107-259-047	---	MO	EVAPORATIVE SEDIMENTARY PRODUCT
120-	GLAUCODOT	078	---	---	020-207-068-116-078	---	OR	PYRITE & COBALTITE COATINGS

MINERAL ASSOCIATIONS & NATURAL OCCURRENCE ENVIRONMENTS

 Copyright 1993 T.D. LaRune

-A-	-B-	-M-	-N-	-P-	-R-	-S-	-T-	-U-
I.D. #	MINERAL NAME (& Field Name)	LOOKS LIKE	ALTERS FROM	ALTERS TO	ASSOCIATED WITH THESE MINERALS	REFRA INDEX	CRYSTAL SYSTEM	PROBABLE ENVIRONMENT OCCURRENCE
121-	GLAUCONITE [Mica]	---	---	---	101	---	MO	MARINE SEDIMENTARY SANDSTONE DEP.
122-	GLAUCOPHANE	---	---	---	144	1.60	MO	LOW TEMP./ HI PRESSURE META-ROCKS
123-	GOETHITE * [*Needle Ironstone ^]	135	207	---	207-161-228-121	---	OR	WEATHERED IRON-RICH FORMATIONS
224-	GOLD	68-207	---	---	213-207-026-116-234-* [*-066-083-230-173-212-^]	---	IS	HYDROTHERMAL QUARTZ VEINS, PLACERS
125-	GRAPHITE	173	---	---	207-127	---	HE	METAMORPHIC ROCKS & SCHISTS
126-	GREENOCKITE	---	---	---	202-177-056-232-234	2.52	HE	WEATHERED ZONE OF ZINC ORES
127-	GROSSULAR * [*Garnet ^]	144	---	---	056-272-265-099-222	1.73	IS	REGIONAL & CONTACT META-CALC ROCKS
128-	GYPSUM	056	016	016	016- 129-056-101-213	1.52	MO	EVAPORATIVE SEDIMENTARY ALTERATION
129-	HALITE * [*Rock Salt ^]	245	---	---	128-016-245-058	1.54	IS	SALINE WATER SEDIMENTARY PRODUCT
130-	HARMOTOME * [*Cross Stone ^]	---	---	---	213	1.50	MO	BASALT ROCK CAVITITES, GNEISS
131-	HAUERITE	---	---	---	128-056-243	---	IS	SULFER-RICH CLAYS & IGNEOUS ROCKS
132-	HAUSMANNITE	161	---	---	204-163-208-123-056	---	TE	METAMORPHIC CONTACT ZONE LIMESTONE
133-	HAUYNITE [Hauyne]	---	---	---	157-166	1.50	IS	EMBEDDED IN BASALTIC ROCKS
134-	HEDENBERGITE	099	---	139-227	255-222-265-251-030	1.71	MO	CONTACT METAMORPHIC SKARNS
135-	HEMATITE * [*Bloodstone^]	92-75	161	139-227	161-158-123-142-228	2.90	HE	SEDIMENTARY BEDS, LAVA ROCKS
136-	HEMIMORPHITE* [*Calamine ^]	202	234	---	232-234-116-015	1.62	OR	OXIDATION ZONE OF Pb.,Zn. SULFIDES
137-	HERDERITE	---	---	---	168	1.59	MO	GRANITE PEGMATITE HYDROTHERMAL DEP
138-	HEULANDITE * [*Lamellar Zeolite ^]	---	---	---	056-202-095-213	1.49	MO	VOLCANIC ROCK CAVITIES, SCHIST
39-	HORNBLENDE	254	42-056	---	221-213-056-106	1.67	MO	CONSTITUENT OF IGNEOUS, META-ROCKS
140-	HYDROMAGNESITE	---	055	---	227-055	---	MO	HYDROTHERMAL BALLS IN SERPENTINE
141-	HYDROZINCITE* [*Zinc Bloom ^]	---	234	---	234-232-136-063-056	---	MO	OXIDIZED ZINC ORE ON SMITHSONITE
142-	ILMENITE * [*Titanoferrite ^]	135	---	---	135-161-219-235-174	---	HE	BASIC IGNEOUS ROCKS, PLACERS
143-	ILVAITE	---	---	---	134-161-014-207	1.91	OR\M	CONTACT METAMORPHIC ROCKS & LAVAS
144-	JADITE * [*Precious Jade^]	265	---	---	227	1.66	MO	SERPENTINE ROCK & SCHIST INSERTS

MINERAL ASSOCIATIONS & NATURAL OCCURRENCE ENVIRONMENTS

 Copyright 1993 T.D. LaRune

-A- -B- I.D. MINERAL NAME # (& Field Name)	-M- LOOKS LIKE	-N- ALTERS FROM	-P- ALTERS TO	-R- ASSOCIATED WITH THESE MINERALS	-S- REFRA INDEX	-T- CRYSTAL SYSTEM	-U- PROBABLE ENVIRONMENT OCCURRENCE
145-JAMESONITE * =[*Brittle Feather Ore ^]	049	---	---	116-234-248-207-240	---	MO	HYDROTHERMAL SILVER-Sb-PB VEINS
146-JOHANNSENITE	---	---	---	134-099	---	MO	QUARTZ VEINS THRU RHYOLITE
147-KAOLINITE	---	---	---	221	1.56	MO	WEATHERED FELDSPAR AS CLAY BEDS
148-KERNITE	128	047	---	259-047-079	1.47	MO	THERMAL SPRING LAKE BASINS
149- KYANITE * ==[*Disthene ^]	---	---	176-211	237-219-153-085	1.72	TR	GRANITE PEGMATITE, SCHIST, GNEISS
150-LABRADORITE * =[*Spectrolite ^]	---	---	011-147	005	1.56	TR	PLAGIOCLASE IN IGNEOUS & META-ROCK
151-LARUNEITE	---	---	---	177	1.77	HE	METAMORPHIC GRANULAR BLUE SCHIST
152-LAUMONTITE* =[*Fibrous Zeolite ^]	---	---	---	224	1.52	MO	IGNEOUS, META & SED-ROCK VEINS
153-LAZULITE * =[*Blue Spar ^]	154	---	---	213-149-013-102-085	1.64	MO	QUARTZ VEINS & PEGMATITES
154-LAZURITE * =[*Lapis Lazuli ^]	233	---	---	207-056	1.50	IS	CONTACT METAMORPHIC LIMESTONE
155-LEPIDOCROCITE	---	207	---	123	---	OR	LEACHED SOIL LATERITE AND GOETHITE
156-LEPIDOLITE * =[*Lithium Mica ^]	176	---	---	213-254-236-010-060	1.57	MO	GRANITE PEGMATITES & TIN ORE DEPS
157-LEUCITE	011	---	147-187	178-011-187	1.51	TE	VOLCANIC ROCK, LAVA & ASHES
158-LIMONITE * =[*Brown Iron Ore ^]	123	---	---	123-155-207	2.05	AM	SURFACE OXIDATION ZONE IRON DEPS
159-LINARITE	034	---	063-162	062-136-052-073-031-063	1.80	MO	UPPER OXIDATION ZONE OF Pb,Cu. DEP
160-MAGNESITE * =[*Bitter Spar ^]	056	---	056-101	052-101-246	1.70	HE	ALTERATION OF SERPENTINE ROCKS
161-MAGNETITE	142	---	135-158	085-235	---	IS	IGNEOUS, METAMORPHIC, SED-ROCKS
162-MALACHITE	117	---	034-092	034-092-158-083-073	1.66	MO	OXIDATION ZONE OF COPPER ORES
163-MANGANITE* =[*Brown Man Ore ^]	208	---	204-208	208-204-132-123-056	---	MO	HYDROTHERMAL VEIN, SEDIMENTARY DEP
164-MARCASITE * =[*White Pyrites ^]	207	---	158-207	116-234-056-101-112	---	OR	HYDROTHERMAL LEAD ZINC VEINS & SED
165-MARGERITE * =[*Brittle Mica ^]	156	085	---	161-085-219	1.65	MO	METAMORPHIC CHLORITE & MICA SCHIST
166-MELILITE* =[Field name for Akermanite ^]	---	---	---	056-192	1.66	TE	VOLCANIC LEUCITE, NEPHELINE BASAL
167-MERCURY * =[*Quicksilver ^]	---	---	---	---	---	HE	THERMAL FIELD CINNABAR DEPOSITS
168-MICROCLINE * =[Amazonite^]	---	---	147	213-005	1.53	TR	PEGMATITES & METAMORPHIC GNEISS

MINERAL ASSOCIATIONS & NATURAL OCCURRENCE ENVIRONMENTS

LaRune's ROCKPECKER Notes — CHART D. Page 8 of 12 — Copyright 1993 T.D. LaRune

-A-	-B-	-M-	-N-	-P-	-R-	-S-	-T-	-U-
I.D. #	MINERAL NAME (& Field Name)	LOOKS LIKE	ALTERS FROM	ALTERS TO	ASSOCIATED WITH THESE MINERALS	REFRA INDEX	CRYSTAL SYSTEM	PROBABLE ENVIRONMENT OCCURRENCE
169-	MICROLITE	---	---	---	156-236-005	1.52	IS	PEGMATITE DIKES,CALCITE INTRUSIONS
170-	MILARITE	---	---	---	056	---	HE	HYDROTHERMAL VEINS & PEGMATITES
171-	MILLERITE *	145	---	---	056-101-112-135	---	HE	CAVITIES IN CALCITE VEINS,SIDERITE
	[*Hair Pyrites ^]							
172-	MIMETITE *	209	---	---	026-209	2.12	HE	OXIDATION ZONE LEAD/ZINC DEPS.
	[*Campylite ^]							
173-	MOLYBDENITE	125	---	---	223-271-026-068-207	---	HE	PEGMATITES, CONTACT META-LIMESTONE
174-	MONAZITE	275	---	---	275-160-020-080-142	1.79	MO	PEGMATITES, LATERITES, PLACERS
175-	MONTMORILLONITE	--	168	---	168	---	MO	IN CLAYS AND VOLCANIC TUFF
176-	MUSCOVITE *	156	13-149	---	213-187-168-005-254	1.60	MO	IGNEOUS & SEDIMENTARY ROCK MICA
	[*Isinglass ^]							
177-	NATUROLITE	189	178-233	---	056-021-095-202	1.48	OR	BASALT CAVITIES LININGS
178-	NEPHELINE	213	---	147-233	221-233-057-275-085	1.54	HE	INTRUSIVE & EXTRUSIVE ROCKS, PEGS
179-	NEPTUNITE	---	---	---	177-178-038	1.70	MO	NEPTHELINE SYENITE PEGMATITES
180-	NICCOLITE *	083	---	17	230-078-212-017-026	---	HE	IN SILVER, NICKEL, COPPER ORE VEIN
	[*Arsenical Nickel ^]							
181-	NICKEL-IRON *	---	---	---	---	---	IS	METEORITES WITH WIDMANSTATTEN LOOK
	[*Taenite^]							
182-	OLIVENITE	---	---	---	026-034-162	---	OR	OXIDATION ZONE OF COPPER SULFIDES
183-	OLIVINE *	---	---	227-158	071-085-227-030-161	1.66	OR	IGNEOUS & META-LIMESTONE, PLACERS
	[*Fosterite,Peridot ^]							
184-	OMPHACITE	---	---	---	213-210-014	---	MO	IGNEOUS ROCKS, MICA SCHIST\GARNETS
185-	OPAL	066	---	066	066	1.44	AM	SANDSTONE, LAVA, RHYOLITE CAVITIES
186-	ORPIMENT *	---	214	---	214-075-056-036-025-240	2.80	MO	HYDROTHERMAL VEINS WITH REALGAR
	[*Tinsel ^]							
187-	ORTHOCLASE	010	---	147	213-176-168-004-221	1.52	MO	PEGMATITES & INTRUSIVE ROCKS
	[*K-Feldspar ^]							
188-	PARAGASITE *	---	---	---	056-106-213-184	1.67	MO	CONTACT METAMORPHIC MARBLES
	[*Hornblende ^]							
189-	PECTOLITE	177	---	---	056-202-095	1.61	TR	BASALT CAVITIES WITH ZEOLITES
190-	PENNINITE *	---	---	---	076	1.62	MO	METAMORPHIC CHLORITE SCHIST ROCK
	[*Green Chlorite ^]							
191-	PENTLANDITE	212	---	---	212-068-200	---	IS	OLIVINE, PYROXENE, AMPHIBOLE ROCK
192-	PEROVSKITE	---	---	---	166-178	2.38	OR	CARBONATITES, LATERITES, PLACERS

MINERAL ASSOCIATIONS & NATURAL OCCURRENCE ENVIRONMENTS

 Copyright 1993 T.D. LaRune

-A- I.D. #	-B- MINERAL NAME (& Field Name)	-M- LOOKS LIKE	-N- ALTERS FROM	-P- ALTERS TO	-R- ASSOCIATED WITH THESE MINERALS	-S- REFRA INDEX	-T- CRYSTAL SYSTEM	-U- PROBABLE ENVIRONMENT OCCURRENCE
193-	PETALITE	---	---	---	236-254-156	1.50	MO	LITHIUM-BEARING PEGMATITES
194-	PHENAKITE * =[*Phenacite^]	213	040	---	040-072-252-254-213	1.65	HE	PEGMATITE VEINS, MICA SCHISTS
195-	PHILLIPSITE * =[*Cross Stone ^]	---	168	---	064	---	MO	BASALT ROCK CAVITY LINING
196-	PHLOGOPITE * =[*Amber Mica ^]	---	---	---	101-227	1.60	MO	PEGMATITES & META-LIMESTONE
197-	PHOSGENITE * =[*Horn Lead ^]	---	116	063	063-116-230-015	2.11	TE	HYDROTHERMAL ALTERATION OF GALENA
198-	PHOSPHORPHYLLITE	--	234	---	234	---	MO	CHEMICAL ALTERATION OF SPHALERITE
199-	PIEDMONTITE * =[*Mangan Epidote ^]	---	139	---	106-122-005-002-223	---	MO	CONTACT METAMORPHISM GREEN SCHIST
200-	PLATINUM	230	---	---	183-161-071-212-191	---	IS	DUNITE & SERPENTINE ROCKS. PLACERS
201-	POLYBASITE	135	---	238-230	206-203-238-024-124	---	MO	HYDROTHERMAL SILVER ORE VEINS
202-	PREHNITE	136	---	---	056-983-095-189-177	1.63	OR	BASALT ROCK CAVITY. META-LIMESTONE
203-	PROUSTITE * =[*Ruby Silver ^]	206	---	024	230-024-206-201-056	2.80	HE	HYDROTHERMAL SILVER SULFIDE VEINS
204-	PSILOMELANE* =[Romanechite ^]	123	216	---	056-215-216-208-213	---	OR	MANGANESE OXIDATION ON LIMESTONE
205-	PURPURITE	---	---	---	RARE	---	OR	PEGMATITE WITH OTHER Mn. PHOSPHATE
206-	PYRARGYRITE * =[*Dark Ruby Silver ^]	203	---	024-230	230-024-201-203-238-* [*238-248-056-213 ^^^]	2.98	HE	SILVER MINERAL HYDROTHERMAL VEINS
207-	PYRITE * =[*Iron Pyrite, Fools Gold^]	124	---	123-158	213-124-068-116-234	---	IS	IGNEOUS, SEDIMENTARY & META-ROCKS
208-	PYROLUSITE	163	216	---	163-204-215-216-056	---	TE	SURFACE ALTERATION Mn. DEPOSITS
209-	PYROMORPHITE	262	116	---	063-158-136-232-262	2.06	HE	OXIDATION ZONE OF LEAD MINERALS
210-	PYROPE * =[*Bohemian Garnet ^]	---	---	---	097-227-183-235	1.71	IS	SERPENIZED ROCKS & PLACERS
211-	PYROPHYLLITE	---	---	---	213-149	1.59	MO	HYDROTHERMAL QUARTZ VEINS, SCHIST
212-	PYRRHOTITE * =[*Magnetic Pyrite ^]	207	---	123-228	068-191-200-124-271	---	HE	BASIC IGNEOUS ROCKS WITH SULFIDES
213-	QUARTZ * =[*Rock Crystal ^]	056	---	066	MANY	1.54	HE	IGNEOUS MAGMA INJECTED IN VEINS
214-	REALGAR	075	---	186	186-124-230-240	2.60	MO	HYDROTHERMAL LEAD, ANTIMONY VEINS
215-	RHODOCHROSITE* =[*Manganese Spar ^]	216	---	163-208	083-230-216-163	1.82	HE	OXIDATION ZONE OF SULFIDE ORES
216-	RHODONITE * =[Fowlerite ^]	215	---	---	056-269-274-113	1.72	TR	SKARNS IN LIMESTONE METAMORPHISM

MINERAL ASSOCIATIONS & NATURAL OCCURRENCE ENVIRONMENTS

 Copyright 1993 T.D. LaRune

-A- -B- I.D. # MINERAL NAME (& Field Name)	-M- LOOKS LIKE	-N- ALTERS FROM	-P- ALTERS TO	-R- ASSOCIATED WITH THESE MINERALS	-S- REFRA INDEX	-T- CRYSTAL SYSTEM	-U- PROBABLE ENVIRONMENT OCCURRENCE
217-RICHTERTITE	---	---	---	056	---	MO	CONTACT METAMORPHIC CALC-ROCKS
218- ROSASITE	---	---	---	158-230-232	---	MO	OXIDATION ZONE ZINC, COPPER ORES
219- RUTILE	060	142	---	213-161-275-174-012	2.61	TE	PEGMATITES, HYDROTHERMAL VEINS
220- SAMARSKITE	---	---	---	080	---	MO	PEGMATITES, LATERITES, PLACERS
221- SANADINE *	213	---	147	168-004-187	1.52	MO	BASALT & EXTRUSIVE ROCKS
[*K Feldspar ^]							
222- SCAPOLITE	---	---	005-106	020-042-099-251-275	1.54	TE	META-LIMESTONE, MICA SCHIST GNEISS
223- SCHEELITE	229	---	110	255-099-106-173-271	1.92	TE	PEGMATITES,HYDROTHERMAL VEINS
224- SCOLECITE *	177	---	---	202-189-056-177	1.52	MO	LAVA & BASALT CAVITY LINING
[*Fibrous Zeolite ^]							
225- SCORODITE	---	026	---	026	1.78	OR	OXIDATION ZONE COPPER-ZINC ORES
226- SEPIOLITE *	---	160	---	160-227	---	OR	ALTERATION OF MAGNESITE-SERPENTINE
[*Meerschaum ^]							
227- SERPENTINE	---	104	---	071-161-160-074	1.55	HOM	ALTERATION OF OLIVINE, PYROXENE
228- SIDERITE *	234	---	123	234-116-112-036-090	1.88	HE	HYDROTHERMAL VEINS BROWN CALCITE
[*Chalybite ^]							
229-SILLIMANITE*	255	---	176-211	085-007-084-213-13-176	1.66	OR	CONTACT METAMORPHIC SCHIST, GNEISS
[*Fibrolite ^]							
230- SILVER	200	---	---	056-213-036-112-83-260	---	IS	SULFIDE REDUCTION LEAD,ZINC ORES
231-SKUTTERUDITE*	026	---	108-017	026-056-213-036-078-43	---	IS	HYDROTHERMAL SILVER,CO.,NI. VEINS
[*Smaltite ^]				[180-230-228^^]			
232-SMITHSONITE*	194	234	---	234-136-116-063-015-34	1.85	HE	OXIDATION ZONE OF ZINC ORES
[*Zinc Spar ^]				[209-172-031-161-141^^]			
233- SODALITE	154	---	057-147	178-157-057-221	1.48	IS	NEPTHELINE SYENITES & VOLCANICS
234-SPHALERITE *	116	---	232-136	116-207-068-056-164-24	2.35	IS	WITH GALENA, HYDROTHERMAL VEINS
[*Zinc Blende ^]				[^^126-136-232]			
235- SPINEL	---	---	227	056-070-030-020-222-85	1.72	IS	METAMORPHOSED DOLOMITES,PLACERS
				[^^125-135-161-265-183]			
236- SPODUMENE	---	---	176-005	254-156-213-221-040	1.67	MO	ONLY IN PEGMATITES
237-STAUROLITE *	---	---	147	149-176-007	1.75	MO	CONTACT METAMORPHIC ROCKS, PLACERS
[*Rood Stone ^]							
238-STEPHANITE *	---	---	201-203	206-203-024-230-036-56	---	OR	SILVER ORE HYDROTHERMAL VEINS
[*Brittle Silver ^]				[^^116-234-068-248-213-101-201]			
239-STIBICONITE	---	240	---	240	---	IS	HYDROTHERMAL ALTERATION STIBNITE
240-STIBNITE *	044	---	239	056-066-075-116-207-36	3.19	OR	HYDROTHERMAL LEAD-SILVER VEINS
[*Antimonite ^]				[^^186-214-164]			

 Copyright 1993 T.D. LaRune

-A- I.D. # (& Field Name)	-D- MINERAL NAME LOOKS LIKE	-M- ALTERS FROM	-N- ALTERS TO	-P- ASSOCIATED WITH THESE MINERALS	-R- REFRA INDEX	-S- CRYSTAL SYSTEM	-T- -U- PROBABLE ENVIRONMENT OCCURRENCE
241- STILBITE *	---	---	---	064-138-056-021-202	1.50	MO	HYDROTHERMAL BASALT CAVITIES
=[Bundle Zeolite ^]							
242-STRONTIANITE	022	---	061	035-056-061	1.67	OR	HYDROTHERMAL VEINS WITH CALCITE
243- SULFUR	186	116	---	128-061-056-022	2.04	OR	SEDIMENTARY EVAPORITE DEPOSIT
244-SYLVANITE *	019	---	124	124-207-112-056-213	---	MO	HYDROTHERMAL VEINS QUARTZ GANGUE
=[*Graphic Tellurium ^]							
245- SILVITE	129	---	---	058-129-016-128	1.49	IS	INTERGROWN WITH HALITE
246- TALC *	211	183	---	183	1.59	MO	MAGNESIA LEACHED FROM DOLOMITE
=[*^ Soapstone,Steatite,Freestone]							
247- TARBUTTITE	---	---	---	031-141-232	---	TR	OXIDATION ZONE OF ZINC ORES
248-TETRAHEDRITE*	---	---	034-162	068-024-048-116-207	---	IS	HYDROTHERMAL VEINS COPPER-SILVER
-TENNANTITE-[*^Gray Copper]				[^^^234-203-206-056-213]			
249-THOMSONITE*	---	---	---	056-202	---	OR	AGATE-LIKE PEBBLES IN BASALT ROCK
=[*Comptonite ^]							
250- THORIANITE	---	---	---	060-026-068-207-116	---	IS	PEGMATITE & CONTACT META-LIMESTONE:
251-TITANITE *	033	---	012-219	178-056	1.84	MO	IGNEOUS ROCKS,MICA SCHIST,GNEISS
=[*Sphene ^]							
252- TOPAZ	213	---	112-147	060-213-254-271-104-	1.62	OR	PEGMATITES, RHYOLITIC CAVITIES
				[^^^173-020-112-110]			
253-TORBERNITE*	---	260	---	032-260	1.59	TE	PEGMATITES. WEATHERED URANIUM MICA
=[*Copper Uranite ^]							
254- TOURMALINE	040	---	042-176	213-175-156-221-040	1.65	HE	PEGMATITES
255- TREMOLITE	272	---	246	056-099-106-272-160	1.62	MO	TALC SCHISTS & DOLOMITIC MARBLES
256-TRIDYMITE *	213	213	087	213-087	---	MO	QUARTZ HI-TEMP STRUCTURE CHANGE
=[*Quartz ^]							
257- TRONA	---	---	---	089-101-056	---	MO	EVAPORATED SALT LAKE DEPOSITS
258-TURQUOISE *	073	---	---	020-068-066-158	1.62	TR	SILICON SCHIST CRACK FILLING
=[*Callaite ^]							
259-ULEXITE *	---	---	128	047-129-128-079-119	1.50	TR	EVAPORITE WITH BORAX IN PLAYA'S
=[*Cotton Balls ^][*T.V. stone]							
260-URANINITE *	---	---	253-032	275-254-174-060-026	---	IS	PEGMATITES, HYDRO ORE VEINS, SED.
=[*Pitchblende ^]				[^^068-207-116-234]			
261-UVAROVITE *	---	---	---	071-227	1.87	IS	CHROMITE-RICH SERPENTINES
=[*Garnet ^]							
262-VANADINITE	209	---	095	063-273-209-172-095	2.3	HE	OXIDATION ZONE LEAD ORES
263- VARISCITE	258	---	---	268-020-158-066	1.58	OR	FISSURES IN SILICA SCHISTS
264-VERMICULITE	---	042	---	042-196	---	MO	WEATHERED BIOTITE [illegible]

MINERAL ASSOCIATIONS & NATURAL OCCURRENCE ENVIRONMENTS

 Copyright 1993 T.D. LaRune

-A-	-B-	-M-	-N-	-P-	-R-	-S-	-T-	-U-
I.D. #	MINERAL NAME (& Field Name)	LOOKS LIKE	ALTERS FROM	ALTERS TO	ASSOCIATED WITH THESE MINERALS	REFRA INDEX	CRYSTAL SYSTEM	PROBABLE ENVIRONMENT OCCURRENCE
265-	VESUVIANITE *	---	---	---	99-127-254-070-251-272	1.71	TE	CONTACT META-LIMESTONE & DOLOMITE
	[*Idocrase ^]							
266-	VIVIANITE *	---	---	---	212-207-158	1.60	MO	OXIDATION ZONE OF SULFIDE ORES
	[*Blue Iron ^]							
267-	VOLBORTHITE	---	---	---	---	---	MO	THERMAL WATER RISING IN SANDSTONE
268-	WAVELLITE	154	---	---	258-020	1.54	OR	PEGMATITES & META-PHOSPHATE ROCKS
269-	WILLEMITE	---	---	---	056-274-113	1.69	HE	METAMORPHOSED LEAD,ZINC, Mn. ORES
270-	WITHERITE	---	---	136	036-116-112	1.68	OR	HYDROTHERMAL VEINS GALENA & BARITE
	[TOXIC^]							
271-	WOLFRAMITE *	080	---	---	213-060-040-252-026	---	MO	HYDROTHERMAL QUARTZ & TIN VEINS
	[*See Also Ferberite ^]				[^^^212-207-068-116-223-043-112]			
272-	WOLLASTONITE*	255	---	---	056-099-265-255-106	1.63	TR	CONTACT META-LIMESTONE & SCHIST
	[*Table Spar ^]							
273-	WULFENITE*	089	---	---	209-262-063-158-116-56	2.40	TE	OXIDIZED ZONE LEAD, MO. DEPOSITS
	[*Yellow Lead Ore ^]							
274-	ZINCITE *	---	---	---	056-269-113	2.01	TE	CONTACT META-CALCITE (WHITE-PINK)
	[*Red Oxide Of Zinc ^]							
275-	ZIRCON	265	---	---	178	1.92	TE	EXTRUSIVE ROCKS, SCHISTS,PLACERS
276-	ZINNWALDITE*	175	---	---	213-236-254-005-010	1.57	MO	PEGMATITES, HYDRO-QUARTZ VEINS
	[*Lithionite ^]							
277-	ZOISITE	---	---	---	077-005-056-106	1.69	OR	META-CALC ROCKS & SCHISTS

THIS AREA RESERVED FOR YOUR SPECIAL REQUIREMENTS.

278-

279-

ABBREVIATIONS:

COLORS= [Red- Rd.][Yellow- Yw.][Green- Gn.][Blue- Be.][Brown- Brn.][Gray- Gy.][Black- Bk.][Orange- Og.] [Pink- Pk][Silver- Si][Violet- Vi.][Gold- Gd.][White- Wh.][Colorless-Cl.]

LUSTERS= [Metallic- MET] [Vitreous- VIT] [Sparkling or Adamantine- SRK] [Silky- SKY] [Submetallic- S-MET] [Pearly-PRY] [Greasy- GSY] [Oily or resinous- RSN] [Earthy- ERY] [Dull-DUL] [Transparent- tp] [Translucent- tl][Opaque- opq]

CRYSTAL SYSTEMS= [Isometric-IS] [Tetragonal-TE] [Orthorombic- OR] [Monoclinic- MO] [Hexagonal- HE] [Triclinic-TR] [Rhombohedral- HE] [Amorphous- AM] [Unknown- UN]

Code numbers refer to Mineral Name I.D numbers.

CHAPTER 68

CHEMICAL ELEMENTS — CHART E

COLUMN **EXPLANATION**

A. – ELEMENT = The chemical elements of the periodic table.

B. – SYMBOL = This is the abbreviation for the chemical element.

C. – ATOMIC WEIGHT = This is the relative weight of the atom on the basis of the element carbon = 12.00.

D. – ATOMIC NUMBER = The number of protons within the atomic nucleus. The electrical charge of these protons determine the number and arrangement of the outer electrons of the atom. This tells you the chemical and physical properties of the element.

J. – HAZARDOUS PROPERTIES = Every element in some form can cause human harm. EXERCISE CARE.

K. – COMMON NATURAL OCCURRENCES = Identifies some minerals in which the element is found.

THE ELEMENTS – HAZARDOUS PROPERTIES & NATURAL OCCURRENCES

 Copyright 1993 by T.D. LaRune

THE ELEMENTS, ALPHABETICAL LIST

A ELEMENT	B SYMBOL	C ATOMIC WEIGHT	D ATOMIC NUMBER	E MELTING POINT [C]	J HAZARDOUS PROPERTIES	K COMMON NATURAL OCCURRENCES
ACTINIUM xx	Ac	*227.00	89	1050.0	Radioactive.	URANIUM ORES- Glows blue in dark
ALUMINUM	Al	26.98	13	660.4	Reacts violently with oxidants.	BAUXITE, ALUM, FELDSPAR
AMERICIUM xx	Am	243.00	95	994.0	Radioactive.Toxic by inhalation.	SYNTHETIC TRANSURANIUM
ARGON	Ar	39.95	18	-189.2	Asphyxiant.	0.94% OF AIR
ARSENIC	As	74.92	33	817.0	Poison. Highly toxic.	REALGAR, ORPIMENT, ARSENOPYRITES
ASTATINE	At	210.00	85	302.0	Radioactive.	URANIUM & THORIUM ISOTOPES
BARIUM	Ba	137.34	56	725.0	Poison. Soluble compounds toxic.	BARITE, WITHERITE
BERKELIUM xx	Bk	247.00	97	986.0	Radioactive.	SYNTHETIC TRANSURANIUM ELEMENT
BERYLLIUM	Be	9.01	4	1278.0	Poison. Carcinogen.	BERYL, CHRYSOBERYL, BERTRANDITE
BISMUTH	Bi	208.98	83	271.3	Poison. Compounds harmful ingestion.	BISMUTHINITE
BORON	B	10.81	5	2300.0	Toxic. Halides can cause burns.	BORAX, ULEXITE, KERNITE
BROMINE	Br	79.91	35	-7.2	Poisonous vapor.	SALT BRINES FROM WELLS
CADMIUM	Cd	112.40	48	321.0	Toxic by ingestion.	SPHALERITE,GREENOCKITE, ZINC ORE
CALCIUM	Ca	40.08	20	839.0	May irritate.	GYPSUM, FLUORITE, APATITE
CALIFORNIUM xx	Cf	249.00	98	899.9	Radioactive.	SYNTHETIC TRANSURANIUM ELEMENT
CARBON	C	12.01	6	3631.9	Carbon Black - Moderately Toxic	GRAPHITE, DIAMOND
CERIUM ++	Ce	140.12	58	799.0	Toxic by inhalation	ALLANITE, MONAZITE, BASTNASITE
CESIUM	Cs	132.91	55	28.5	Reacts explosively with water	LEPIDOCITE, POLUCITE
CHLORINE	Cl	35.45	17	-100.9	Toxic Gas. Irritant.	SODIUM CHLORIDE, CARNALLITE
CHROMIUM	Cr	52.00	24	1856.0	Toxic. Irritant.	CHROMITE
COBALT	Co	58.93	27	1495.0	Compounds may be toxic.	COBALTITE, ERYTHRITE, SMALTITE

THE ELEMENTS – HAZARDOUS PROPERTIES & NATURAL OCCURRENCES

 Copyright 1993 by T.D. LaRune

THE ELEMENTS, ALPHABETICAL LIST

A ELEMENT	B SYMBOL	C ATOMIC WEIGHT	D ATOMIC NUMBER	E MELTING POINT [C]	J HAZARDOUS PROPERTIES	K COMMON NATURAL OCCURRENCES
COPPER	Cu	63.54	29	1063.0	Metal dust poison by inhalation.	CUPRITE, BORNITE
CURIUM xx	Cm	245.00	96	1340.0	Radioactive	MINUTE AMOUNTS IN URANIUM ORE
DYSPROSIUM ++	Dy	162.50	66	1411.0	Metal readily ignites & burns.	FERGUSONITE, GADOLINITE
EINSTEINIUM xx	Es	252.00	99	860.0	Radioactive	SYNTHETIC TRANSURANIUM ELEMENT
ERBIUM ++	Er	167.26	68	1529.0	Metal readily ignites & burns.	GADOLINITE
EUROPIUM ++	Eu	151.96	63	822.0	Moderately toxic (ingestion)	MONAZITE, SOME ZINC ORES
FERMIUM xx	Fm	255.00	100	[?]	Radioactive	SYNTHETIC TRANSURANIUM ELEMENT
FRANCIUM	Fr	223.00	87	21.0	Radioactive	ALKALI METAL GROUP ELEMENT
FLUORINE	F	18.99	9	219.6	Flammable gas-toxic by inhalation.	FLUORITE, CRYOLITE
GADOLINIUM ++	Gd	157.25	64	1313.0	Metal ignites & burns rapidly.	GADOLINITE, MONAZITE
GALLIUM	Ga	69.72	31	29.8	Reacts violently with chlorine.	DIASPORE, SPHALERITE
GERMANIUM	Ge	72.59	32	975.3	Reacts violently with oxidants.	ZINC ORES, COAL
GOLD	Au	196.97	79	1064.0	Can react with oxidants.	AS ELEMENT & IN TELLURIDES
HAFNIUM	Hf	178.49	72	2150.0	Toxic – metal burns readily.	ZIRCONIUM MINERALS
HELIUM	He	4.00	2	-272.2	Asphyxiant	ATMOSPHERE HAS 1 PART IN 200,000
HOLMIUM ++	Ho	164.93	67	1461.0	Can react violently with halogens.	MONAZITE, GADOLINITE
HYDROGEN	H	1.00	1	-259.1	Flammable gas. Explosive.	WATER, COAL, PETROLEUM
INDIUM	In	114.82	49	156.6	Compounds toxic(ingestion)	ZINC MINERALS. IRON,LEAD,Cu.
IODINE	I	126.90	53	113.5	Vapor harmful.	CALICHE, SEA WATER & SALT BRINES
IRIDIUM	Ir	192.22	77	2410.0	Metal ignites & burns rapidly.	PLATINUM & NICKEL ORES
IRON	Fe	55.85	26	1535.0	Metal dust flammable.	HEMATITE, MAGNETITE, TACONITE
KRYPTON	Kr	83.80	36	-156.6	Asphyxiant	AIR – 1 PART PER MILLION

THE ELEMENTS - HAZARDOUS PROPERTIES & NATURAL OCCURRENCES

 Copyright 1993 by T.D. LaRune

THE ELEMENTS, ALPHABETICAL LIST

A ELEMENT	B SYMBOL	C ATOMIC WEIGHT	D ATOMIC NUMBER	E MELTING POINT [C]	J HAZARDOUS PROPERTIES	K COMMON NATURAL OCCURRENCES
LANTHANUM xx	La	138.91	57	921.0	Metal ignites & burns rapidly.	ALLANITE, MONAZITE
LAWRENCIUM xx	Lr	257.00	103	[?]	Radioactive.	SYNTHETIC TRANSURANIUM ELEMENT
LEAD	Pb	207.19	82	327.5	Poison (cumulative)	GALENA, CERUSSITE, ANGLESITE
LITHIUM	Li	6.94	3	180.5	Flammable-Can ignite in air.	PEGMATITES, MINERAL SPRINGS
LUTETIUM ++	Lu	174.97	71	1663.0	Compounds moderately toxic.	MONAZITE, (Yttrium oxide)
MAGNESIUM	Mg	24.31	12	649.0	Flammable-Violent reaction oxidants.	DOLOMITE, MAGNESITE
MANGANESE	Mn	54.94	25	1244.0	Chronic poison.	PYROLUSITE, RHODOCHROSITE
MENDELEVIUM xx	Md	256.00	101	[?]	Radioactive.	SYNTHETIC TRANSURANIUM ELEMENT
MERCURY	Hg	200.59	80	-38.8	Cumulative poison. Toxic by inhalation.	CINNABAR
MOLYBDENUM	Mo	95.94	42	2610.0	Compounds toxic(inhalation)	MOLYBDENITE, WULFENITE
NEODYMIUM ++	Nd	144.27	60	1024.0	Metal ignites & burns readily.	MONAZITE, BASTNASITE
NEON	Ne	20.18	10	-248.7	Asphyxiant.	AIR - 1 PART IN 65000
NEPTUNIUM xx	Np	237.00	93	640.0	Radioactive, radiotoxic.	TRACE IN URANIUM ORES
NICKEL	Ni	58.71	28	1455.0	Poison, dust carcinogenic.	PENTLANDITE, PYRRHOTITE
NIOBIUM * (Columbium^*)	Cb	92.91	41	2468.0	Compounds may irritate skin.	COLUMBITE, PYROCHLORE
NITROGEN	N	14.00	7	-209.9	Low toxicity.	78 % OF AIR
NOBELIUM	No	253.00	102	[?]	Radioactive.	SYNTHETIC TRANSURANIUM ELEMENT
OSMIUM	Os	190.20	76	3945.0	Osmium tetroxide highly toxic.	PLATINUM SANDS, NICKEL ORE
OXYGEN	O	15.99	8	-218.4	Oxidizer.	21 % OF AIR
PALLADIUM	Pd	106.4	46	1554.0	Metal dust combustible.	PLATINUM & COPPER-NICKEL ORES
PHOSPHOROUS	P	30.97	15	44.1	Highly flammable solid. Toxic to inhale.	APATITE ROCK
PLATINUM	Pt	195.09	78	1772.0	Compounds toxic by ingestion.	NATIVE, SPERRYLITE

THE ELEMENTS - HAZARDOUS PROPERTIES & NATURAL OCCURRENCES

 Copyright 1993 by T.D. LaRune

THE ELEMENTS, ALPHABETICAL LIST

A ELEMENT	B SYMBOL	C ATOMIC WEIGHT	D ATOMIC NUMBER	E MELTING POINT [C]	J HAZARDOUS PROPERTIES	K COMMON NATURAL OCCURRENCES
PLUTONIUM xx	Pu	242.06	94	641.0	Radioactive, toxic.	SYNTHETIC TRANSURANIUM ELEMENT
POLONIUM	Po	210.00	84	254.0	Radioactive, highly toxic.	URANIUM ORES
POTASSIUM	K	39.10	19	63.3	With water give off flammable gas.	SYLVITE, CARNALLITE
PRASEODYMIUM ++	Pr	140.91	59	931.0	Metal ignites & burns readily.	MONAZITE, BASTNASITE
PROMETHIUM ++	Pm	147.00	61	1168.0	Poison, radioactive, radiotoxic.	A RARE EARTH METALLIC
PROTACTINIUM xx	Pa	231.00	91	1552.0	Radiotoxic, radioactive, carcinogen.	PITCHBLENDE
RADIUM	Ra	226.05	88	700.0	Radiotoxic-carcinogen by inhalation.	URANANITE,CARNOTITE
RADON	Rn	222.00	86	-71.2	Radiotoxic, carcinogen by inhalation.	SOIL, 1 Gm/Sq.MILE
RHENIUM	Re	186.20	75	3180.0	Radioactive.	MOLYBDENITE, COLUMBITE, PLATINUM
RHODIUM	Rh	102.91	45	1966.0	Toxic by inhalation.	PLATINUM ORES, Cu/Ni SULFIDE ORE
RUBIDIUM	Rb	85.47	37	38.9	Highly flammable, ignites in air.	ZINNWALDITE, LEUCITE
RUTHENIUM	Ru	101.07	44	2310.0	Compounds may be toxic.	PLATINUM ORES, PENTLANDITE
SAMARIUM ++	Sm	150.35	62	1077.0	Metal ignites & burns readily.	MONAZITE, BASTNASITE
SCANDIUM	Sc	44.96	21	1541.0	Flammable, reacts with air.	TRACE IN 800 MINERALS
SELENIUM	Se	78.96	34	217.0	Cumulative poison.	SELENITE
SILICON	Si	28.09	14	1410.0	Pulmonary inhalation.	SAND, QUARTZ, AGATE, CLAY, MICA
SILVER	Ag	107.87	47	961.9	Compounds irritate.	NATIVE, GOLD, LEAD, COPPER, ZINC ORE
SODIUM	Na	22.99	11	97.8	Flammable as a solid.	CRYOLITE, HALITE, ZEOLITES
STRONTIUM	Sr	87.62	38	769.0	Metal ignites & burns readily.	STRONTIANITE, CELESTITE
SULFUR	S	32.06	16	112.8	Irritant.	PYRITE, GALENA, SPHALERITE
TANTALUM	Ta	180.95	73	2996.0	Metal ignites in air.	COLUMBITE, TANTALITE
TECHNETIUM	Tc	99.00	43	2172.0	Radioactive.	OBTAINED BY FISSION OF URANIUM

THE ELEMENTS – HAZARDOUS PROPERTIES & NATURAL OCCURRENCES

 Copyright 1993 by T.D. LaRune

THE ELEMENTS, ALPHABETICAL LIST

A ELEMENT	B SYMBOL	C ATOMIC WEIGHT	D ATOMIC NUMBER	E MELTING POINT [C]	J HAZARDOUS PROPERTIES	K COMMON NATURAL OCCURRENCES
TELLURIUM	Te	127.60	52	449.5	Poison	CALAVERITE
TERBIUM	Tb	158.92	65	1360.0	Flammable.	GADOLINITE, MONAZITE
THALLIUM	Ti	204.37	81	303.5	Cumulative poison.	PYRITE, CROOKSITE, HUTCHINSONITE
THORIUM xx	Ac	232.04	90	1750.0	Poison, radioactive, carcinogen	MONAZITE
THULIUM ++	Tm	168.93	69	1545.0	Dust flammable, compounds can be toxic.	MONAZITE
TIN	Sn	118.69	50	231.9	Organic compounds highly toxic.	CASSITERITE
TITANIUM	Ti	47.90	22	1660.0	Solid is flammable.	RUTILE, ILEMITE, TITANITE
TUNGSTEN	W	183.85	74	3410.0	Toxic dust, inhaled or eaten.	WOLFRAMITE, SCHEELITE
URANIUM xx	U	238.03	92	1132.3	Poison, radiotoxic.	URANINITE, CARNOTITE, PITCHBLENDE
VANADIUM	V	50.94	23	1890.0	Flammable, possible carcinogen.	VANADINITE, CARNOTITE
XENON	Xe	131.30	54	-111.9	Asphyxiant.	AIR - 1 PART IN 20,000,000
YTTERBIUM ++	Yb	173.04	70	819.0	Dust flammable. Possible carcinogen.	MONAZITE
YTTRIUM	Y	88.91	39	1522.0	Dust flammable. Compounds toxic.	MONAZITE, BASTASITE
ZINC	Zn	65.37	30	419.5	Reacts violently with oxidants.	SPHALERITE, SMITHSONITE
ZIRCONIUM	Zr	91.22	40	1852.0	Reacts violently with oxidants.	ZIRCON + 30 MINERALS

xx ACTINIDES
++ LANTHANIDES

TRANSURANIUM ELEMENT - Any element having an atomic number greater than 92 (the atomic number of uranium). ALL SUCH ELEMENTS ARE RADIOACTIVE.
(They are synthesized by bombarding a heavy element with a light particle or element.)

S//Q ™

Memo

to: MINERAL PROSPECTORS, ROCKHOUNDS, STUDENTS

from: T. D. LaRune

" ROCKPECKER ™ v.1 DOS PROFESSIONAL "

DESCRIBE, IDENTIFY & CHEMICAL TEST UNKNOWN ORE

Computerized Versions

of PART THREE & PART 4 ARE AVAILABLE

NOW FOR PC's

DOS 3.3+, 400K+ RAM,

XT\486 COMPUTERS.

277 MINERALS ARE DETAILED.

Prospector can identify unknown mineral / ore;

then PRE-ASSAY chemical test to confirm finding.

WRITE TO: Skill//Quest Co.

675 Fairview Drive, Suite 246,

Carson City, Nevada 89701 U.S.A.

CHAPTER 69

PART FOUR

LaRUNE'S

COOKBOOK CHEMISTRY

NOTES

PRE-ASSAY CHEMICAL TESTS

CHEMICAL TESTING RECIPES

COPYRIGHT NOTICE

COPYRIGHT 1991 BY T.D. LaRUNE.

All rights reserved

From the book:

LaRune's ROCKPECKER Notes

A Mineral Prospector's Primer

No part of this publication may be copied or distributed, transmitted, transcribed, stored in a retrieval system, or translated into any human or computer language, in any form or by any means, electronic, mechanical, magnetic, manual, or otherwise, or disclosed to third parties without express written permission of: T. D. LaRune, P.O. Box 620995, San Diego, California 92162 U.S.A.

PRELUDE TO PART FOUR

CHEMICAL TESTING

ACID BURNS -- CAUTIONS,

SAFETY TIPS & FIRST AID

EYES — Wearing safety glasses or a face mask should prevent any problems of acid getting into the eyes.

FIRST AID — Hold your head under running tap water and generously flush out the eyes with water; then immediately get to your doctor or an emergency room. DON'T TAKE CHANCES!

ACID BURNS — ANTIDOTES

ANTIDOTES FOR ACID BURNS DEPEND ON THE KIND AND STRENGTH OF ACID.

FIRST AID —

1. Flush the burn area thoroughly with water.

2. Cover the burn area for about 15 minutes with a paste made of bicarbonate of soda and water.

3. Wash the paste off and cover the burn area with a mixture of lime water and linseed oil.

P.S. WATER ALONE IS NOT RECOMMENDED.

CAUTIONS

A - Keep acids out of reach of children. Store in a safe place.

B - Wear rubber gloves and rubberized apron.

C - Avoid breathing acid fumes; work in a well ventilated area.

D - ACID IS ALWAYS POURED INTO WATER. REVERSING THE PROCESS PRODUCES HOT SPATTERING AND SPLASH HAZARD.

FIRST SAFETY RULE - WHEN TESTING, ALWAYS WEAR PROTECTIVE CLOTHING, GLOVES AND GLASSES.

LaRUNE'S COOKBOOK CHEMISTRY NOTES

PRE-ASSAY CHEMICAL TESTS

THESE TESTS ARE QUALITATIVE - THIS MEANS THAT THE ELEMENT TESTED FOR IS EITHER PRESENT OR ABSENT. THIS IS IMPORTANT TO KNOW, FOR IN BUYING AN "ASSAY", YOU NEED TO SPECIFY AN ELEMENT CERTAIN. THE ASSAYER WILL DETERMINE THE PERCENTAGE OR VALUE OF THAT SPECIFIED ELEMENT IN YOUR ROCK OR ORE. IF YOU HAD CALLED FOR GOLD AND SILVER AS THE SEARCH ELEMENTS, YOUR ASSAY REPORT MIGHT READ: 0.25 oz. Au and 9.5 oz. Ag/ton. FROM THE CURRENT MARKET PRICES YOU CAN THEN DETERMINE THE VALUE OF YOUR ORE OR ROCK.

IF YOU DON'T WANT TO TEST FOR THE ELEMENTS YOURSELF, AND CAN AFFORD A "QUANTATIVE ANALYSIS", YOU CAN ORDER A SEARCH OF EVERY ELEMENT PRESENT AND THE PERCENTAGE OR VALUE OF EACH. HOWEVER, IF YOU DRAW A BLANK, YOU MUST BE PREPARED FOR A "NO VALUES" REPORT.

"SHOTGUN" & "IT AIN'T THERE" TESTS

VISUAL IDENTIFICATION OF NATIVE ELEMENTS IS POSSIBLE FOR A FEW ELEMENTS WITH THE AID OF A HAND-LENS. GOLD, SILVER, PLATINUM, IRON, COPPER, MERCURY, BISMUTH, ANTIMONY ARE YOUR PRACTICAL POSSIBLES. HOWEVER, YOU WILL MISS THESE ELEMENTS WHEN THEY ARE CHEMICALLY COMBINED AND LOOK LIKE ORDINARY COUNTRY ROCK. FOR THIS REASON THE "SHOTGUN" & "IT AIN'T THERE" TESTS ARE VALUABLE.

TEST EVERY HEAVY, COLORED, COUNTRY ROCK FOR BLIND ELEMENTS.

ROCK & ORE COLORS

ROCK & ORE COLORS	MINERALS THAT MAY BE PRESENT
Black	COLUMBITE, TANTALITE, PITCHBLENDE
Brown	MICROLITE, WOLFRAMITE
Red or Brown	CASSITERITE, BISMUTH
Black Specks	COLUMBITE, CASSITERITE, TANTALITE, SAMARSKITE
Silver-White specks	SYLVANITE, TELLURIUM
Yellow	CARNOTITE
Green	GARNIERITE
Gray	COBALTITE, SKUTTERUDITE, NICCOLITE
Bronze	GOLD, COPPER, PYRRHOTITE, PLATINUM

Test colors are designed to show POSITIVE when the element tested-for is present in COMMERCIAL GRADES. Test color density can sometime point to the grade of ore. The deeper the color, the richer the ore.

PRE-ASSAY TEST -- DON'T GUESS !

FOR OUR PURPOSES

LIQUID QUANTITY MEASURES --

1 MEDICINE DROPPER DROP	=	0.05 ml	=	0.05 cc
20 MEDICINE DROPPER DROPS	=	1.0 ml	=	1 cc
1 TEASPOON	=	5.0 ml	=	5 cc
1 TABLESPOON	=	15.0 ml	=	15 cc
1 FLUID OZ.	=	30.0 ml	=	30 cc

DRY QUANTITY MEASURES -- 2BB's = 1 GRAM.
------ We use BB sizes as dry quantity measures. ------
POWDERED TEST MINERAL MIXTURES -- It is always best to reduce the test mineral to dust particle size. If you can see grains, pulverize further. (A 3 inch cast iron end cap with an iron rod works fine as a mortar & pestle.)

WHAT IS A TEST MINERAL ?

REGARDLESS OF ITS APPEARANCE, IT IS ANY ROCK OR ORE YOU WISH TO TEST AFTER YOU HAVE GROUND IT TO POWDERED FORM.

ROCK - A solid material composed of a mix of minerals.

ORE - In the classic sense, any mineral from which one or more metallic elements may be extracted at a profit. WE DEFINE AN ORE HEREIN AS CONTAINING METALLIC OR NON-METALLIC ELEMENTS that can be extracted at a profit.

MINERAL - An inorganic chemical composition of elements occurring in nature as natural combinations.

THREE TEST PHASES

FIRST PHASE - AS YOU FOUND IT, take your rock or crude ore and powder it down to a tablespoonful. Run your tests.

SECOND PHASE - Take another FIRST PHASE SAMPLE MATERIAL spoonful and wash it in your goldpan to concentrate the heavies. RUN YOUR TEST ON THE HEAVIES. (If you skip the first phase, you wash out some of the lighter elements. You will miss elements that are oxidized or in chemical combination with sulfur, arsenic or tellurium. If you find sulfides, arsenic or tellurium, first roast the material. TO ROAST - place on charcoal. With blowpipe heat to red heat until any evolving smoke ceases.)

THIRD PHASE - Where your ore shows as nodules in a rock, or as a strata layer of similar minerals, pick out the suspected ore particles and/or separate the gangue from the test mineral. (This is another concentration method.) Test for arsenic, sulfides and tellurium. If any found, roast the material as in phase two. Then run the rest of your tests.

NOTE: YOU ONLY NEED A SPOONFUL OF FINELY POWDERED MATERIAL.

FIRST SAFETY RULE - WHEN TESTING, ALWAYS WEAR PROTECTIVE CLOTHING, GLOVES AND GLASSES.

SOLUTION & SATURATED SOLUTION -- Liquids, such as water, acids, or other chemicals are used to dissolve test minerals. Distilled water is best, but for many minerals acids are needed. If some of your mineral remains undissolved, you have a saturated solution. Gently shake for more complete dissolution. Don't worry if some stays in the solid state. The liquid may not be able to hold any more solid. If solid does not dissolve, change solvents.

DILUTION OF ACIDS, HYDROGEN PEROXIDE, ETC. -- Concentrated hydrochloric acid is known as 12 MOLAR HCL. (WRITTEN AS "12 M HCL") To dilute 12 M HCL to half strength (6 M HCL), you take one part concentrated acid added to one part distilled water. Similarly, if you have 30% hydrogen peroxide solution, a mix of one part peroxide to one part pure water gives you a 15% solution.
(Drugstore shelf peroxide is a 3 % solution.)

MOLARITY OF CONCENTRATED ACIDS --

ACETIC ACID	-----	17 M
HYDROCHLORIC ACID	-----	12 M
NITRIC ACID	-----	16 M
SULFURIC ACID	-----	18 M

CONCENTRATED ACIDS ARE USED AS REAGENTS. (Unless otherwise directed) A reagent is a material used to bring about a change.

IT IS IMPERATIVE THAT REAGENTS BE PRESERVED AGAINST CONTAMINATION. (Droppers of reagent bottles should always be held ABOVE the test tube and not be allowed to touch them.)

PREPARING FRESH SOLUTIONS PRIOR TO TESTING -- USE WITHIN 4 HOURS.

AQUA REGIA -- (For totally dissolving GOLD, PLATINUM, PALLADIUM) To prepare aqua regia - use one part nitric acid to 3 parts hydrochloric acid.

STANNOUS CHLORIDE -- (Used as reagent in GOLD and PLATINUM GROUP) To make a small amount take 1 ml of 12 M hydrochloric acid, drop 2 small pieces of metallic tin into the acid and boil SLIGHTLY.

SAFETY RULE - SPLASH HAZARD -- POUR ACID INTO WATER.

NEVER ADD AMMONIA TO A HOT SOLUTION.

WATER -- As used in ALL these tests should be DISTILLED water. Impurities in water, particularly CHLORINE contamination from well-salt or city water purifications, can upset tests. This is especially true for silver and lead tests. (If you don't have pure distilled water, then you need to pre-test the water for chlorine content with a silver nitrate solution. CHLORINE shows up as cloudy water. You cannot use it for lead or silver tests.)

FIRST SAFETY RULE - WHEN TESTING, ALWAYS WEAR PROTECTIVE CLOTHING, GLOVES AND GLASSES.

CLEAN EQUIPMENT MEANS ACCURATE TESTING.

For flame tests always start with a clean wire. After each use, dip the wire in concentrated hydrochloric acid and bring wire to the flame. Repeat until the wire ceases to color the flame. A bottle cork makes a suitable holder. Multiple wires are helpful.

You should have a separate medicine dropper for each acid and reagent. Acids should be concentrated strength, you can always dilute if need be. Reagents should be used freshly made.

Test tubes in the 4" size are used for boiling solutions over the alcohol lamp. (Pipe cleaners make adequate test tube holders.)

IMPORTANT -- HEATING SPEEDS UP REACTION TIME.

BOILING means passing the tube back and forth over the flame to obtain the boilpoint. SLOW HEATING uses the same action only further from the flame. WARMING and KEEPING WARM is accomplished by heating the test tube in water and leaving it in warm water until needed. CLEAN YOUR GLASSWARE AFTER EACH TEST USE. (Having 6 or 8 test tubes on hand speeds your work.)

FILTER PAPERS -- Use the small filter with a glass funnel. (Used for separating solution from precipitate and for SPOT TESTS.)

~~~~~~~~~~~ CALIFORNIA PROPOSITION 65 WARNING ~~~~~~~~~~

BE CAREFUL WHEN WORKING WITH ACIDS.

ACIDS & MOST ELEMENTS CAN BE HARMFUL TO YOUR HEALTH.

ALWAYS WEAR SAFETY EQUIPMENT & USE COMMON SAFETY SENSE.

~~~~~~~~~~~~~~~ READER AND/OR USER WARRANTIES ~~~~~~~~~~~~

~~ The author believes that each of these tests, and ~~
~~ simplified test directions is reasonably accurate.~~

NO EXPRESS OR IMPLIED REPRESENTATION OF SAFETY IS MADE. In using these tests, the READER AND/OR USER agrees to accept full responsibility FOR ALL KNOWN OR UNKNOWN HAZARDS, and expressly agrees to hold the author harmless from any loss or damage, of whatever kind or nature, whether direct, indirect or consequential.

FIRST SAFETY RULE - WHEN TESTING, ALWAYS WEAR PROTECTIVE CLOTHING, GLOVES AND GLASSES.

TEST # 1 -- FLAME TESTS --

MATERIALS NEEDED: 1. ALCOHOL LAMP 2. NICHROME WIRE & HOLDER
3. TEST TUBE 4. HYDROCHLORIC ACID (12 M)
5. POTASSIUM IODIDE 6. COBALT BLUE BRAZING LENS

TEST ACTION --
1. FINELY POWDER YOUR TEST MINERAL.
2. MOISTEN A LITTLE OF THE SOLID SAMPLE WITH CONCENTRATED 12 M HCL INTO A PASTY MIXTURE.
3. DIP CLEAN WIRE IN MIXTURE, THEN BRING IT TO OUTER FLAME CONE TO OBTAIN FLAME COLOR.
4. WHERE FLAME COLOR IS YELLOW, USE COBALT LENS TO MASK STRONG YELLOW THAT MAY HIDE A COLOR.

EXPECTED FLAME TEST RESULTS --

	TEST CODE	FLAME COLOR	MINERAL ELEMENT	
TEST # 1.	[1 A]	STRONG YELLOW --------	SODIUM	(Na)
	[1 B]	WEAK VIOLET ----------	POTASSIUM	(K)
	[1 B]	RED THRU COBALT LENS-	POTASSIUM	(K)
	[1 C]	STRONG ORANGE-RED ---	CALCIUM	(Ca)
	[1 D]	CARMINE RED ----------	STRONTIUM	(Sr)
	[1 E]	YELLOW-GREEN ---------	BARIUM	(Ba)
	[1 F]	GREEN ----------------	COPPER	(Cu)
	[1 G]	BLUISH-WHITE * -------	ARSENIC	(As)
	[1 H]	BLUISH-WHITE ** ------	ANTIMONY	(Sb)
	[1 J]	PALE BLUE *** --------	LEAD	(Pb)
	[1 K]	DARK BLUE ------------	SELENIUM	(Se)
	[1 L]	BLUE-GREEN ------------	PHOSPHOROUS	(P)
	[1 M]	BLUE-GREEN **** ------	ZINC	(Zn)
	[1 N]	GREEN-YELLOW ---------	BORON	(B)
	[1 P]	GRASS-GREEN ----------	TELLURIUM	(Te)
	[1 R]	WHITE-GREEN ----------	BISMUTH	(Bi)
	[1 S]	PALE YELLOW-GREEN ---	MOLYBDENUM	(Mo)
	[1 T]	CRIMSON-RED ----------	LITHIUM	(Li)

TIP -- Always start with a clean wire.

DISTINGUISH -- ARSENIC * ---- ANTIMONY ** ---- LEAD ***
* ARSENIC - emits a garlic odor. (Smell test)
** ANTIMONY from *** LEAD minerals. Dissolve powdered mineral in HCL acid solution. ADD a few CRYSTALS of potassium iodide (KI).

TEST CODE
TEST RESULT -- [1 U] [P] ANTIMONY -- BRIGHT ORANGE PRECIPITATE
[1 W] [P] LEAD -- LEMON YELLOW PRECIPITATE

DISTINGUISH -- [1 X] [P] **** ZINC -- from PHOSPHOROUS minerals. Moisten mineral fragment with a few drops of cobalt nitrate solution, then heat intensely. BRIGHT GREEN SOLUTION indicates ZINC. (Roast zinc silicates on charcoal with sodium carbonate before heating with cobalt nitrate.)

FIRST SAFETY RULE - WHEN TESTING, ALWAYS WEAR PROTECTIVE CLOTHING, GLOVES AND GLASSES.

TEST # 2 -- BEAD TESTS --

MATERIALS NEEDED --
1. NICHROME OR PLATINUM WIRE WITH MATCH-HEAD SIZE LOOP & WIRE HOLDER.
2. ALCOHOL LAMP (For heat.)
3. BORAX - ($Na_2B_4O_7.10H_2O$ - used dry.)
4. SALT OF PHOSPHOROUS (Microcosmic salt) ($HNaNH_4PO_4.4H_2O$ - used dry.)
5. TINY BIT OF FINELY POWDERED TEST MINERAL.

TEST ACTION --
1. HEAT NICHROME WIRE IN FLAME.
2. DIP HOT LOOP INTO DRY FLUX. (Borax or Salt of phosphorous) CONTINUE DIPPING AND REHEATING UNTIL THE LOOP IS FILLED WITH A CLEAR BEAD.
3. DIP HOT CLEAR BEAD INTO A FEW OF THE FINELY POWDERED TEST MINERAL CRYSTALS.
4. FLAME TEST -- IN OUTER FLAME CONE (oxidizing flame) HEAT THE TEST MINERAL LOOP. OBSERVE BEAD COLOR IN BOTH HOT AND COLD BEADS.
5. CLEAN WIRE AND REPEAT WITH NEW TEST MINERAL SAMPLE. HEAT IN INNER FLAME CONE. (Reducing (flame) OBSERVE BOTH HOT & COLD BEAD COLOR.

FALSE TEST RESULTS -- If Borax and SoP beads are BLACK, it means that too much test mineral was used --OR-- FIRST roast test mineral, then run test.

A CORRECT TESTING PROCEDURE WILL GIVE THE FOLLOWING RESULTS.

TEST #2 ------------ COLD BEAD COLORS

TEST CODE	FLUX	OUTER FLAME CONE	INNER FLAME CONE	MINERAL ELEMENT
[2 A]	Borax	Green	Green	CHROMIUM
[2 B]	Borax	Blue	Blue	COBALT
[2 C]	Borax	Blue-Green	Red-Opaque	COPPER
[2 D]	SoP	Colorless	Red-Brown	COLUMBIUM
[2 E]	Borax	Yellow	Pale Green	IRON
[2 E]	SoP	Yellow	Colorless	IRON
[2 F]	B & SoP	Violet	Colorless	MANGANESE
[2 G]	Borax	Colorless	Dark Brown	MOLYBDENUM
[2 H]	Borax	Red-Brown	Gray (opaque)	NICKEL
[2 H]	SoP	Yellow-Brown	Yellow-Brown	NICKEL
[2 J]	SoP	Colorless	Violet	TITANIUM
[2 K]	Borax	Colorless	Yellow	TUNGSTEN
[2 K]	SoP	Colorless	Blue	TUNGSTEN
[2 L]	Borax	Yellow	Pale Green	URANIUM
[2 L]	SoP	Colorless	Green	URANIUM
[2 M]	Borax	Green-Yellow	Green	VANADIUM

Distinguish IRON from URANIUM by the Salt of Phosphorous test.

TIP -- Only a bit of test mineral is required for these tests.

FIRST SAFETY RULE - WHEN TESTING, ALWAYS WEAR PROTECTIVE CLOTHING, GLOVES AND GLASSES.

TEST # 3 --- BLOWPIPE ON CHARCOAL

FLUX FOR BLOWPIPE CHARCOAL TESTS -- (Level teaspoon measures) Mix well, 1 tsp. POWDERED CHARCOAL with 2 tsps. ANHYDRATED SODIUM CARBONATE with 2 tsps. POWDERED BORAX. Quantity can be doubled.

PREPARING THE TEST SAMPLE -- 6 BB's OF WELL-POWDERED TEST MINERAL MIXED WITH 12 BB's OF CHARCOAL FLUX. MOISTEN WITH WATER DROPS INTO A PASTY MIX. PLACE IN SMALL CHARCOAL BLOCK DEPRESSION. WITH BLOWPIPE FUSE INTO LIQUID STATE WITH INNER CONE OF FLAME. TURN SAMPLE OVER AND CONTINUE FUSION. IF NO OBSERVABLE RESULTS, HEAT IN OUTER FLAME CONE. IF STILL NO RESULTS, REDO TEST WITHOUT THE CHARCOAL FLUX. SEE BELOW FOR EXPECTED REACTIONS.

These tests employ the use of :

1. A blowpipe to direct the flame of the alcohol lamp;
2. Onto a square of charcoal with a depression;
3. Into which a paste of test mineral mixed with flux;
4. Is heated to fusion, after or during which;

5. A TELL-TALE REACTION IN THE FORM OF:

   a. A coating on the charcoal; (Sublimate)
   b. A smell of garlic or horseradish; (Arsenic or Selenium)
   c. Fumes smart the eyes; (Osmium)
   d. Heavy white smoke; (Antimony)
   e. Metallic buttons that are malleable, cuttable, magnetic;
   f. The color of the metallic button both hot and cold.

6. REVEALS THE ELEMENT EITHER POSITIVELY OR AS AN INDICATOR.

--- POSITIVE ELEMENT IDENTIFICATION --- [P]

TEST CODE			TEST # 3 RESULTS
[3 A]-	[P] --	ARSENIC	-Garlic odor, a light WHITE coating painted away from the fusion.
[3 B]-	[P] --	CADMIUM	-RED, BROWN coating with BLUISH edges.
[3 C]-	[P] --	ANTIMONY	-WHITE smoke very heavy, coating is a dense WHITE with BLUISH EDGING.
[3 D]-	[P] --	SELENIUM	-Horseradish odor, GRAY coating with RED edges. (Odor is positive.)
[3 E]-	[P] --	SULFUR	-Strangling, sharp odor. (Odor positive)
[3 F]-	[P] --	OSMIUM	-Fumes sting eyes; sharp acrid odor.

--- ELEMENT INDICATED BUT SHOULD BE CONFIRMED --- [I]

[3 G]-	[I] --	LEAD	-Cuttable metallic bead; YELLOW coating.
[3 H]-	[I] --	BISMUTH	-Hard-to-cut metallic bead; YELLOW coat.
[3 J]-	[I] --	TELLURIUM	-BLUISH-YELLOW edge on a dense WHITE coating with BLACKISH shading near rim.
[3 K]-	[I] --	ZINC	-Hot YELLOW coat turns WHITE-BLUE cold.
[3 M]-	[I] --	COPPER	-RED metallic mass. Malleable.
[3 N]-	[I] --	SILVER	-WHITE metallic, malleable button.
[3 P]-	[I] --	GOLD	-Soft, metallic YELLOW, malleable button

SHOTGUN TEST # 4 --

COLORED ROCK BLINDFOLD TESTS -- BLACK, RED, BROWN, GRAY, YELLOW, GREEN, RED-BROWN - THAT HEFT AS "HEAVY".

DIRECTIONS -- Start with TEST ACTION # 1 and work continuously thru to TEST ACTION # 7 D.

ORDER IN WHICH BLIND ELEMENTS SHOW UP:

1. SULFUR, 2. SULFIDES, 3. COPPER, 4. COBALT, 5. NICKEL, 6. LEAD, 7. BISMUTH, 8. TELLURIUM, 9. MOLYBDENUM, 10. ARSENIC, 11. TUNGSTEN 12. COLUMBIUM (Niobium) 13. SILVER.

MATERIALS NEEDED --
1. NITRIC ACID (16 M) - Mix 1 part concentrated acid with 2 parts water.
2. HYDROCHLORIC ACID (12 M) concentrated.
3. DIMETHYLGLYOXIME - Used wet, so dissolve 1 BB in 3 tbsp. 95% grain alcohol.(45 ml)
4. POTASSIUM IODIDE (KI) - Used dry.
5. AQUEOUS AMMONIA (15 M NH_3) - strong.
6. METALLIC TINFOIL - small pieces.
7. DISTILLED WATER - (Must be pure)
8. TEST TUBES, EVAPORATING DISHES, FILTER PAPER, ALCOHOL LAMP, GLASS SHEET.

BEGIN SHOTGUN TESTS -- Results Code: [P] Positive for element. [I] Element indicated, but confirm.

TEST ACTION # 1.

Test solution preparation: NITRIC ACID 2 cc's with 6 BB's TEST MINERAL in clean test tube. Boil for 5 minutes, cool for 5 minutes.

TEST CODE RESULTS - TEST # 4 --

[4 A]-[I] SULFUR, SULFIDE -- a rising BLACK or YELLOW, spongy mass in tube. Confirm test [3.E]

[4 B]-[I] COPPER, NICKEL (OR BOTH) -- GREEN COLORED SOLUTION. Confirm Test [1.F][2.H]

[4 C]-[I] COBALT -- ROSE COLORED SOLUTION. Confirm test [2.B]

TEST ACTION # 2 --SPOT TEST --

Test preparation:
1. Place filter paper on 6" x 6" glass sheet.
2. Add 1 drop DIMETHYLGLYOXIME to paper.
3. Add 1 drop TEST SOLUTION on top of dimeth.
4. Add 1 drop AMMONIA on top of test solution.

TEST CODE RESULT --

[4 D]-[P] NICKEL -- RED OR ROSE COLORED SPOT.

TEST # 4 CONTINUED

TEST ACTION # 3 --

Test preparation:
1. In a clean evaporation dish;
2. Add 2 drops of TEST SOLUTION.
3. Dilute with 8 drops of water.
4. Add a few crystals of POTASSIUM IODIDE.
5. Allow solution to stand and settle out.

TEST CODE RESULT --

[4 E]-[I] LEAD -- YELLOW OR TAN-ISH PRECIPITATE. Confirm with test [1.J]

[4 F]-[I] BISMUTH -- BLACK SOLUTION. Confirm with test [1.R]

[4 G]-[I] BOTH LEAD & BISMUTH -- BLACK SOLUTION DISAPPEARS to leave YELLOW or BROWNISH PRECIPITATE. Confirm tests [1.J][1.R]

TEST ACTION # 4 --

Test preparation:
1. In an evaporating dish add 10 drops of TEST SOLUTION and boil until dish is dry.
2. Remove dish from flame. Add 1 cc 12 M HCL.
3. Save for FOLLOW-UP use in TEST ACTION #5.

TEST CODE RESULT --

[4 H]-[P] COBALT -- SOLUTION TURNS GREEN.

TEST ACTION # 5 --
1. Take TEST ACTION # 4 END-UP solution.
2. Slowly BOIL-OFF 50% of test solution # 4.
3. Remove from heat source. Drop small piece of tin metal into HOT solution.
4. Slowly agitate dish with circular motion.
5. Save for follow-up use in TEST ACTION # 6.

TEST CODE RESULTS --

[4 J]-[P] TELLURIUM -- SOLUTION TURNS BLACK (or solution around tin metal darkens)

[4 K]-[P] MOLYBDENUM -- BLUE STAIN ON DISH.

[4 L]-[I] ARSENIC -- REDDISH-BROWN COATING ON DISH. Confirm test [1.H]

TEST ACTION # 6 -- BOIL TEST ACTION # 5 SOLUTION UNTIL SOLUTION CHANGES COLOR.

TEST CODE RESULT --

[4 M]-[I] TUNGSTEN & COLUMBIUM -- BLUE solution turns BROWN upon cooling. (Same for both) Confirm test [2.K][2.D]

TEST ACTION # 7 -- SILVER - LEAD TESTS

Test preparation:
1. From TEST ACTION # 1 use remaining solution
2. Bring to BOIL and FILTER into a clean tube.
3. When solution is COLD, divide equally into test tubes # 1 & # 2.
4. ADD 2 DROPS 12 M HCL TO TEST TUBE # 1.
5. HOLD test tube # 2 for later use.

TEST CODE RESULT --

[4 N]- [I] SILVER & LEAD (both) - WHITE, MILKY, LUMPY PRECIPITATE (Silver & Lead is in precipitate)

TEST ACTION # 7 A --

Test preparation:
1. FILTER-OFF SOLUTION. SAVE WHITE PRECIPITATE.
2. Divide precipitate into test tubes # 3 & # 4
3. ADD 2 cc's of PURE WATER to test tube # 3
4. Heat to a BOIL. (Hold tube # 4 for later)

TEST CODE RESULT --

[4 R]-[P] LEAD - WHITE PRECIPITATE DISSOLVES IN WATER.

[4 S]-[I] SILVER - WHITE PRECIPITATE DOES NOT DISSOLVE IN WATER.

TEST ACTION # 7 B --

Test preparation:
1. TAKE TEST TUBE # 4, ADD 2 cc's pure water.
2. ADD 1 cc 15 M AMMONIA (strong)
3. Shake tube gently to thoroughly mix.

TEST CODE RESULT --

[4 T]-[P] SILVER - WHITE PRECIPITATE DISSOLVES.

TEST ACTION # 7 C --

Test preparation: 1. TO TEST TUBE # 4 ADD 2 DROPS NITRIC ACID.

TEST CODE RESULT --

[4 V]-[P] SILVER - SILVER IN SOLUTION FROM TEST ACTION # 7 B REPRECIPITATES.

TEST ACTION # 7 D -- SILVER SPOT TEST

Test preparation: 1. Place filter paper on glass sheet. From TEST TUBE #2 ADD 1 drop of test solution to paper. Then add 1 drop of hydrochloric acid to the test solution drop. On the drop of acid add another drop of test solution.

TEST CODE RESULT --

[4 W]-[P] SILVER - SPOT TURNS DARK AFTER SETTING A WHILE.

END OF SHOTGUN TEST # 4

FIRST SAFETY RULE - WHEN TESTING, ALWAYS WEAR PROTECTIVE CLOTHING, GLOVES AND GLASSES.

TEST # 5 -- BLIND TEST

-- THE LaRUNE " It ain't there" TEST --

IF NO PRECIPITATE FORMS -- THESE METALS ARE ABSENT

TEST CODE--

[5 A] [I] -- SILVER
[5 B] [I] -- MERCURY
[5 C] [I] -- BISMUTH
[5 D] [I] -- COPPER
[5 E] [I] -- COBALT
[5 F] [I] -- CADMIUM
[5 G] [I] -- IRON
[5 H] [I] -- MANGANESE
[5 J] [I] -- NICKEL
[5 K] [I] -- MAGNESIUM

TEST ACTION #1 --

Test preparation:

1. Place 6 BB's of well-powdered test mineral in a clean test tube and test for solubility. First in water, then in whichever acid will dissolve your mineral. (The pur pose is to completely dissolve the mineral.)
2. ADD THE WATER OR ACID DROP BY DROP (SHAKE TUBE AFTER EACH DROP) until a solution is reached. Use new tube for each solvent change. Solvents are used cold. You may need to heat to a BOIL.
3. TEST MINERAL IS NOW IN SOLUTION.
   a. Put 2 drops of test solution in a clean test tube. (Add a few drops of water to get a volume.)
   b. ADD SODIUM HYDROXIDE 1 DROP AT A TIME. (Mix well by flicking the tube with your finger.)
   c. SOLUTION MUST BE STRONGLY BASIC. TEST WITH RED LITMUS PAPER WHICH SHOULD TURN BLUE. (Dip your glass rod into solution, then touch litmus paper. Avoid sides of tube when wetting rod.)

TEST RESULTS --

NO PRECIPITATE FORMS ----- SEE *TEST CODE* [5 A TO 5 K] ABOVE.

COLORED PRECIPITATE FORMS -----

TEST CODE -- [5 M] [I] COPPER or COBALT - BLUE PRECIPITATE
[5 N] [I] IRON - RED-BROWN PRECIPITATE
[5 P] [I] MERCURY - YELLOW PRECIPITATE
[5 R] [I] SILVER - BROWN PRECIPITATE

Results are INDICATORS of metals PRESENT.
If PRECIPITATE formed, do TEST # 4. (BUCKSHOT TEST)

GOLD AND PLATINUM GROUP ELEMENTS

ROCKPECKER TIP # 218 -- If your ore tests positive for GOLD, have it tested for PLATINUM GROUP METALS. If it tests positive for platinum & palladium, have the ore assayed for OSMIUM, RHODIUM, and RUTHENIUM. Two or more of these metals are usually alloyed together as PLATINUM GROUP METALS.

ROCKPECKER TIP # 219 - ASSAYS can pick up low GOLD values ranging to .008 oz. Au/ton with good repeatability and acceptable limits. With low gold values a check assay can be ordered where INQUARTS of gold are added to the sample. (This is a spiking procedure where an ore of known gold value is added to the sample.) For example: If you have 1 gold coin and place it in a glass of water, then add 2 more gold coins, you expect to count 3 coins.

THE ROCKPECKER TIP PROBLEM -- What if you only count 2 coins ?

IT HAS BEEN FOUND THAT CERTAIN NATURAL, PLATINUM GROUP ELEMENT ORES, WHEN COMBINED WITH GOLD, BARELY SHOW THE GOLD VALUE AND SOMETIMES NONE OF THE PLATINUM GROUP ELEMENTS. In fact, the more spiking or inquarting of gold into the sample, the lower the GOLD values reported. The reason for this strange behavior is unknown. It is as if the gold is being eaten without trace. Your tip-off to look more closely, will be a fire assay which produces a gold bead that looks unusual, oftentimes indicating the probable presence of other elements. THESE MAY BE PGM's. CHECK !

FOR THESE REASONS, GOLD TEST # 6 AND GOLD TEST # 7 ARE PRESENTED.

IF YOUR PRE-ASSAY TESTS INDICATE GOLD, PLATINUM, PALLADIUM AS PRESENT, AND THE FEE-ASSAY REPORT READS "NONE REPORTABLE", YOU MAY HAVE THIS STRANGE ORE.

If test # 6 GOLD SPOT TEST shows the presence of GOLD, you should retest for antimony, sulfides, and tellurium as combining elements in the GOLD ORE before ROASTING. Too high a roasting heat makes ores containing those elements REFRACTORY. (If you consider using the rosebud on your acetylene torch, you must be extremely careful in heating.) HEATING temperatures OVER 1000 C (1800 F) lock gold, platinum and palladium into a refractory so that chemical identity is retained in spite of high temperatures. THIS MEANS THE ORE SHOULD NEVER EXCEED "VERY DULL RED HEAT".

ROASTING ORE -- Ore is roasted to get rid of sulfides, tellurides, antinomides, and to reduce elements to the metallic state.

PROCEDURE -- The idea is to check for heavier elements in your crude ore. You'll need a few pounds of ore. Crush it, bring it to a dull red heat, stirring, until all the gases such as smoke and fumes are released. Then pulverize the sample. Pan it to concentrate the "heavies" and wash off the lighter material. Use a magnet to remove the magnetics. USE 1 SPOONFUL FOR TESTS.

TEST # 6 ---
GOLD, PLATINUM, PALLADIUM

MATERIALS NEEDED -- a) Prepare test solution
1. 4 BB's powdered test mineral.
2. Make AQUA REGIA - In clean test tube mix 1 cc 16 M nitric acid with 3 cc's 12 M hydrochloric acid. ADD test mineral; bring to BOIL for 5 to 6 minutes.

For Gold Spot Test -- b) Prepare fresh stannous chloride (SnCl)
1. In clean test tube dissolve 1 small piece of METALLIC TIN in 1 cc 12 M hydrochloric acid.
2. Boil SLIGHTLY. (Use within 4 hours)

OTHER MATERIALS NEEDED -- 2 Filter papers, 2 small evaporating dishes, 4 cc's pure water, 1 glass sheet, 2 BB's POTASSIUM IODIDE (KI), alcohol lamp.

TEST ACTION # 1 -- GOLD SPOT TEST --
1. Place filter paper on glass sheet
2. ADD 1 drop test solution on paper.
3. ADD 1 drop STANNOUS CHLORIDE to test solution drop. LET PAPER DRY.

TEST RESULT --- *TEST CODE*

[6 A] [P]-GOLD -- ROSE COLORED SPOT.

TEST ACTION # 2 -- PLATINUM TEST --
1. Put 1 cc (20 drops) of test solution in evaporating dish. BOIL UNTIL DRY.
2. When cool, add 4 cc's pure water.
3. Bring dish to BOIL.
4. FILTER solution into clean dish.
5. ADD 2 BB's POTASSIUM IODIDE to solution
6. HEAT dish to SLIGHT WARM.

TEST RESULT --- *TEST CODE*

[6 B] [P]-PLATINUM -- ROSE-PINK SOLUTION (The speedier the color reaction, the greater the platinum content.)

[6 C] [P] - PLATINUM PRESENT IN ORE -- CONFIRM ROSE COLOR - BOIL OFF most of the TEST ACTION # 2 solution, then let dish air-dry. ADD 5 drops of water to dish. RESULT - BRIGHT ROSE COLOR.

[6 D] [P] - PALLADIUM TEST - When conducting the TEST ACTION # 2 (Platinum Test) step # 3 & # 4, if the solution turns DARK COLORED plus leaves a BLACK PRECIPITATE on the dish bottom, PALLADIUM IS PRESENT. (The blacker the solution, the higher the palladium content of your ore.)

TEST # 7 -- GOLD — PLATINUM — PALLADIUM — RHODIUM

THIS TEST PROVES THE PRESENCE OR ABSENCE OF THESE ELEMENTS. CONFIRMATION TESTS ARE NECESSARY.

Roasting to remove arsenic, antimony, tellurium is NOT necessary, and may lock our tested-for elements in a refractory should those elements be present.

1. Crush and pulverize your test ore containing silver-gray particles.

2. Pan out the heavy concentrates and place 2 BB's of the silver-gray concentrate in a clean test tube. Over heat, dry out the concentrates.

3. ADD 3 ml's (60 drops) of AQUA REGIA to dissolve the concentrates. (Aqua regia = 1 part nitric acid to 3 parts hydrochloric acid.)

4. When test tube reaction stops, cap the tube and shake well to agitate the contents.

5. HEAT the tube in a water bath to body temperature and keep it there for about 2 hours, or longer if your test particles are of the larger sizes. (THIS IS A LEACHING PROCESS.)

6. After the leaching period, filter off the solution and DIVIDE THE SOLUTION into 4 clean test tubes and rack them.

7. ADD 2 drops of pure alcohol to each of the 4 tubes. (This is to stabilize the color reactions. It may be omitted.)

8. PREPARE FRESH STANNOUS CHLORIDE REAGENT.
(2 cc's 12 M HCL + 2 small pieces metallic tin, boil slightly.)

9. With a medicine dropper ADD 1 drop of stannous chloride to each of the 4 tubes, cap and shake well.

10. IF NO COLOR REACTION IN 60 SECONDS, REPEAT STEP # 9.

11. NOTE SOLUTION COLORS IN EACH TEST TUBE.

TEST RESULTS --

TEST CODE
[7 A] [I] - RHODIUM - Lemon-Yellow solution.
[7 B] [I] - PLATINUM - Orange solution.
[7 C] [I] - PALLADIUM - Blue solution.
[7 D] [I] - GOLD - Rose to Black solution.
[7 F] [I] - MERCURY - Silver-White silky threads.
[7 G] [I] - RHODIUM, PLATINUM & PALLADIUM - Brown solution.

(QUESTIONABLE RESULTS? RETRY, BUT ADD 3 ml WATER TO AQUA REGIA.)

TEST # 8 -- SULFIDES & SULFUR

MATERIALS NEEDED --

1. Test tube
2. Hydrochloric acid (12 M HCL)
3. Alcohol lamp
4. Powdered test mineral

TEST ACTION # 1 --

1. Place 1 BB test mineral in tube.
2. ADD 2 ml hydrochloric acid
3. Waft tube under nose.

TEST RESULT --

TEST CODE - [8 A] [P] SULFIDE - ROTTEN EGG ODOR (Smell Test)

TEST ACTION # 2 --

1. BOIL test action solution # 1

TEST RESULT --

TEST CODE - [8 B] [P] SULFUR - A rising sponge-like YELLOW or BLACK mass in tube.

TEST # 9 - VANADIUM, MANGANESE, CARBONATES (CO_3) -- SULFIDES

MATERIALS NEEDED --

1. 6 BB's powdered test mineral.
2. Cold hydrochloric acid (12 M HCL)
3. Clean evaporating dish
4. HCL eye dropper.

TEST ACTION # 1 --

1. Place powdered test mineral in dish.
2. ADD 4 to 5 drops HCL to mineral.

TEST RESULT --

TEST CODE - [9 A] [P] SULFIDE - Rotten egg smell.

[9 B] [P] CARBONATE - Effervescence with various degrees of bubbling.

(IF NO BUBBLING - DILUTE HCL with 4 or 5 drops of water & retest. or HEAT dilute acid and retest.

[9 C] [P] MANGANESE - GREEN-BLACK solution.

[9 D] [P] VANADIUM - RED or BROWN solution and residue.

TEST # 10 -- IRON -- CALCIUM -- MAGNESIUM

MATERIALS NEEDED --
1. CLEAN TEST TUBES (3)
2. NITRIC ACID CONCENTRATED (16 M)
3. HYDROCHLORIC ACID (12 M HCL)
4. AMMONIA (Strong 15 M)
5. AMMONIUM OXALATE (0.2 M)
6. SODIUM PHOSPHATE (used dry)
7. DISTILLED WATER
8. FILTER PAPER
9. ALCOHOL LAMP FOR HEAT
10. WELL-POWDERED TEST MINERAL

ROCKPECKER TIP # 220 - If test mineral is a silicate, ROAST with sodium carbonate on charcoal before powdering.

PREPARE TEST SOLUTION --
1. In a large test tube combine 3 cc's NITRIC ACID with 2 drops HYDROCHLORIC ACID and 6 cc's pure water. (tube #1)
2. ADD 6 BB's powdered test mineral.
3. HEAT to BOILPOINT, set aside until COLD
4. This test solution is used in all tests.

TEST ACTION #1 -
1. To test tube # 1 TEST SOLUTION ADD 2 cc's AMMONIA to COLD solution.

TEST RESULT -- *TEST CODE* --
[10 A] [P] - IRON - RED-BROWN PRECIPITATE.

TEST ACTION # 2 -
1. FILTER test action #1 solution into tube # 2 (Leaves iron on filter, calcium & magnesium in solution.)
2. ADD 12 BB's AMMONIUM OXALATE to solution.
3. HEAT to BOILPOINT, set aside until COLD.

TEST RESULT -- *TEST CODE* --
[10 B] [P] - CALCIUM - WHITE PRECIPITATE.
(The precipitate is crystalline - like sugar)

TEST ACTION # 3 -
1. FILTER test action #2 solution into tube #3. (Leaves calcium on filter, magnesium in solution.)
2. ADD 2 BB's SODIUM PHOSPHATE.

TEST RESULT -- *TEST CODE* --
[10 C] [P] - MAGNESIUM - WHITE PRECIPITATE.
(The precipitate is crystalline - like sugar)

TEST # 11 —— ZINC

All zinc SILICATES should be ROASTED before making any tests.

MATERIALS NEEDED -
1. CHARCOAL BLOCK AND CHARCOAL POWDER
2. SODIUM CARBONATE
3. HYDROCHLORIC ACID (12 M) CONCENTRATED
4. NITRIC ACID (16 M) CONCENTRATED
5. STRONG AMMONIA (15 M)
6. SODIUM SULFIDE - 6 BB's IN 1 TBSP. WATER
7. FILTER PAPER, BLOWPIPE, ALCOHOL LAMP,
8. WELL-POWDERED TEST MINERAL
9. PURE WATER
10. COBALT NITRATE - 2 BB's IN 1 TBSP. WATER.

TEST ACTION # 1 --
1. Make a paste of 4 BB's SODIUM CARBONATE, 4 BB's POWDERED BORAX, 2 BB's POWDERED CHARCOAL, 3 BB's OF WELL-POWDERED TEST MINERAL. Add 1 or 2 drops WATER to make it pasty.
2. Place on CHARCOAL BLOCK and with BLOWPIPE use INNER FLAME of alcohol lamp. ROAST INTENSELY to thoroughly fuse the sample.

TEST RESULT - *TEST CODE*
[11 A] [I] ZINC - WHITE SUBLIMATE COATING ON CHARCOAL EDGED IN BLUE NEAR SAMPLE.

TEST ACTION #2 -- CONFIRMING TEST. Place 1 drop of COBALT NITRATE on the white coating and heat drop with INNER FLAME.

TEST RESULT - *TEST CODE*
[11 B] [I] ZINC - COATING TURNS BLUISH-GREEN WHEN COLD.

TEST ACTION # 3 -- CONFIRMING TEST. --
1. Pulverize the mass and divide equally in test tube # 1 and # 2.
IN TEST TUBE # 1 --
2. ADD 10 drops of hydrochloric acid.
3. ADD 1 drop of nitric acid.
4. Slowly heat to boilpoint. LET COOL.
5. ADD 1 drop cobalt nitrate.
6. INTENSELY HEAT.

TEST RESULT - *TEST CODE*
[11 C] [I] -ZINC - SOLUTION TURNS GREEN.

TEST ACTION # 4 --
1. IN TEST TUBE # 2 REPEAT STEPS 2, 3, & 4 FROM TEST ACTION # 3. WHEN COLD --
2. ADD 2 cc's water and 2 cc's of ammonia. (Zinc is in solution.) Filter into test tube # 3 and add 3 drops sodium sulfide.

TEST RESULT - *TEST CODE*
[11 D] [P] ZINC - WHITE LUMPY MASS ON BOTTOM.

TEST # 12 — SULFATES --(SO_4)

MATERIALS NEEDED -
1. NITRIC ACID - CONCENTRATED 16 M
2. BARIUM NITRATE - CONCENTRATED 0.1 M
3. PURE WATER
4. WELL POWDERED TEST MINERAL

TEST ACTION #1 --
1. In clean test tube ADD 4 BB's test mineral.
2. ADD 10 drops of 16 M nitric acid.
3. ADD 10 drops of water.
4. ADD 3 to 4 drops of barium nitrate.
5. Mix contents with finger flick action.

TEST RESULT - *TEST CODE*

[12 A] [P] -SULFATE (SO_4) - WHITE PRECIPITATE.
(No other test is required.)

If you wish to confirm the presence of SULFATE, filter the white precipitate into a clean test tube and add concentrated acid.

THE FORMATION OF SULFATE IS DETECTED BY THE INSOLUBILITY OF THE BARIUM SALT IN ACID SOLUTION.

TEST # 13 -- IS IT A DIAMOND ? ———OR CRYSTAL QUARTZ ?

GEMSTONE TRIVIA -- Gemstones are valued for their beauty and durability or hardness. A DIAMOND has a hardness of 10 on the MOHS scale compared with QUARTZ at 7, RUBY and SAPPHIRE at 9 (corundum), and GARNET AT 7. This tells us that these gems are highly resistant to abrasion due to their hardness quality.

As natural weathering releases these crystals from their rock vault, erosional processes move them intact into watercourses. Oftentimes, these stones reveal themselves in a ROCKPECKER'S goldpan.

The overall class of gem crystal quartz is known as ROCK CRYSTAL, which includes COLORED QUARTZ with names such as amethyst, citrine, smoky, morion, cat's eye, tiger's eye, falcon's eye, blue, rose, adventurine and prase. OUR TARGET IS TO IDENTIFY THE ROCK CRYSTAL THAT MOST RESEMBLES A DIAMOND. I call it "CRYSTAL QUARTZ".

As you can appreciate, discarding a DIAMOND, thinking it CRYSTAL QUARTZ, can be hazardous to your pocketbook. In the real world, you keep both and do nothing else. (It could be a diamond.)

HOW CAN WE BE SURE IT IS "CRYSTAL QUARTZ" AND NOT A DIAMOND ?

The test is simple and involves the "INDEX OF REFRACTION" concept shown in CHART F, column "S". In checking the refractive index of a diamond and of quartz, we learn:

DIAMOND - 2.42
QUARTZ - 1.54

REFRACTIVE INDEX BACKGROUND - As light speeds through air into a medium of higher optical density, it is both slowed down and forced to change it's path slightly. (The trout in the stream idea.) The angular change in the light path can be measured. That angle measurement is the "index of refraction" number, which is always greater than 1. (The law of refraction was devised in 1621.)

THE TEST BACKGROUND - If you have 2 glasses of water each half full, then fill one to the top by emptying the other into it, you cannot identify the last-poured water. This is because light travels through the same medium (water) at the same speed. The I.R of water is 1.33. However, if you drop an uncut diamond and a piece of crystal quartz in the glass of water, both stones are clearly and completely visible.
(water IR = 1.33, quartz IR = 1.54, diamond IR = 2.42.)

SOLUTION: IF WE COULD FIND A LIQUID MEDIUM WITH AN *INDEX OF REFRACTION AT OR NEAR* QUARTZ'S I.R. OF 1.54, IMMERSION OF QUARTZ IN THAT MEDIUM WOULD PERMIT IT TO DISAPPEAR. (Or nearly so.)

Kerosene, carbon oil, lamp oil, coal oil, K-1 are all the same thing.... KEROSENE.

For our purposes it is not important, but you should know that there is the old kerosene and K 1, a newer, lighter product with more of the sulfur removed so to comply with the pollution laws

THE INDEX OF REFRACTION OF EACH IS:
KEROSENE (old) = 1.455
K1 KEROSENE = 1.458

THE SOLUTION APPLIED AS A TEST

TEST ACTION --
1. Into a clear glass ADD enough old kerosene or new K 1 to cover your test piece of crystal quartz.
2. Hold the glass up to the light and inspect for the crystal outline.

TEST RESULT - *TEST CODE*
[13 A] [P] - ROCK CRYSTAL - THE EDGES OF CRYSTAL QUARTZ DISAPPEAR, AND THE OUTLINE OF THE CRYSTAL BECOMES ALMOST INVISIBLE.

(K1 index of refraction 1.458 is near enough to quartz's I.R. of 1.548 to make the test POSITIVE.)

THEREFORE, A DIAMOND WOULD BE CLEARLY VISIBLE IN KEROSENE.

TEST # 14
CHLORINE & CHLORIDES

MATERIALS NEEDED --
1. Clean test tube
2. Pure chlorine-free water
3. Alcohol lamp
4. Concentrated nitric acid (16 M)
5. Powdered test mineral
6. Silver nitrate solution

TEST ACTION --

Test preparation:
1. Place 6 BB's of well-powdered test mineral in clean test tube.
2. Add 40 drops (2 cc's) of chlorine-free water to test tube.
3. Add 6 to 7 drops of 16M nitric acid.
4. Heat tube just to the point of boiling.
5. Set aside tube until cold.
6. To COLD tube, add 3 to 5 drops of silver nitrate solution.

TEST RESULTS -- *TEST CODE*

[14 A] [P] CHLORIDE OR CHLORINE
A whitish, milky or cloudy color AND/OR a white mass of curds is a positive test for chlorine or chloride minerals.

[14 B] [P] CHLORINE IN WATER -
Repeat test #14 but reduce silver nitrate to 2 drops. OMIT THE TEST MINERAL.

ROCKPECKER TIP # 221 - Test 14-B should be performed anytime you suspect that water contains salt (Sodium Chloride) or any chlorine-bearing material.

This is particularly important if you also suspect the presence of SILVER in your test mineral. ....

FIRST SAFETY RULE - WHEN TESTING, ALWAYS WEAR PROTECTIVE CLOTHING, GLOVES AND GLASSES.

SOLUBILITIES

MANY MINERALS DISSOLVE IN EITHER WATER, AN ACID OR BOTH.

SOLUBILITY GUIDE

USUALLY WATER SOLUBLE

1. NITRATES
2. CHLORIDES - NOT LEAD, SILVER OR MERCURY CHLORIDES.
3. HYDROXIDES - BUT ONLY SOME.
4. BORATES - WHEN TERMED "HYDROUS".
5. CARBONATES - OF SODIUM, POTASSIUM AND AMMONIUM.
6. SULFATES - NOT LEAD, BARIUM OR CALCIUM SULFATES.
7. PHOSPHATES - OF SODIUM, POTASSIUM AND AMMONIUM.
8. ARSENATES - ONLY A FEW.

SOLUBLE IN ACIDS - GENERALLY.

1. METALS - BUT GOLD & PLATINUM IN AQUA REGIA.
2. SULFIDES
3. HYDROXIDES
4. FLUORIDES
5. OXIDES - BUT ONLY SOME.
6. SILICATES - WITH DIFFICULTY IN BOILING HYDROFLUORIC ACID.
7. SULFATES
8. ARSENATES
9. NITRATES
10. VANADATES
11. MOLYBDATES
12. PHOSPHATES
13. ZEOLITES - GENERALLY IN HYDROCHLORIC ACID.

GOLD DISSOLVES IN AQUA REGIA WHICH IS A MIXTURE OF ONE PART NITRIC ACID TO THREE PARTS HYDROCHLORIC ACID.

PLATINUM DISSOLVES IN AN ACID MIXTURE OF TWO PARTS NITRIC ACID TO NINE PARTS HYDROCHLORIC ACID. (AS WELL AS AQUA REGIA.)

HYDROFLUORIC ACID IS A NASTY PRODUCT. STAY AWAY FROM IT! IF IT SPILLS ON YOU, IT CAN BURN DOWN TO THE BONE; YOU MAY NOT FEEL ANY WARNING PAIN.

WHEN WORKING WITH ACIDS, KEEP A NEUTRALIZING BASIC SOLUTION HANDY TO MINIMIZE AN ACID BURN. ADDING A BASE TO AN ACID NEUTRALIZES THE ACID.

SOLUBILITY PRODUCTS

IF YOU FIND:	TEST FOR THESE ELEMENTS
ARSENATES	Silver, aluminum, barium, bismuth, calcium, cadmium, cobalt, chromium, copper, iron, magnesium, manganese, nickel, lead, zinc, strontium.
CARBONATES	Silver, barium, cadmium, calcium, copper, iron, mercury, lead, magnesium, manganese, nickel, strontium, zinc.
CHLORIDES	Silver, copper, mercury, lead.
CHROMATES	Silver, barium, calcium, copper, mercury, lead, strontium.
HYDROXIDES	Silver, aluminum, bismuth, cadmium, cobalt, chromium, copper, iron, magnesium, nickel, manganese, lead, tin, zinc.
OXALATES	Silver, barium, calcium, copper, iron,lead, magnesium, manganese, strontium, zinc.
PHOSPHATES	Silver, barium, aluminum, calcium, lead, chromium, iron.
SULFATES	Silver, barium, calcium, mercury, lead, strontium.
SULFIDES	Silver, bismuth, cadmium, cobalt, copper, iron, mercury, manganese, nickel, lead, tin, zinc.

SECOND SAFETY RULE -- NEVER LICK THE SPOON !

FIRST SAFETY RULE - WHEN TESTING, ALWAYS WEAR PROTECTIVE CLOTHING, GLOVES AND GLASSES.

COLORS OF TRANSITION METALS

TRANSITION METALS, are the elements with atomic numbers 21 to 30 inclusive. Generally, they are metallic, and, when clean, are shiny and lustrous, good conductors of heat and electricity. Some when finely divided, such as cerium and lanthanum oxidize immediately when exposed to air. However, these transition elements form brightly colored simple compounds and more complex substances. THEIR SOLUTIONS ARE ALSO BRIGHTLY COLORED.

Solid compounds of transition metals, and their aqueous solutions are colored. Copper, chromium and iron are some of our most useful metals. They are transition elements. The COLOR depends on the metal, its oxidation state, and the anion with which it is combined. THE COLOR IS DIAGNOSTIC.

A compound is a substance that can be decomposed into two or more simpler substances by ordinary chemical means. Water can be decomposed into hydrogen and oxygen elements. Thus, water is also a compound.

WHEN THE FOLLOWING TRANSITION ELEMENT COMPOUNDS ARE PRESENT IN A MINERAL DISSOLVING SOLUTION, THE COLOR SHADES OF BOTH THE SOLUTION AND PRECIPITATE INDICATE THE ELEMENT PRESENT. A CONFIRMING TEST IS NEEDED.

COMPOUND	SOLUTION & PRECIPITATE COLOR SHADES
COPPER CARBONATE	YELLOW
COPPER OXIDE	RED OR BLACK
COPPER SULFATE (AND NITRATE)	BLUE
NICKEL HYDROXIDE (AND NITRATE)	GREEN
NICKEL SULFATE	YELLOW
COBALT CHLORIDE (AND NITRATE)	RED
COBALT SULFATE	PINK OR BLUE-GREEN
IRON CHLORIDE	GREEN OR BROWN-YELLOW
IRON SULFATE	BLUE-GREEN OR YELLOW
MANGANESE CHLORIDE	PINK
MANGANESE SULFATE	GREEN
MANGANESE OXIDE	BLACK
CHROMIUM ACETATE	RED
CHROMIUM SULFATE	BLUE
CHROMIUM CHLORIDE	VIOLET
VANADIUM CHLORIDE	PINK OR VIOLET OR GREEN
VANADIUM SULFATE	VIOLET
TITANIUM CHLORIDE	YELLOW OR VIOLET
TITANIUM SULFATE	GREEN

FIRST SAFETY RULE - WHEN TESTING, ALWAYS WEAR PROTECTIVE CLOTHING, GLOVES AND GLASSES.

TEST CODES

ELEMENT or RADICAL		TEST CODE FOR ELEMENT or RADICAL
1. ANTIMONY	(Sb)	---- 1-H, 1-U-P, 3-C-P.
2. ARSENIC	(As)	---- 1-G, 3-A-P, 4-L-I,
3. BARIUM	(Ba)	---- 1-E
4. BISMUTH	(Bi)	---- 1-R, 3-H-I, 4-F-I, 4-G-I, 5-C-I.
5. BORON	(B)	---- 1-N.
6. CADMIUM	(Cd)	---- 3-B-P, 5-F-I.
7. CALCIUM	(Ca)	---- 1-C, 10-B-P.
8. CARBONATES	(CO_3)	---- 9-B-P.
9. CHLORINE	(Cl)	---- 14-A.
10. CHLORIDES		Test for chlorine
11. CHROMIUM	(Cr)	---- 2-A.
12. COBALT	(Co)	---- 2-B, 4-C-I, 4-H-P, 5-E-I, 5-M-I.
13. COLUMBIUM	(Ni)	---- 2-D, 4-M-I. (NIOBIUM)
14. COPPER	(Cu)	---- 1-F, 2-C, 3-M, 4-B-I, 5-D-I, 5-M-I.
15. DIAMOND	(C)	---- 13-A-I. (VERSUS QUARTZ)
16. GOLD	(Au)	---- 3-P-I, 6-A-P, 7-D-I.
17. IRON	(Fe)	---- 2-E, 5-G-I, 5-N-I, 10-A-P.
18. LEAD	(Pb)	---- 1-J, 1-W-P, 3-G-I, 4-E-I, 4-G-I, 4-N-I, 4-R-P.
19. LITHIUM	(Li)	---- 1-T.
20. MAGNESIUM	(Mg)	---- 5-K-I, 10-C-P.
21. MANGANESE	(Mn)	---- 2-F, 5-H-I.
22. MERCURY	(Hg)	---- 5-B-I, 5-P-I, 7-F-I.
23. MOLYBDENUM	(Mo)	---- 1-S, 2-G, 4-K-P.
24. NICKEL	(Ni)	---- 2-H, 4-B-I, 4-D-P, 5-J-I.
25. NIOBIUM	(Nb)	---- Use columbium tests.
26. OSMIUM	(Os)	---- 3-F-I.
27. PALLADIUM	(Pd)	---- 6-D-P, 7-C-I, 7-G-I.
28. PHOSPHOROUS	(P)	---- 1-L.
29. PLATINUM	(Pt)	---- 6-B-P, 6-C-P, 7-B-I, 7-G-I.
30. POTASSIUM	(K)	---- 1-B
31. QUARTZ	(SiO_2)	---- 13-A-P (ROCK CRYSTAL)
32. RHODIUM	(Rh)	---- 7-A-I, 7-G-I.
33. SELENIUM	(Se)	---- 1-K, 3-D-P.
34. SILVER	(Ag)	---- 3-N-I, 4-N-I, 4-S-I, 4-T-P,4-V-P, 4-W-P, 5-A-I, 5-R-I.
35. SODIUM	(Na)	---- 1-A
36. STRONTIUM	(Sr)	---- 1-D
37. SULFIDES	(S)	---- 4-A-I, 8-A-P, 9-A-P.
38. SULFATES	(SO_4)	---- 12-A-P.
39. SULFUR	(S)	---- 3-E-P, 4-A-I, 8-B-P.
40. TELLURIUM	(Te)	---- 1-P, 3-J-I, 4-J-P.
41. TITANIUM	(Ti)	---- 2-J.
42. TUNGSTEN	(W)	---- 2-K, 4-M-I.
43. URANIUM	(U)	---- 2-L.
44. VANADIUM	(V)	---- 2-M, 9-D-P.
45. ZINC	(Zn)	---- 1-M, 1-X-P, 3-K-I, 11-A-I, 11-B-I, 11-C-I, 11-D-P.

SECOND SAFETY RULE - NEVER, NEVER, LICK THE SPOON !

S//Q ™

Memo

to: MINERAL PROSPECTORS, ROCKHOUNDS, STUDENTS

from: T. D. LaRune

" ROCKPECKER ™ v.1 DOS PROFESSIONAL "

DESCRIBE, IDENTIFY & CHEMICAL TEST UNKNOWN ORE

Computerized Versions

of PART THREE & PART 4 ARE AVAILABLE

NOW FOR PC's

DOS 3.3+, 400K+ RAM,

XT\486 COMPUTERS.

277 MINERALS ARE DETAILED.

Prospector can identify unknown mineral / ore;

then PRE-ASSAY chemical test to confirm finding.

WRITE TO: Skill//Quest Co.

675 Fairview Drive, Suite 246,

Carson City, Nevada 89701 U.S.A.